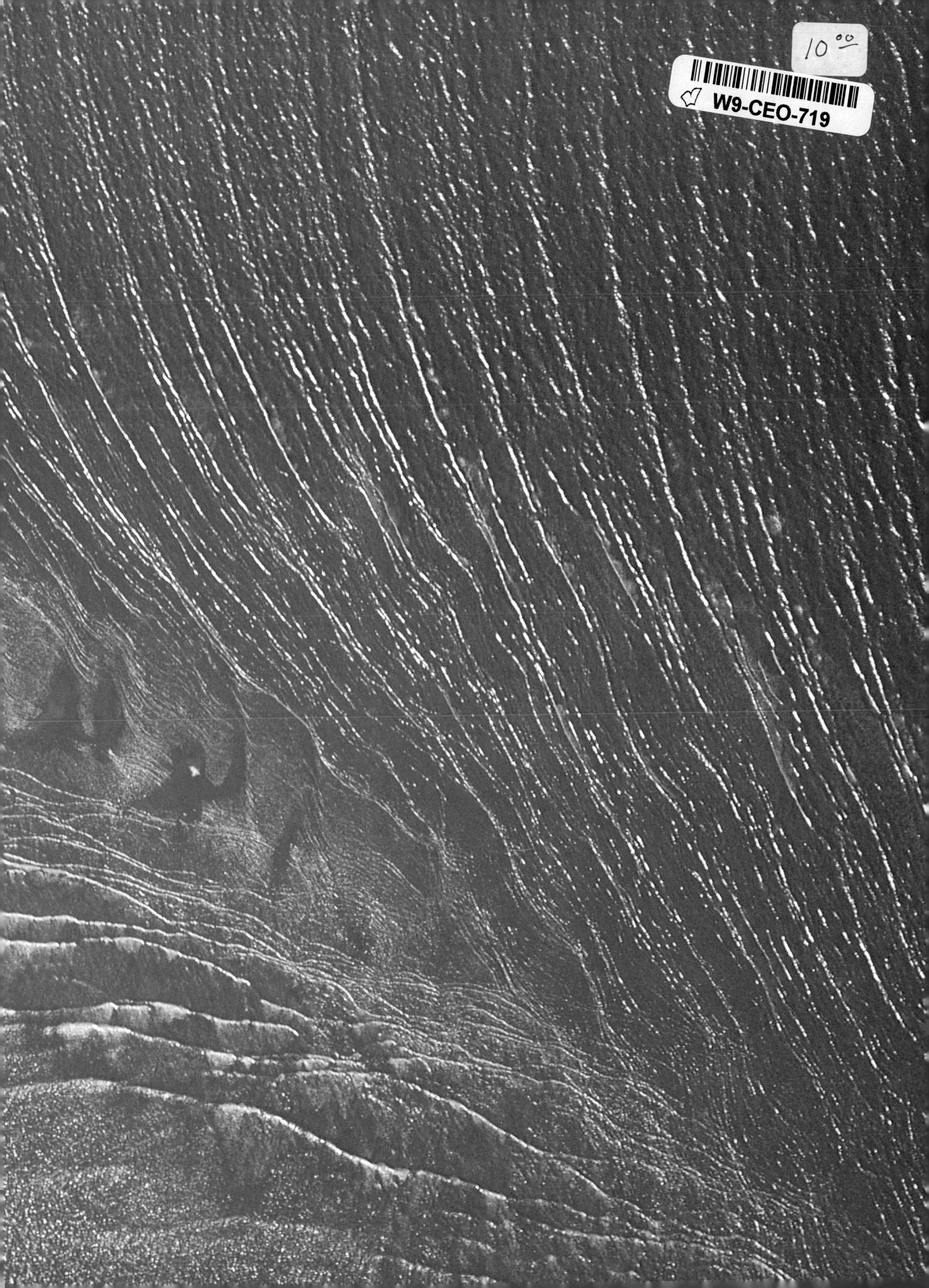

To Dad;

Happy Fathers' DAY 1992

From Ray and Fyrne Lantz

THE ENCYCLOPEDIA OF THE EARTH

# OCEANS AND ISLANDS

THE ENCYCLOPEDIA OF THE EARTH

# OCEANS AND ISLANDS


CONSULTANT EDITORS

DR FRANK H. TALBOT

Director, National Museum of Natural History, Smithsonian Institution, Washington DC, USA

DR ROBERT E. STEVENSON

Secretary-General, International Association for the Physical Sciences of the Ocean, California, USA

ILLUSTRATIONS BY

MIKE GORMAN

First published in the United States in 1991 by SMITHMARK Publishers Inc.,
112 Madison Avenue, New York, NY 10016

SMITHMARK books are available for bulk purchase for sales promotion and premium use. For details write or telephone the Manager of Special Sales, SMITHMARK Publishers, Inc, 112 Madison Avenue, New York, NY 10016. (212) 532 6600

Produced by Weldon Owen Pty Limited
43 Victoria Street, McMahons Point, NSW 2060, Australia
Telex 23038; Fax (02) 929 8352
and
Weldon Owen Inc
90 Gold Street, San Francisco, California 94133, USA
Fax (415) 291 8841

A member of the Weldon International Group of Companies
Sydney • London • Paris • San Francisco

Chairman: Kevin Weldon
President: John Owen
General Manager: Stuart Laurence
Publisher: Alison Pressley
Project Coordinator: Sheena Coupe
Foreign Editions Editor: Derek Barton
Copy Editor: Sheena Coupe
Editorial Assistant: Veronica Hilton
Picture Research: Brigitte F. Zinsinger
Captions: Margaret Atkinson
Index: Dianne Regtop
Designer: Warren Penney
Production Manager: Mick Bagnato


ISBN 0 8317 2813 2



Typeset by Midland Typesetters, Maryborough, Victoria
Printed by Singapore National Printers Ltd
Printed in Singapore

A WELDON OWEN PRODUCTION

*Endpapers*: Aerial seascape, Keppel Bay, Queensland, Australia. (Leo Meier/Weldon Trannies)
*Page 1:* The beauty and unharnessed power of a breaking wave. (Frans Lanting/Minden Pictures)
*Pages 2–3:* Kiritimati, the world's oldest known atoll.
*Page 6:* The Rarotongan Express, Cook Islands.
*Pages 8–9:* Dawn breaks over the Polynesian island of Moorea.
*Pages 10–11:* Bigeye soldier fish, Seychelles, Indian Ocean.
*Right:* The island of Raiatea in French Polynesia.

M. Claye/Explorer

FLAG ME DOWN
SARAO MOTO S INC.
Rarotongan Express
RARO
RARO
MN 3703
PGS·242
PILIPINAS

## CONSULTANT EDITORS


DR FRANK H. TALBOT
Director, National Museum of Natural History,
Smithsonian Institution, Washington DC, USA

DR ROBERT E. STEVENSON
Secretary-General, International Association for the
Physical Sciences of the Ocean, California, USA


## CONTRIBUTORS


MARGARET ATKINSON
Research Assistant,
University of Sydney, Australia

MICHEL A. BOUDRIAS
Scripps Institution of Oceanography,
University of California, USA

DR ALEC C. BROWN
Professor of Marine Biology,
University of Cape Town, South Africa

DR MICHAEL BRYDEN
Professor of Veterinary Anatomy,
University of Sydney, Australia

M. G. CHAPMAN
Professional Officer, Institute of Marine Ecology,
University of Sydney, Australia

DR SYLVIA A. EARLE
President and Chief Executive Officer,
Deep Ocean Engineering Inc., USA

DR RICHARD S. FISKE
Research Geologist, National Museum of Natural History,
Smithsonian Institution, Washington DC, USA

SCOTT C. FRANCE
Scripps Institution of Oceanography,
University of California, USA

DR CHRISTIAN D. GARLAND
Senior Lecturer,
University of Tasmania, Australia.

DR STEPHEN GARNETT
Environmental and Scientific Consultant and Writer,
Victoria, Australia

DR ALISTAIR J. GILMOUR
Professor of Environmental Studies,
Macquarie University, Sydney, Australia

DR RICHARD W. GRIGG
Professor of Oceanography,
University of Hawaii, USA

DR RICHARD HARBISON
Senior Scientist, Woods Hole Oceanographic Institution,
Massachusetts, USA

DR HAROLD HEATWOLE
Associate Professor, Department of Zoology,
University of New England, Australia

DR ROBERT R. HESSLER
Professor of Biological Oceanography,
Scripps Institution of Oceanography,
University of California, USA

STUART INDER MBE
formerly Editor *Pacific Islands Monthly*
Sydney, Australia

DR ANGELA MARIA IVANOVICI
Senior Project Officer,
Australian National Parks and Wildlife Service,
Canberra, Australia

DR DAVID JOHNSON
Associate Professor in Sedimentology,
James Cook University, Townsville, Australia

GRAHAM JOYNER
Lecturer in History
Macquarie University, Sydney, Australia

DR E. ALISON KAY
Professor of Zoology,
University of Hawaii, USA

DR JAMES C. KELLEY
Dean of Science and Engineering,
San Francisco State University, California, USA

DR KNOWLES KERRY
Senior Research Scientist,
Australian Antarctic Division, Hobart, Australia

DR G. L. KESTEVEN
formerly FAO Fisheries Biologist,
Sydney, Australia

DR M. J. KINGSFORD
Lecturer, School of Biological Sciences,
University of Sydney, Australia

DR JOHN E. McCOSKER
Director, Steinhart Aquarium,
California Academy of Sciences, USA

DR KENNETH McPHERSON
Executive Director,
Center for Indian Ocean Peace Studies,
Curtin University and the
Unversity of Western Australia, Australia

DR ALEXANDER MALAHOFF
Professor of Geological Oceanography,
University of Hawaii, USA

DR COLIN MARTIN
Reader, Department of Scottish History,
University of St Andrews, Scotland

DR SIDNEY W. MINTZ
William L. Strauss Jr Professor of Anthropology,
John Hopkins University, Baltimore, USA

DR STORRS L. OLSON
Curator of Birds, National Museum of Natural History,
Smithsonian Institution, Washington DC, USA

DR JOHN R. PAXTON
Senior Research Scientist,
Australian Museum, Sydney, Australia

DR VICTOR PRESCOTT
Professor of Geography,
University of Melbourne, Australia

DR PATRICK G. QUILTY
Assistant Director, Science,
Australian Antarctic Division, Hobart, Australia

WILLIAM REED
Pearling Consultant
Perth, Australia

DR PAUL SCULLY-POWER
Research Associate, Scripps Institution of Oceanography,
California, USA

DR J. R. SIMONS
formerly Associate Professor of Biology
and Dean of the Faculty of Science,
University of Sydney, Australia

DR ROBERT E. STEVENSON
Secretary-General, International Association for the
Physical Sciences of the Ocean, California, USA

DR A. J. UNDERWOOD
Director, Institute of Marine Ecology,
University of Sydney, Australia

DR DIANA WALKER
Lecturer in Marine Botany,
University of Western Australia, Australia

DR G. M. WELLINGTON
Associate Professor of Biology,
University of Houston, USA

# CONTENTS

PART THREE

# THE FUTURE OF OCEANS AND ISLANDS

# INTRODUCTION

In our world of jet travel and space exploration, oceanic islands still remain magical and mysterious. They range from tropical havens of great beauty, fringed with coconut palms and growing exotic fruits, to huge craggy ice-topped bastions where seabirds breed in their hundreds of thousands. We may be lucky enough to visit a few islands, but no one could see them all in a lifetime of searching. Most will remain islands of our imagination, known from our readings or from pictures, or perhaps seen through an aircraft window as a speck in an immense blue sea. Their animals and plants are often intriguingly different, because isolation has allowed their evolution to take a separate course from those of other island or mainland habitats.

The great oceans that surround most islands are the predominant feature of our watery, life-sustaining planet. In this book, you will discover that one ocean current transports 60 times more water than all the rivers in the world together; that the highest undersea mountains dwarf Everest; and that the forces that move our restless continents can best be seen in the spreading of the volcanic mid-oceanic ridges and the great trenches at the rim of the oceans. Though linked, the oceans are not uniform masses of water, and each has a marked internal structure. Their great currents are determined by areas of varying temperature and saltiness, by the Earth-girdling winds, and by the spin of the Earth itself.

In the oceans there are areas of amazing richness, where deep nutrient-laden water wells up to the surface and supports huge quantities of plankton that feeds millions of fish, mammals, and birds. There are other areas of clear blue water that sustain few living things except for the rich oases formed by coral reefs, or swirling beds of floating *Sargassum* weed with a host of strange inhabitants.

Since our emergence as tool makers we have created rafts and boats of wood, reeds, bark, or skin to cross water. Tales of the exploring of oceans and the colonizing of islands are among the most stirring in human history. We are now also starting to search the ocean depths themselves, and have briefly visited their deepest point. On the deep-sea floor whole new living systems have been found based on chemical, not solar, energy. But in spite of these findings, the sea cloaks the largest unexplored portion of our globe.

I hope this book begins, or perhaps continues, your personal exploration of oceans and islands. There is much to discover!

FRANK H. TALBOT
Consultant Editor

Pam & Willy Kemp/Oxford Scientific Films

Hans-Jürgen Burkard/Bilderberg

Dolphins in Dusky Sound, in the South Island of New Zealand.

# PART ONE
# THE MIRACLE OF THE SEA

# 1 THE UNDERSEA WORLD

NASA

▲ The African continent and the frozen Antarctic—as seen from space—swirled by cloud and surrounded by oceanic blue.

◄ A meeting of elemental forces as molten lava flows into the pounding waves, producing clouds of steam.

Our "Blue Planet", so named because of its distinctive appearance from outer space, derives its brilliant coloration from the vast oceans that cover 70 percent of its surface. From this vantage point the Earth appears as a blue and white ball totally surrounded by the blackness and infinity of space. One can begin to appreciate the immense size of the oceans by considering that they contain over one billion cubic kilometers (300 million cubic miles), which is equivalent to nearly 300 billion liters (over 60 billion gallons) for every inhabitant of the Earth. The world ocean, a vital natural resource, is the result of ancient geological processes that have created extraordinary undersea landscapes of ridges and trenches, mountain chains, valleys, and plains.

## THE FORMATION OF THE OCEANS

Hunting and gathering gave way to incipient farming in Mesopotamia perhaps as early as 6000 BC, but there was no concept of the global expanse even until 400 BC when Aristotle was puzzled by marine fossils in rocks high above sea level. In the centuries that followed, the history of the Earth continued to puzzle scholars to the extent that in 1654 Archbishop James Ussher decided to settle the question once and for all by declaring that Heaven and Earth had been created "upon the entrance of the night preceding" Sunday October 23, 4004 BC.

Ussher's dogmatic pronouncement was not acceptable to the scientists of the day, but the power of the Church was such that it would take another 200 years to overcome this Church-dominated concept. In the mid-nineteenth century James D. Dana, an American geologist, declared that the Earth had originated in a molten state, that it had contracted upon cooling, and that it continued to do so. The continents and ocean basins, he declared, were permanent features, having stabilized at the beginning of geologic time. The continents had cooled first, and further contractions had lowered the crust to form the oceans. As the cooling continued, the Earth's interior shrunk, causing great forces that created the upheaval of mountain ranges.

Dana's concept was quickly adopted and geologic research into the twentieth century was based on his theory. Then, after half a century of speculation, and an extraordinary decade of oceanographic expeditions in the mid-century, the origin of the Earth's present continents and ocean basins was finally known. The long and controversial efforts to prove or disprove that moving, crustal plates formed the ocean basins and continents finalized in 1967; the process is known as plate tectonics.

Giraudon, Paris

◄ One of the earliest world maps was drawn by the Greek scientist Ptolemy in AD 140. Subsequently lost, Ptolemaic maps were rediscovered in 1400 and, despite limited information on the oceans, they profoundly influenced European geographic thought until 1516, when Ptolemy's ideas were rejected following the production of a new map based on discoveries by hydrographers from Lisbon and Seville.

## CONTINENTAL DRIFT

At the beginning of the twentieth century there were no topographical maps of the sea floor. Geologists believed the Earth was slowly cooling and shrinking, and that continents and oceans had been in their respective places since the beginning of time. With this accepted prevailing thought, we can understand why Dr Alfred Wegener's theories describing continents that drift or spread apart were not taken well by his fellow German colleagues when in January 1912 Wegener presented his ideas on "The Formation of Major Features of the Earth's Crust (Continents and Oceans)". In 1915, his theory of continental drift was published in a small but comprehensive book entitled *The Origin of Continents and Oceans.*

**STAGES IN CONTINENTAL DRIFT**

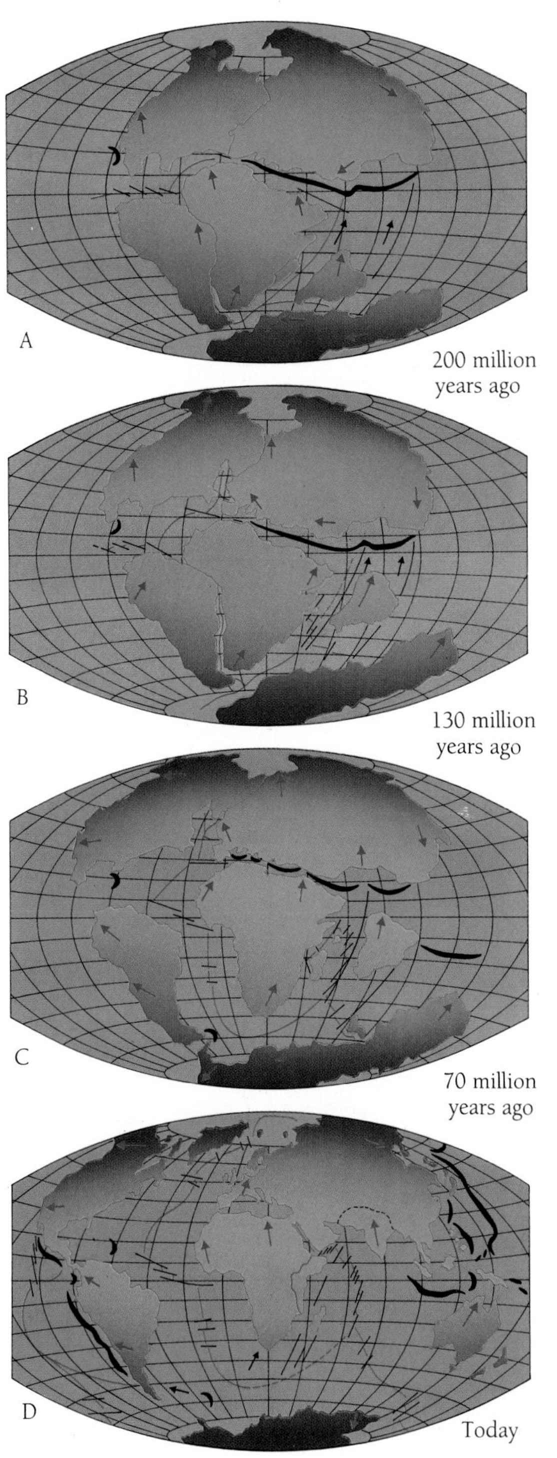


▸ Some 200 million years ago (A), the landmass of Pangaea began to break up to create the supercontinents of Laurasia in the north and Gondwana in the south. This southern landmass was itself breaking up: India, South America, and Africa were moving apart, although Australia was still part of Gondwana. By 130 million years ago (B) the north Atlantic and Indian oceans had widened, and Africa was moving away from South America. Seventy million years ago (C), South America had separated from Africa, but Australia was still joined to Antarctica, and North America and Europe were linked in the north. In the last 70 million years (D) the Atlantic Ocean has widened, North America and Europe have separated, and North America has curved to join South America. Australia has separated from Antarctica, and India has collided with Asia to form the Himalayas.

Wegener described how continents had once been united, only to drift to their present locations in the past millions of years. It was a well-conceived hypothesis. He pointed out that continental boundaries lay at the edge of the surrounding shelves rather than the present coastlines. The oceanic crust was similar to pitch, he suggested, which flows when put under lengthy pressure, yet breaks into brittle pieces when struck a blow. Fossil and geological evidence indicated former connections between continents. He concluded that the major landmasses had been joined in the geologic past and had since drifted apart.

Wegener named his supercontinent Pangaea, from the Greek root meaning "all land". He assumed Pangaea existed 300 million years ago, and began to separate about 200 million years before the present, with the continents slowly, imperceptibly moving to the positions in which we know them today.

Geologists around the world were stung by the impertinence of Wegener's proposition. From a special meeting of the Royal Geographical Society in London in January 1923, he learned that no one who "valued his reputation for scientific sanity" would advocate a wild theory like continental drift. Despite his unpopularity, Wegener continued to pursue the theory until his death in 1930, when it became dormant. The data and technology needed to prove, or disprove, the concept of continental mobility would remain the secret of the ocean basins for another thirty years.

## A DECADE OF OCEAN EXPLORATION

In 1950, ocean-testing equipment was more sophisticated than in Wegener's day. Acoustic depth recorders enabled scientists to determine depths for mapping the ocean floor; sensitive heat probes and radioactive carbon-dating techniques were used to analyze age and polarity of the layers of the sea floor; and core sea-floor samples told its biological history. After the Second World War, surplus naval vessels equipped with these new instruments were made available to United States oceanographers. Scientists from Scripps Institution of Oceanography, California, sailed into the Pacific Ocean, and other exploring cruises were mounted from Woods Hole Oceanographic Institution, Massachusetts, and the Lamont-Doherty Geological Laboratory, New York.

Assumptions based on the prevailing geologic thought of the day were challenged by the data collected on the expeditions of the 1950s. By 1960, the three major ocean basins and their adjacent seas had produced some confirmations, some disproofs, and an extraordinary amount of

**A DECADE OF OCEANIC EXPLORATION**

| Assumptions about the Earth in 1950 | Facts about the Earth in 1960 |
|---|---|
| The ocean basins were old; the Pacific was older than the Atlantic. | The ocean basins are relatively young, no older than the Cretaceous Period: about 140 million years. |
| Considering the supposed age of the ocean basins, the sediments should be some 5 km (3 miles) thick. The sea floor was expected to be smooth and gently sloping. | There are no thick layers of sediments. On the ocean ridges sediments are either absent or thinly deposited in valleys between parallel ridges. |
| Nothing was known of the oceanic crust nor the depth of the discontinuity between the crust and the mantle — the Mohorovicic Discontinuity or "Moho". | The oceanic crust is far thinner than that of the continents, averaging some 5 km (3 miles) in contrast to the 35 km (20 miles) of the major landmasses. The rocks making up the ocean's crust are quite different from those of the continents. |
| Except for the margins and the Hawaiian Islands, the Pacific Basin seemed free of earthquakes. There were probably no faults because all the earthquakes would be of volcanic origin. | The crests of the ocean ridges are seismically active. In fact, the most active are in the Pacific Ocean. |
| The seamounts and guyots scattered across the floor of the central Pacific were probably drowned Precambrian islands, more than 500 million years old. | All the large volcanoes of the Pacific Basin, including seamounts, guyots, and the oceanic islands, are Cretaceous in age. |
| Since continental rocks have greater amounts of radioactive minerals than the basalts of the sea floor, heat coming from the oceanic crust would be far less than from the continents. | Convection currents exist in the Earth's mantle beneath the crust, accounting for the mid-ocean ridges, the high oceanic heat flow over the ridges, the island arcs and trenches, and the great fracture zones that cut the ridges in each ocean. |

new data leading to new avenues of thought and controversy among scientists. If we compare the data in the accompanying table, we can see the remarkable amount of information learned in less than ten years.

SEA-FLOOR SPREADING

The implication that the young sea floor was moving, and had been doing so for the past 130 million years, presented a puzzling enigma to most geologists of the day. Not so to Dr Robert S. Dietz in California, nor to Professor Harry Hess at Princeton. Simultaneously, they considered the new knowledge and proposed that the ocean's crust is continuously spreading from the oceanic ridges. Dietz called the concept "sea-floor spreading". After many expeditions to the Atlantic and Pacific oceans, and arduous years of data analysis, Dietz and Hess established the reality of convection currents—the powerful force behind the movement of the continents. As Dietz pointed out: "Wegener was correct in his sequence of the break-up of Pangaea. He just didn't have the data to describe the mechanism."

Sea-floor spreading in 1960 was as incredible to the scientific community as continental drift was in 1912. But with the overwhelming data, even non-believers sailed forth to address the Dietz/Hess claim. From Cambridge University came Dr Drummond H. Matthews and his student, Frederick J. Vine, who tested the hypothesis by measuring the magnetic anomalies in the oceanic crust on either side of the ridges in the Atlantic and Indian oceans. By 1963, they had confirming data: the sea floor is spreading from the ridge crests and at rates that agree with the 130-million-year period starting with the break-up of Pangaea.

With sea-floor spreading accepted, the next question arose: do the oceans' bottoms spread as two massive units in each basin? Not likely, thought Dr J. Tuzo Wilson, a Canadian geologist on sabbatical at Cambridge University, in 1965. He wrote, "These huge transverse faults are not isolated features, but connect with the ridges and trenches to divide the oceanic crust into large,

▼ Convection currents bring partly molten material from the mantle into the magma chambers. Some sinks back into the mantle; the rest is ejected to the surface where, over time, a ridge develops.

**STAGES IN SEA-FLOOR SPREADING**

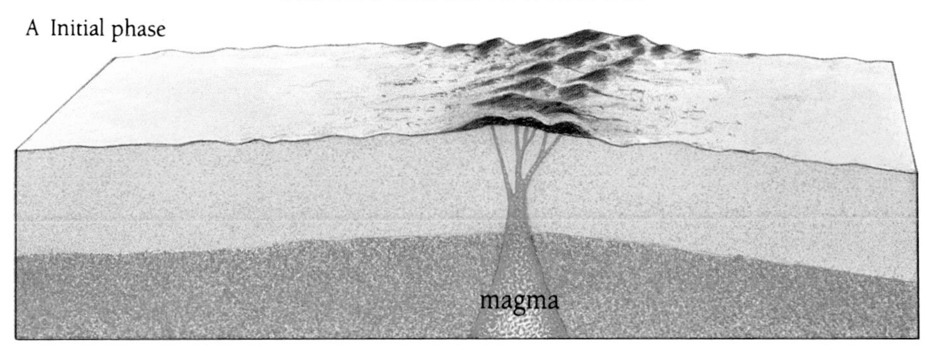


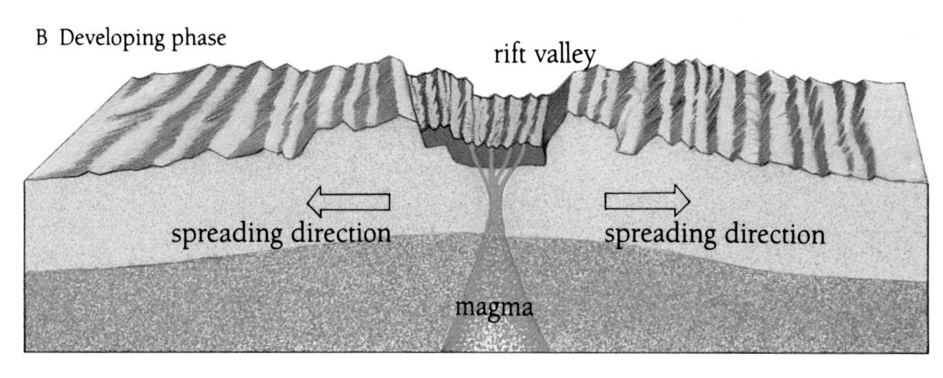


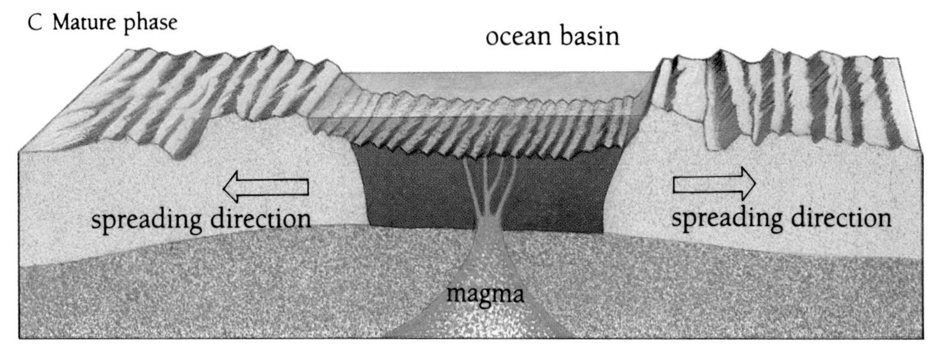

# DISCOVERING THE OCEAN'S SECRETS

In the early 1970s, global plate tectonics seemed to make sense, but important questions remained about how the process actually worked. The only way to find out was to examine the formations themselves in detail. A century before, it would not have been possible, but from 1972 to 1975, teams of scientists from France and the United States cooperated in the appropriately named "Project Famous" (for French American Mid Ocean Undersea Study) using submersibles to descend 2,000 to 3,000 meters (6,500–10,000 feet) below the surface for direct observation and sampling. Meter by meter, scientists were able to examine rock on the floor and walls of the median valley of the mid-ocean ridge. British and Canadian teams provided preliminary topographic mapping of the study area using precision echosounding, while magnetic patterns were surveyed using an airborne magnetometer. In addition, huge side-scan sonars operated from a British research ship mapped 15-kilometer (9-mile) swaths of the sea floor, and remotely operated cameras were towed back and forth over the study areas.

In 1975 the submersibles made numerous visits to the median valley and associated transform faults. Geologists were able to view fissures and cracks that run parallel to the median valley, and from within the submersibles operated special drilling and coring devices to sample from selected sites. Dark, cloud-shaped lumps of rock, aptly termed pillow lava, were found lying right over the area where it was thought injections of new crust were received from below.

Evidence provided by magnetic patterns and direct observations of rift areas gave strong support to the idea that plate movements have been going on for millions of years—and are still going on. But further important clues were derived from what some have called one of the most successful scientific experiments ever conducted. From 1968 to 1975, during the Deep Sea Drilling Project, more than 400 cores were drilled into the structure of the sea floor from every ocean during 44 cruises of the deep-sea drilling ship *Glomar Challenger*. Technically, this was a tremendous accomplishment, especially since few holes had been previously drilled in very deep water. Scientifically, the results made history.

On one cruise, drilling was directed at a location whose age—38 million years—had been predicted based on the mid-ocean spreading theory and the magnetic patterns on the sea floor. Dating of a core sample indicated the true figure to be remarkably close, geologically speaking—39 million years. Analysis of cores taken from the Atlantic, Pacific, Indian, and Antarctic oceans all gave important new and supportive evidence favoring the sea-floor spreading concept, and plate tectonics moved into the realm of general acceptance by Earth scientists. In the years since the heady "early days" of developing and confirming theories of plate tectonics, there has been much refinement and additional evidence that supports the basic theory. Some suggest that little now remains for future generations of scientists to do in this area of research, except for tying up some loose ends, but as with other exciting revelations, one mystery solved leads to numerous other profound questions.

SYLVIA A. EARLE

Al Giddings/Ocean Images

Exploring the Galapagos Rift from the research submersible *Alvin*.

rigid plates." This made sense to Dr Jason Morgan, a young geologist, who in 1967 laid the last piece of the puzzle before his colleagues in Washington, DC by demonstrating how all the Earth's plates are spreading and rotating around their individual poles.

TECTONIC FORCES

Three major structural (or "tectonic") features form within the Earth's crust: mountains and trenches, oceanic ridges, and faults with large horizontal movements. The oceanic ridges extend continuously through the great ocean basins. Deep trenches edge the basins where the oceanic crust meets the continents. Horizontal shear faults, called "transform faults" connect all the trenches and ridges around the world to form large, rigid crustal plates.

All these plates are moving away from "spreading centers" at an average rate of 6 centimeters (2.5 inches) per year—the growth rate of a fingernail. The main spreading centers are the oceanic ridges: the Mid-Atlantic Ridge, the East Pacific Rise, and the Mid-Indian Ridge. There are also smaller areas of spreading, including the Gulf of California where the peninsula of Baja California is moving away from Mexico, and the Red Sea–Gulf of Oman where Africa is moving away from the Middle East lands. As the crust spreads away from the ridges, new crust is formed by lava erupting from the Earth's mantle. This new crust of hot, molten rock creates the mountainous mid-oceanic ridges.

At the boundary opposite the ridges, the moving sea floor collides with continental landmasses, and is forced downward to form deep ocean trenches which in turn cause the formation of continental mountains. An example of this huge ocean-bordered mountain range is the Andes of South America.

Since 1967, with the explosion of space-age technologies, scientists can predict what the Earth will look like millions of years from now. Computer models suggest that in 250 million years, all the continents will once again be united as one connected landmass. Can this be the second (or third, or fourth) coming of Pangaea?

ROBERT E. STEVENSON

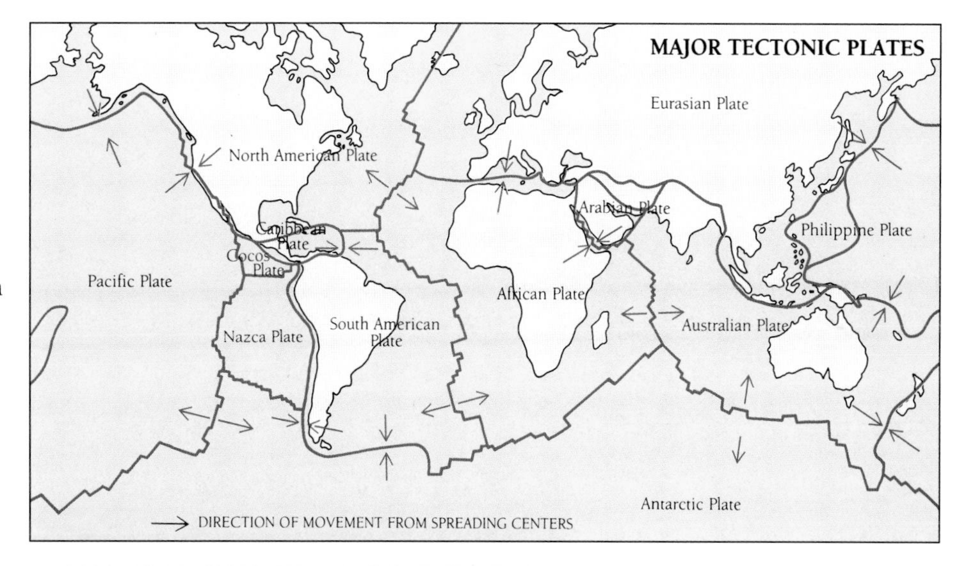


▲ The Earth's surface consists of six major tectonic plates, and a number of smaller ones. Most earthquakes and volcanoes occur at the boundaries of these plates, the result of either the destruction of old material at trenches, or the upwelling of new material at ridges.

NASA

◄ Looking south, a satellite view of the Red Sea. The twin forks of the Gulf of Suez on the right and the Gulf of Aqaba on the left are separated by the rugged mountains of the Sinai peninsula. Five million years ago the Red Sea was a shallow basin. Now widening, as the sea floor spreads to separate the Saudi peninsula from continental Africa, it is considered to be an embryonic ocean.

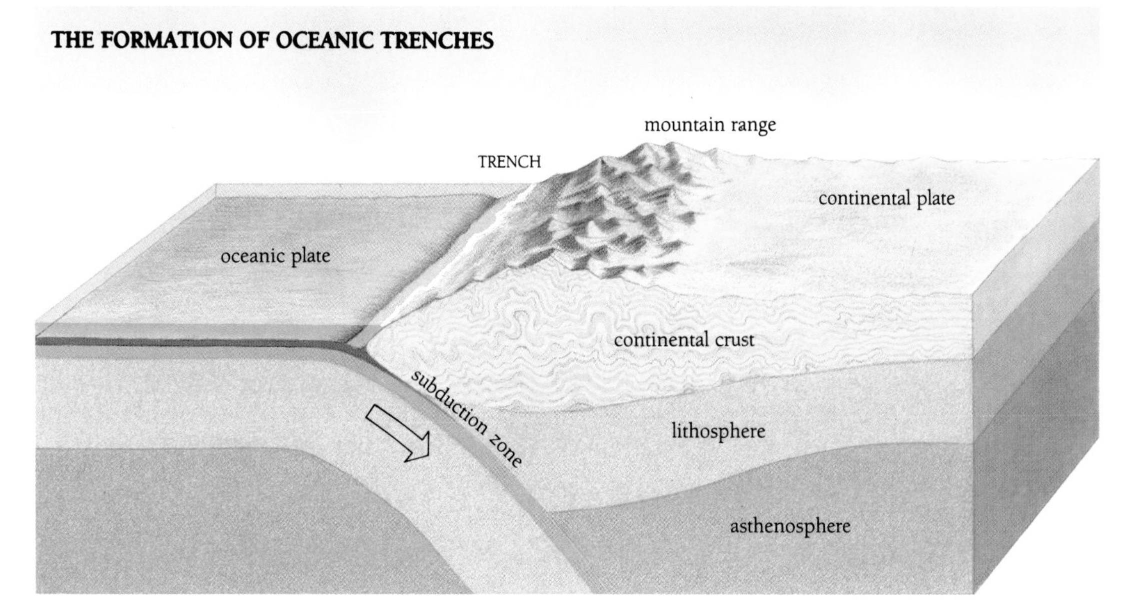


◄ Trenches can occur either when two oceanic plates converge, or, as here, when an oceanic plate meets a continental plate. As the two plates collide, the oceanic plate slides underneath the continental plate, and is carried downward to form a trench. The lighter continental plate is pushed upward to create a mountain range. The Peru-Chile Trench at the conjunction of the Nazca and South American plates is an example of these forces at work: the Andes mountain range is the result.

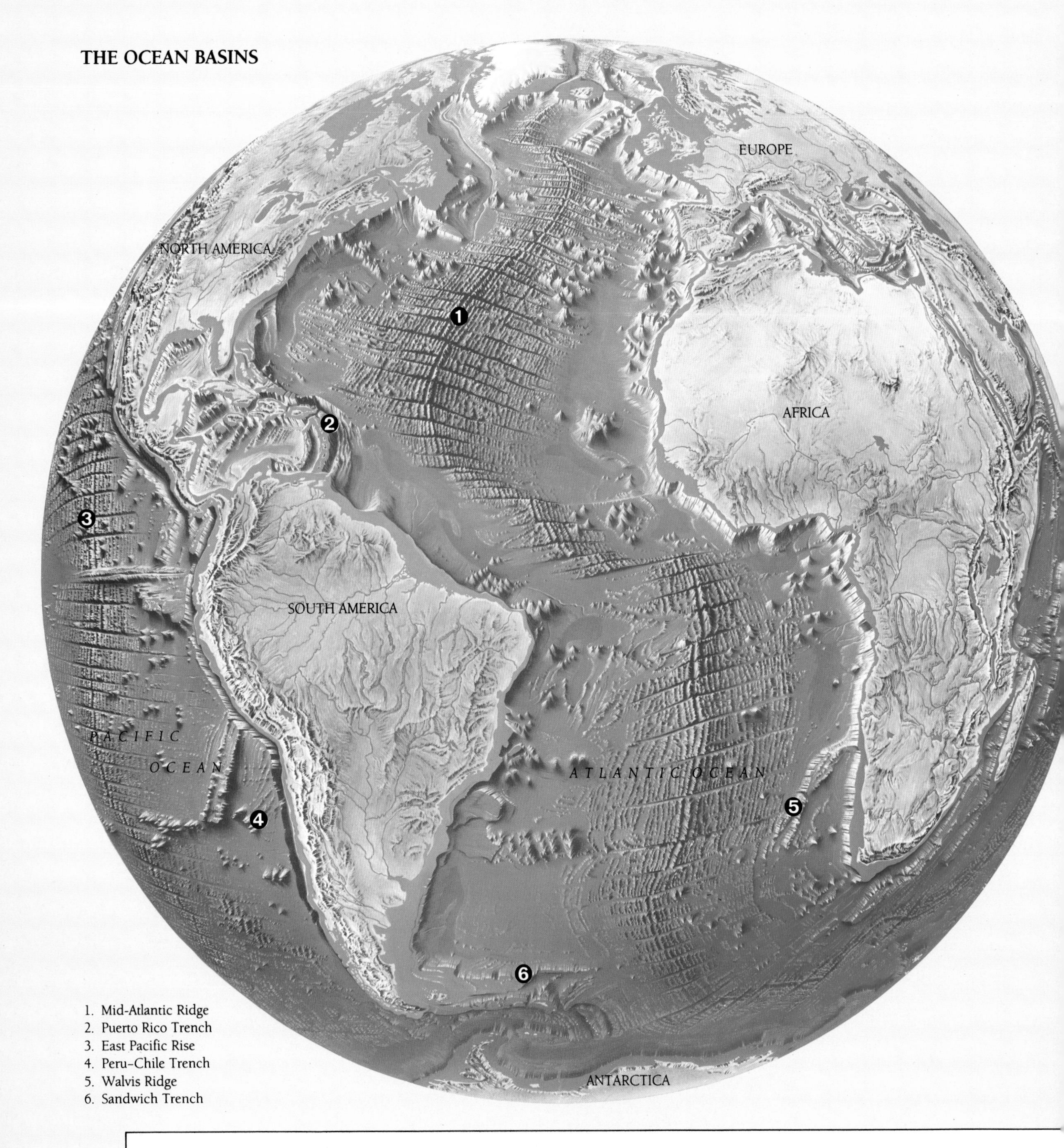


1. Mid-Atlantic Ridge
2. Puerto Rico Trench
3. East Pacific Rise
4. Peru-Chile Trench
5. Walvis Ridge
6. Sandwich Trench

THE ATLANTIC OCEAN

The most prominent feature of the Atlantic Basin is the Mid-Atlantic Ridge—the longest mountain range in the world—which curves through the basin, almost parallel to the coast of America on one side, and the coasts of Europe and Africa on the other. The ridge is an area of great geological stress; along its length new oceanic crust is being created as the sea floor spreads away from it. The ridge is bisected by a multitude of fractures, which are the center of intense earthquake and volcanic activity. Other, smaller ridges divide the east and west basins of the ocean, chief of which is the Walvis Ridge off the coast of Africa. The British Isles are continental islands, still linked under the sea to the continent of Europe. The Azores Islands, on the other hand, are the results of the rise of the Mid-Atlantic Ridge above sea level during a period of volcanism. There are two trenches in the Atlantic Basin, the deeper of which is the Puerto Rico Trench.

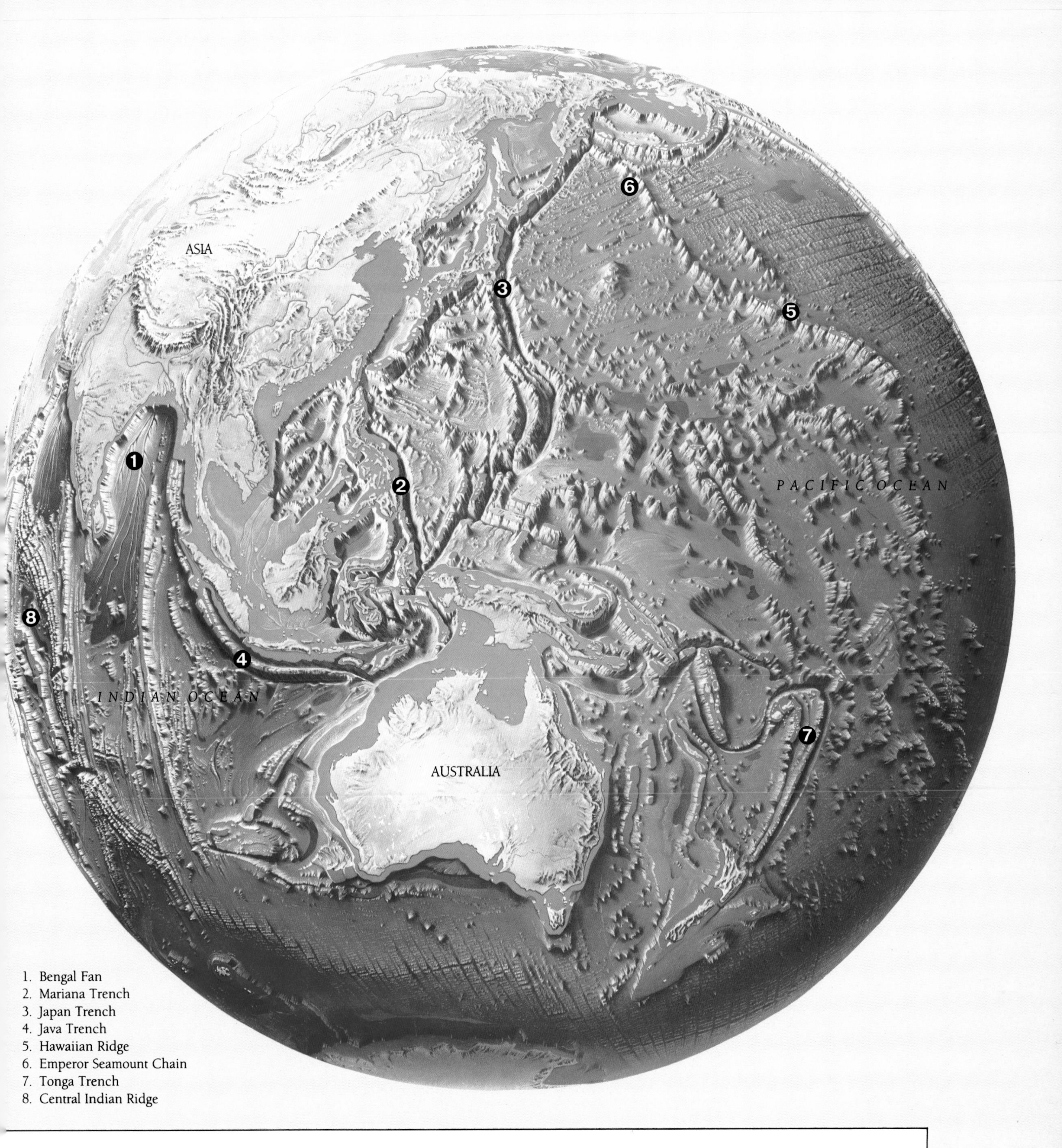


1. Bengal Fan
2. Mariana Trench
3. Japan Trench
4. Java Trench
5. Hawaiian Ridge
6. Emperor Seamount Chain
7. Tonga Trench
8. Central Indian Ridge

THE PACIFIC AND INDIAN OCEANS

The Pacific is the largest and deepest of the oceans, and the western Pacific—particularly the northwestern section—is notable for the depth of the sea floor: the Mariana Trench, the deepest part of the world ocean, is found here. Great fracture zones dissect the northern part of the Pacific Basin. The ridge known as the East Pacific Rise is the dominant feature of the eastern section of the basin. The Pacific rim is ringed with a circle of volcanoes—both active and dormant—which are the result of geological stresses at the conjunction of lithospheric plates. Almost all the oceanic islands of the Pacific are of volcanic origin. The Indian Ocean is the smallest of the three major oceans, and is largely confined to the Southern Hemisphere. Its greatest depths are in troughs near the Sunda Strait and the Andaman Islands. East of the Indian subcontinent lies the Bengal Fan, a huge deposit of sediment on the sea floor.

Nicholas Devore/Photographers Aspen

▲ The atoll of Tetiaroa, with its semicircular reef dotted with irregular islands, lies to the east of the volcanic chain of the Society Islands. Typically, atolls form when an extinct volcano (once in the center of the lagoon) sinks beneath the waves, through either erosion or subsidence of the crustal plate beneath it.

## THE OCEAN BASINS

Although they are interconnected, the three major ocean basins of the world—the Pacific, the Atlantic, and the Indian—have distinctive outlines. The smaller marginal seas of the Gulf of Mexico, the Caribbean Sea, the Mediterranean Sea, and the North Polar Sea, as well as the marginal basins landward of the island arcs off eastern Asia and Australia, all have pronounced barriers that partially separate them from the major oceans.

### THE PACIFIC BASIN

The Pacific, the largest and deepest of the oceans, covers 166 million square kilometers (64 million square miles) and has a mean depth of 4,188 meters (13,737 feet). The circular Pacific lacks the symmetry of the Atlantic, which curves proportionally with the landmasses that surround it. In contrast, the eastern Pacific, north of the tip of Baja California, is dominated by long east-west fracture zones that intersect North America at right angles. South of Baja, the East Pacific Rise and its fractures bend gently westward, cross the south Pacific and merge with the Indian Ocean Ridge south of Australia.

The sea floor of the central Pacific, both north and south, is made up of a number of ridges that trend northwest-southeast. These seamount chains are crowned in many places with islands: Hawaii in the north Pacific, and the coral atoll belts of the tropical and southwest Pacific. The island arcs and their associated trenches ring the Pacific Basin from southernmost New Zealand to the tip of Tierra del Fuego. The largest gap in this ring of trenches is between British Columbia and the southern end of the Gulf of California.

The East Pacific Rise is part of the mid-ocean ridge system of the major ocean basins. Its crest rises 2-3 kilometers (1-2 miles) above the sea floor, but in contrast with the Mid-Atlantic Ridge, the East Pacific Rise is thousands of kilometers wide with gradual slopes. The ridges and troughs parallel to the crest do not have the relief of those in the Atlantic, the flanks are covered with volcanoes, and the crest is almost devoid of sediments.

The Gulf of California began forming about 6 million years ago as Baja California split away from the mainland of Mexico. The splitting continues today as Baja moves west at about 6 centimeters (2.5 inches) per year. The East Pacific Rise, the splitting Gulf of California, and the San Andreas Fault in California are interconnected. The East Pacific Rise interacts with the North American continent by the same mechanism that has created the rift valleys of Africa and Israel's Dead Sea and Sea of Galilee!

From the international date line to the Americas, and from Punta Gorda south nearly to the equator, the crust of the Pacific sea floor is ripped by the greatest fractures on earth. The Clipperton Fracture Zone, for example, extends past the Line Islands, south of Hawaii, and through the Phoenix Islands across the equator. The fracture zones are basically parallel to each other, evenly spaced, with ridges and depressions along their length. Where the fractures cross the East Pacific Rise, the sea floor is displaced in an east-west direction and it is here that great earthquake activity occurs. The Mendocino Fracture Zone has the largest escarpment, the sea floor on the south side being some 1,200 meters (3,935 feet) deeper than on the north.

Although at first glance the seamounts, coral reefs, and volcanic islands of the central Pacific seem to be scattered like so many dobs of mud, they are actually aligned on north-south and northwest-southeast volcanic ridges. The longest is the Hawaiian Ridge, which extends northwest for 4,500 kilometers (2,790 miles) from the island of Hawaii, connecting there with the Emperor Seamount Chain which continues north until it disappears at the junction of the Kuril and Aleutian trenches. South of Hawaii are the Line Islands, including Kiritimati, the oldest known atoll. South and west of Kiritimati are the great series of coral reefs that make up the Marshall and Caroline islands and Kiribati. South of the equator, atop linear ridges, lie the Tuamotos, Samoa, the Society Islands, and the Austral Ridge. The high islands of all the ridges are volcanic, and borings confirm that volcanic material underlies the coral reefs.

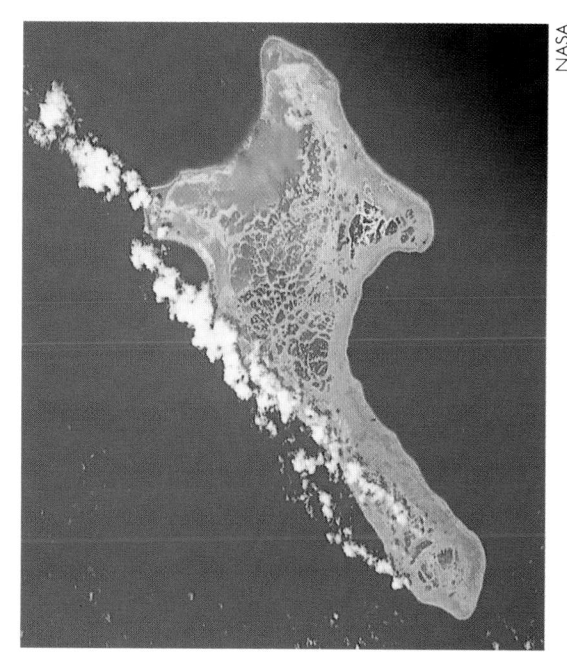

NASA

◄ Kiritimati—one of the Line Island group south of Hawaii—is the oldest known atoll. Evidence of its geological age was clearly seen when the space shuttle *Discovery* passed overhead—the lagoon is almost entirely filled by coral growth.

The islands and submerged reefs of the Hawaiian Ridge extending to Midway are all volcanic, and detailed studies of the ages of the eruptive rocks indicate that the vulcanism has progressed southeast at a rate of about 15 centimeters (6 inches) per year. The implication is that there is a "hot spot" in the oceanic crustal layer and that the sea floor has moved to the northwest over this spot to create the volcanic peaks. The "Big Island" of Hawaii currently sits on a hot spot, evidenced by Kilauea's recent

Krafft-Explorer

◄ Dawn on Mauna Loa, Hawaii, reveals glowing trails, many kilometers long—one of the frequent and famous spurts of flaming lava from the Mokuaweoweo caldera.

eruptions. In the past decade, scientists have encountered sea-floor eruptions south of the "Big Island" which may indicate that Hawaii is moving off the center of this eruptive hot spot.

To the south and west of Hawaii is a vast area of coral reefs and volcanic peaks. Many of the peaks have flat tops, suggesting erosion by surface-wave action during an earlier geologic time when they were at the sea's surface. These flat-topped peaks, called guyots after the Swiss geologist Arnold Guyot, have average depths of 1,200 meters (3,935 feet). If the guyots were eroded by waves, then the whole central Pacific must have subsided by 1,200 meters (3,935 feet) too! But we have no evidence of any such subsidence, or how it may have happened.

The most geologically active structures in the world are the island arcs and their accompanying trenches in the western Pacific, the great trench off Indonesia and that bordering Peru and Chile. They make up the deepest parts of the ocean basins, and experience the greatest amount of earthquake activity. There are gaps in the trenches adjacent to North and South America, especially along the United States and Canada. The southern termination of the trench off Chile has a structural relationship to the only other trench in the Atlantic: the South Sandwich Arc.

The western Pacific trenches are much deeper than those along the borders of the eastern Pacific. Starting near the Kamcatka peninsula on the eastern fringe of the Soviet Union, they form a disconnected series south to the Kermadec Islands north of New Zealand. The similarity of the depths in the trenches is striking. Except for that off New Britain, which bottoms at 8,320 meters (27,300 feet), they are all deeper than 9,000 meters (29,520 feet), with the Mariana Trench taking the "depth prize" at 10,860 meters (35,620 feet), a bare distinction over the Tonga Trench at 10,800 meters (35,425 feet) deep.

## THE ATLANTIC BASIN

The elongated Atlantic, with its curving parallel shores, has an area of 86 million square kilometers (33 million square miles) and a mean depth of 3,736 meters (12,254 feet). The borders of the ocean are remarkably symmetrical, a symmetry mirrored by the Mid-Atlantic Ridge which bisects the basin. The ridge forms half of the abyssal depths; the rest are basins that lie on its flanks. The mean relief of the ridge above the basins is close to 700 meters (2,300 feet), but in some places it stands as much as 3,000 meters (9,840 feet) above an adjacent plain.

Scientists discovered a central ridge in the north Atlantic in 1873. After the First World War, and the development of acoustic depth finders, the ridge was mapped to the south Atlantic. Later, with the use of precision depth recorders, it became clear that the Mid-Atlantic Ridge was part of a world-encircling, submarine mountain system. Today we can view this dynamic ocean ridge system using radar altimetry.

The central rift valley of the ridge is the focus of numerous earthquakes, indicative of the tensile forces that pull the ocean crust apart. The central portion of the rift contains recent lavas, and at many places, hot gases and magma pour out onto the sea floor, constantly forming new rocks, highly mineralized deposits, and nutritive bottom water.

The mid-oceanic ridge is cut by a great number of transverse fractures with deep, elongated valleys that offset the ridge crest. These are epicenters for nearly continuous earthquake activity adjacent to the ridge, and have long

▼ The sea floor is studded by seamounts and mighty peaks, dissected by ridges, and cut by plunging trenches. The range and magnitude of ocean formations place the world's highest land mountain, Mt Everest, and one of its deepest canyons, the Grand Canyon, into perspective.

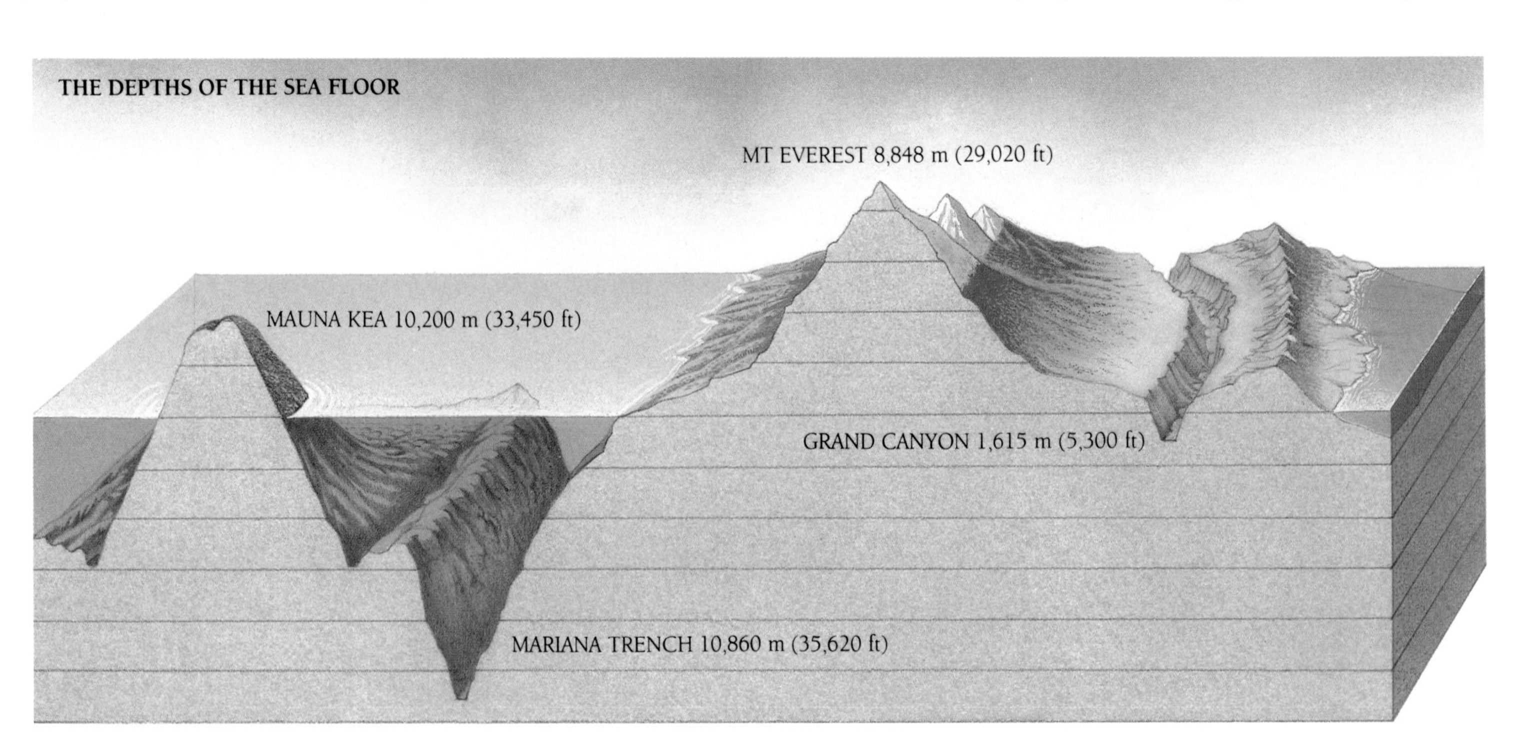

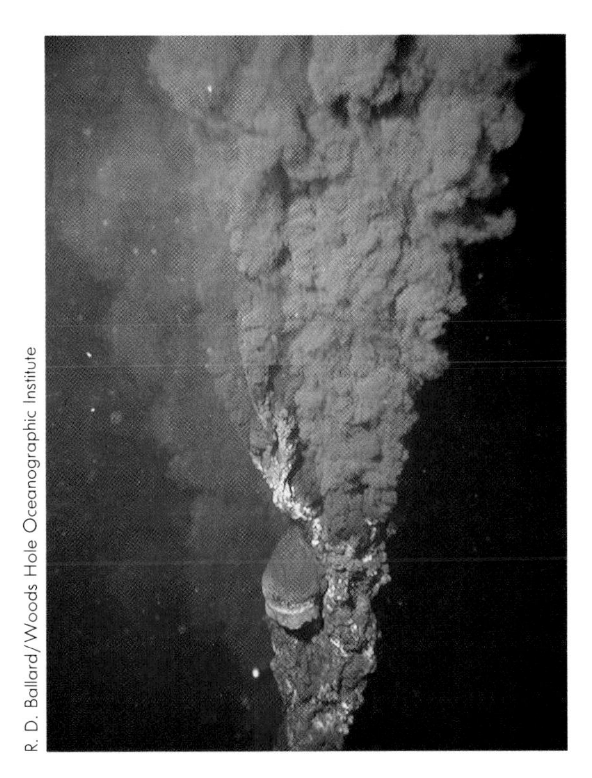

R. D. Ballard/Woods Hole Oceanographic Institute

◀ The mineralized stack of this black smoker (hydrothermal vent) on the East Pacific Rise spews out an inky stream of steaming water and hydrogen sulfide. Material from the fountain is utilized by the sulfur-oxidizing bacteria that are found in many of the animals inhabiting hydrothermal vent areas.

histories of vulcanism, some of which has produced seamount chains, as off New England, or punched islands to sea level, as in the Azores and St Helena. The fractures sometimes extend into the continents bordering the ocean. The Romanche Fracture, cutting the ridge at the equator, is the deepest at 7,856 meters (25,768 feet). The Falkland Fracture Zone is the longest, extending to the very tip of South Africa and intersecting the continental margin off Argentina through the Falkland Plateau.

Of the two trenches in the Atlantic Ocean, the Puerto Rico Trench is the deeper: some 9,200 meters (30,175 feet). Scientists have traced it completely around the volcanic and seismic arc of the West Indies. To the south, the deep-sea boundary of the arc of the Sandwich Trench connects South America with Antarctica. This trench is continuously active with volcanic eruptions and earthquakes, and has a depth of 8,264 meters (27,106 feet).

THE INDIAN BASIN

The Indian Ocean covers an area of 73 million square kilometers (28 million square miles), and has a mean depth of 3,872 meters (12,700 feet). Its most striking sea-floor feature is an inverted Y-shaped ridge, the counterpart to the Mid-Atlantic Ridge and the East Pacific Rise. One arm of the "Y", the Southwest Indian Ridge, extends around Africa to connect with the Mid-Atlantic Ridge. The other arm, the Mid-Indian Ridge, bends around Australia to join the East Pacific Rise. North of the "Y", the Central Indian Ridge continues north toward the Gulf of Aden, and bends into the Gulf, as the Carlsberg Ridge, to join with the rift valleys of Africa and the spreading Red Sea.

Bounding the east side of the Bay of Bengal and its huge sedimentary deposit, the Bengal Fan, is the strikingly straight, north-south, Ninetyeast Ridge. To the west of India is another linear ridge, more or less north-south, atop which lie the Laccadive and Maldive islands. In the western Indian Ocean, northeast of the island of Madagascar, is the unusual Mascarene Plateau. The rocks of this plateau, including the Seychelles, are granitic and of continental origin, giving the plateau the appearance of being a mini-continent amid an ocean basin. South of Madagascar, coral reefs lie on top of volcanic rocks which are clearly part of the ocean basin. Yet further south, the Agulhas Plateau off South Africa, another mini-continent, is made up of rocks similar to those of the Seychelles.

The Bengal Fan in the eastern part of the Indian Ocean is unique. In no other place has the sea floor been so influenced by deposition from the adjacent land—in this case, from the monumental discharges of the Ganges-Brahmaputra rivers. Extending for 2,000 kilometers (1,240 miles) from their great delta to the south of Sri Lanka, this massive deposit on the sea floor slopes at a nearly even gradient of 1.5 meters per kilometer (8 feet per mile). Numerous branching channels criss-cross the fan, indicating the continuous slow currents that carry the land-derived sediments over its entire length. It is without question the most gigantic pile of sediments in the world!

ROBERT E. STEVENSON

# WAVES, TIDES, AND CURRENTS

The oceans were formed over the millennia from steam given off from the interior of the Earth by the action of volcanoes, and they acquired their salty composition by the continual weathering and leaching of the rocks through countless cycles of evaporation and precipitation. This has resulted in the composition of sea water being 3.5 percent by weight of dissolved salts, a percentage that remains constant within very narrow limits throughout the world's oceans.

The oceans act as giant heat reservoirs that capture and then slowly release the radiated energy from the sun, and are the most important controlling factor in the long-term weather patterns, or climate, around the world. This huge global thermostat strongly moderates the range of temperatures and seasonal fluctuations, thereby facilitating the existence of many forms of life. Even small changes in the temperature of the oceans, or their currents, or even in their height, can have dramatic effects on the remaining one-third of the Earth's surface, the land.

Eric Crighton/Bruce Coleman Limited

▲ At any point in time, each water molecule in the ocean travels only a short distance in a circular motion. However, this movement transmits energy between molecules, and enables the surface water oscillations (visible waves)—set up by wind sweeping the open ocean—to be transmitted great distances. These wind-generated waves are hurled against the land, causing erosion and thus shaping our coastlines.

## AN OCEAN IN MOTION

Whether you go to sea in a small boat on the Sea of Galilee as recounted in the Bible or travel across the north Atlantic in a luxury liner, you are forever aware that the ocean is in motion. From the lightest breeze through to the wildest gale, the ocean's surface changes accordingly, from a gentle ballet of ruffles to an awe-inspiring and sometimes frightening vista of waves. These wind-created waves, known to mariners as the "sea", are measured by a scale called the "sea state" and logged in numeric values from zero through nine, depending on the wave height: zero sea state corresponds to a calm mirror-like surface, while sea state nine signifies waves over 14 meters (45 feet) in height. In addition, distant storms telegraph their presence by another, more regular, set of waves called "the swell". All these motions can produce the feared *mal de mer* or sea-sickness which is not unlike, but is certainly different from, both air-sickness and space-sickness.

And if this is not enough, the traveler, having reached the safe haven of a port, can go ashore to celebrate, only to return to find that his or her ship is not where it was left; it can be as much as 12 meters (40 feet) higher or lower. This confusing syndrome is not brought about by the onshore celebration, but is a manifestation of the tide, which is difficult to observe when at sea but is quite obvious at or near the shore.

All these kinds of oceanic movement fall into the general category of waves, which are measured by their height (amplitude), frequency (period), and direction. Added to these are the huge rivers of the ocean, the major current systems, which transport water horizontally over great distances. And finally there is the lesser, vertical, motion of water throughout the volume of the ocean.

## SURFACE WAVES

As anyone who has been to sea will surely testify, almost as soon as the wind begins to blow the ocean responds by creating waves; and as the winds continue to blow these waves build up to such an extent that the ship will roll and pitch and heave. The extent of this build-up is governed by three factors: the speed of the wind; the length of time it has been blowing; and the fetch or distance over which it has traveled. The larger any of these three factors is, the higher the waves that will be generated. Ultimately, however, an equilibrium is reached where, for a given wind speed, the ocean surface can absorb no more energy and the waves break to dissipate any further energy.

Ocean "swell" refers to surface waves that propagate beyond both the region where they were generated and the local influence of the wind. In traveling a considerable distance these waves undergo an internal ordering so that they become far more regular in appearance and direction. This sorting process results from the spreading of the waves from the generating area under the influence of a phenomenon called dispersion, which causes the longer waves to travel faster than the shorter ones. Hence it is possible to find waves at some distance from a storm, where there may be no wind at all. But these waves are usually relatively smooth with rather long periods, in contrast to the confused waves with different periods that are typical within the area of the storm. Thus the direction of the swell can point the way to the location of a distant storm.

## TIDES: THE OCEAN'S PULSE

Tides are the pulse of the ocean. Their regular rhythm is caused by that most universal of all forces, gravity, and is due to the presence of both the moon and the sun. The force of gravity controls the motion of the planets in orbit, but it is the small differences in the gravitational attraction of the moon (and also of the sun) from one side of the Earth to the other that gives rise to the tide-producing forces. These differences result from the varying distances of points on the Earth's surface from the moon (and the sun). The net result is a slight bulging of the ocean both directly in line with the moon (and the sun) and also on the opposite side of the Earth.

This twin effect is a result of the difference in gravity being positive on the "near" side and negative on the "far" side; hence water on each side bulges outward relative to the orbital path of the center of mass of the Earth. The combination of these bulging motions, with the sun contributing about half that of the moon, together with the rotation of the Earth about its own axis, produces the typical semidiurnal or half-daily tides with which we are familiar.

Of course the local shape of the coastline, the water depth, and whether the body of water is enclosed or not can also play a large part in determining the size of the tides. This came as a great surprise to Julius Caesar when he first sailed to Britain in 55 BC, since the tidal range in the Mediterranean with which he was familiar is only some 15 centimeters (6 inches) compared with a range of over a meter along the English coast. Moreover, in confined areas, the ebb and flow of the tides can produce significant tidal currents which may reach values of several knots.

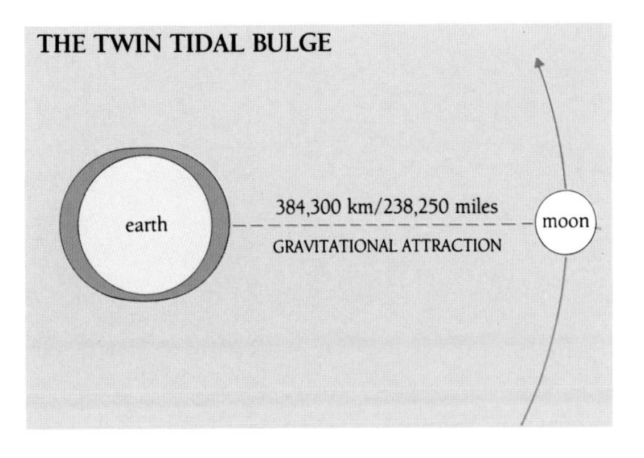


◂ The gravitational attraction between the Earth and the moon is greater on the side of the Earth facing the moon, thus creating a tidal bulge in the ocean. On the opposite side of the Earth, the corresponding centrifugal force forms another tidal bulge.

It should be noted that tides have nothing to do with tidal waves. These are a misnomer, and really refer to the waves set up in the ocean by underwater earthquakes: waves that can travel undiminished for thousands of kilometers across an ocean basin and which sometimes cause massive damage when they come ashore on a distant coastline. To differentiate, tidal waves are now more commonly called "tsunamis", the Japanese term for "tidal wave".

## OCEAN CURRENTS

Navigators have known for thousands of years that the ocean is characterized by variable winds and variable currents. However, it is only in the last half century that a reasonably cogent picture has emerged of the patterns of ocean currents and their underlying causes.

The major force setting up and maintaining the oceanic current system is that of the prevailing

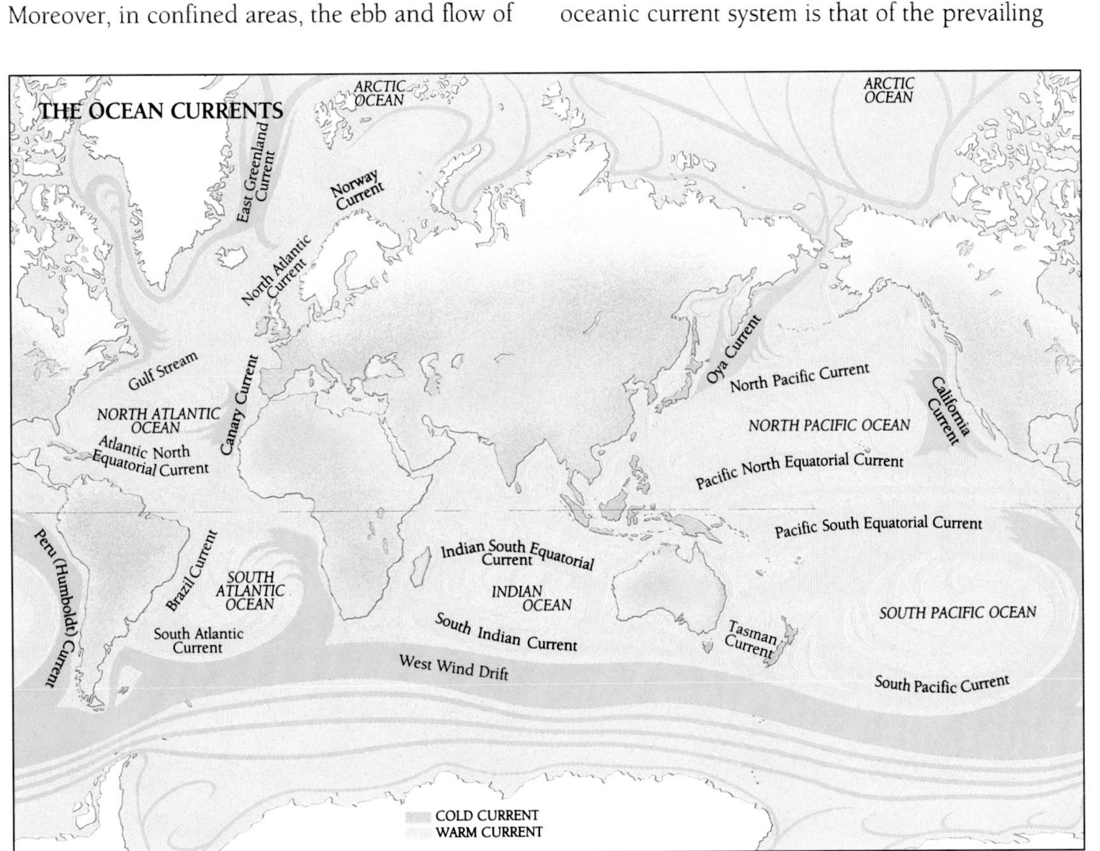


◂ Ocean currents result partly from the action of wind on the sea surface, and partly from convection processes deep within the ocean. The world's surface currents together make up five major gyres or broadly circular systems: the North Atlantic, South Atlantic, South Indian, North Pacific, and South Pacific.

Bill Wood/Planet Earth Pictures

▲ As this wave approaches a reef crest, the decreasing depth of water shortens the wave length and increases its height. Instability follows and the water spills over as a crashing breaker, causing considerable underwater turbulence.

winds, and the direction of current flow is governed by these winds and the fact that the Earth is itself rotating. However, the way in which currents manifest themselves is quite subtle.

Surface winds, especially over the oceans, tend to fall into a regular pattern: near the equator are the doldrums, followed, as you move north or south from the equator, by the trade winds, then the westerlies, and finally the polar easterlies. Basically these are zonal (east-west) bands of wind that alternate in direction: the trade winds blow essentially from the east in the band from the doldrums to about 30° of latitude, the westerlies from there to about 60°, and the polar easterlies from 60° of latitude to the poles.

The effect of these three bands of winds in each hemisphere is to create three regimes of ocean currents: the equatorial current system, the subtropical gyre, and the subpolar gyre. However, these current regions are not superimposed on the wind regions as you would at first expect, since the ocean does not respond directly to the stress of the wind but rather to the changes in the average winds with latitude. This is because a uniform zonal wind over an ocean would not produce a corresponding current, but would only cause water to pile up on the opposite side of the ocean basin. This results in the rather interesting phenomenon that the major dividing lines between the current systems are not at the latitudes where the winds are zero, but rather at the latitudes of maximum zonal winds. Hence the current regimes are bounded by the latitude bands of 0–15° for the equatorial current system, 15–45° for the subtropical gyre, and above 45° for the subpolar gyre.

Added to this, the currents are flowing on a rotating Earth. This introduces another force, called Coriolis, which deflects the currents to the right in the Northern Hemisphere and to the left in the Southern Hemisphere.

The consequence of the winds being zonally banded and hence changing with latitude, together with the rotation of the Earth, is twofold: firstly, that currents are indeed induced in the underlying ocean; and secondly, that these currents tend to join together and recirculate within each band, thus forming gyres (circular formations) across their domain of an ocean basin. Hence in the equatorial region, there is a gyre made up of the westerly-flowing equatorial

▶ Beaches form at the foot of cliffs (such as here in California), or between headlands, as a result of the transitory accumulation of debris from the erosion of rock or shells. There is usually some seasonal variation in the sediment cover on such beaches—depending on the strength or type of onshore wave patterns in the area.

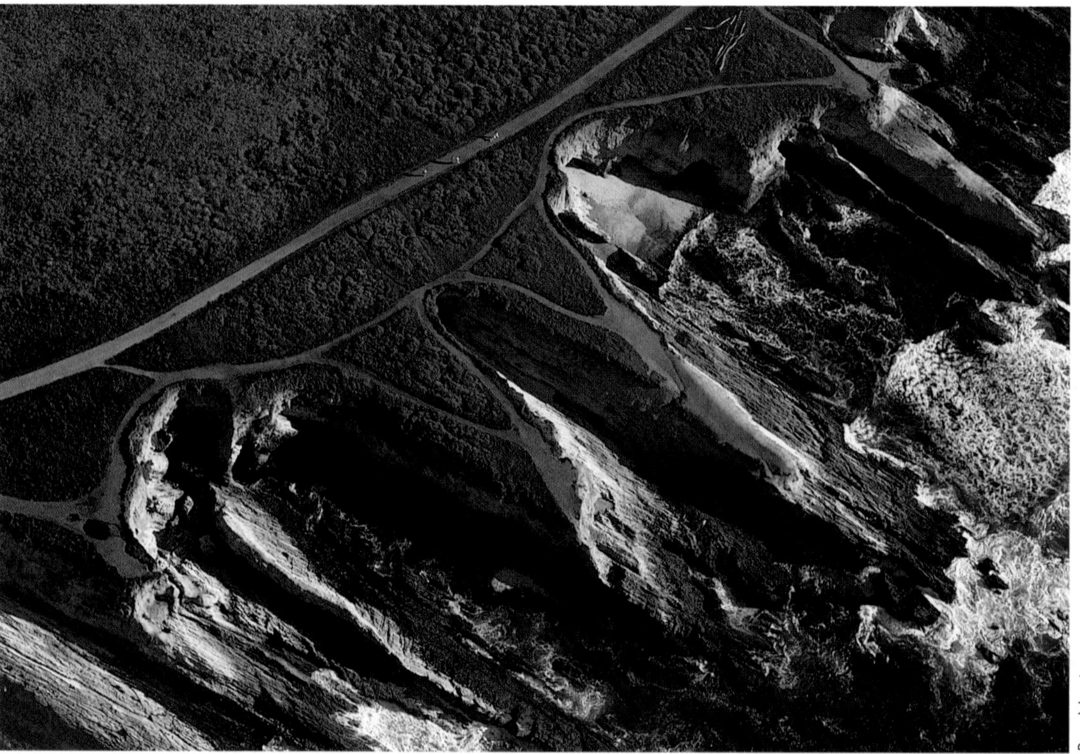

Reg Morrison

**THE FORMATION OF A GULF STREAM RING**

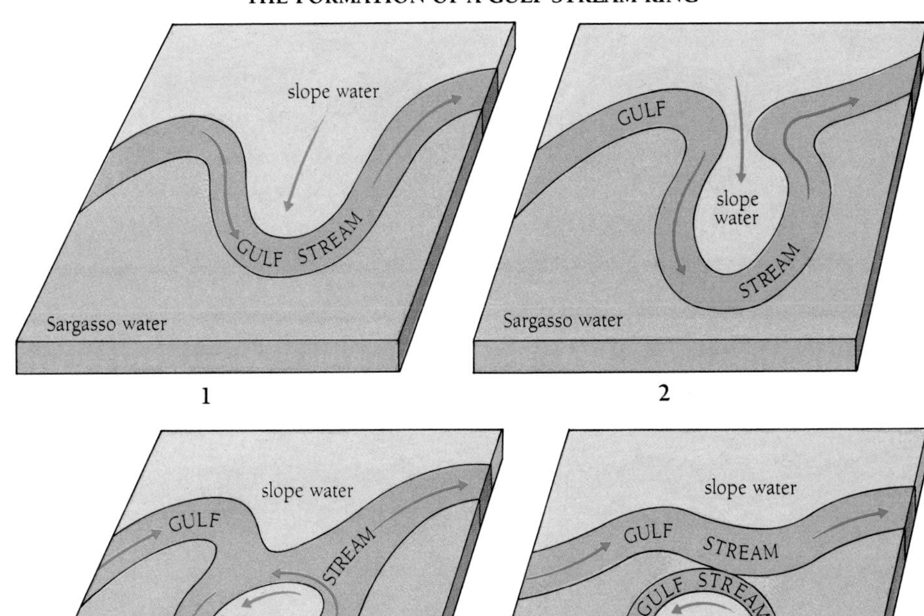


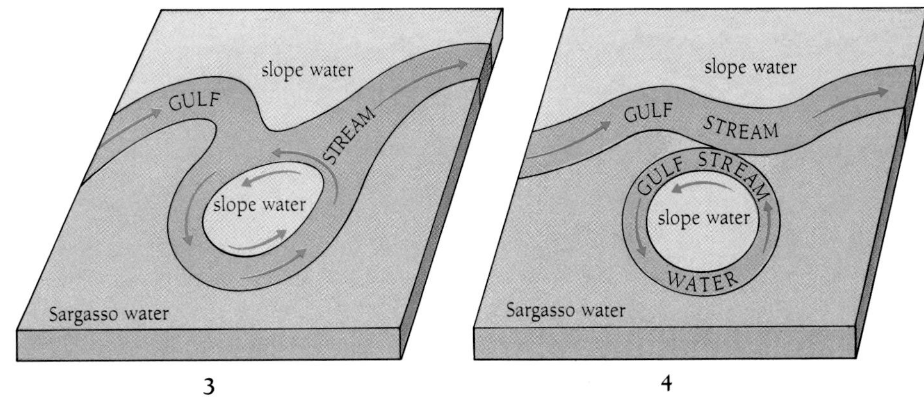


▲ The Gulf Stream, the strongest current in the North Atlantic Gyre, is characterized by meanders, eddies, and self-contained and sometimes long-lived loops or rings, which break off from the main current as a result of its speed and intensity.

current centered at about 15° latitude and a countercurrent flowing in the opposite (easterly) direction near the equator. Poleward (north and south), the subtropical gyre circulates in the opposite direction, being formed by the westerly equatorial current at 15° latitude and the easterly-flowing "west wind drift" at about 45° latitude. Further poleward, the circulation is again in the opposite direction, at least in the Northern Hemisphere where the ocean basins are closed off by the landmasses. In the Southern Hemisphere, the ocean basins are not similarly closed off, and the "west wind drifts" in each ocean join together to form the Antarctic Circumpolar Current, the only current in the world that completely circumnavigates the globe.

There is an additional, but most important, aspect of the Earth's rotation, which is that the amount of local rotation changes with latitude. To understand this, imagine yourself standing at the north pole, in which case you would be rotating in a counterclockwise direction; if you were standing at the south pole, however, you would be rotating in a clockwise direction. The local rotation changes direction at the equator, and indeed is zero there. This change in the direction of rotation causes the currents in the gyres to flow in opposite directions on each side of the equator; they are mirror images of each other.

Furthermore, the increasing local value of the rotation as you move from the equator to the poles causes the currents on the western sides of ocean basins to be much stronger than those on the eastern sides. This "western intensification" explains the existence of such mighty currents as the Gulf Stream off the east coast of North America and the Kuroshio off the east coast of Japan. An appreciation of the magnitude of these currents can be obtained by considering that the Gulf Stream has current speeds ranging from about half a knot to more than 3 knots, and carries about 135 billion liters (30 billion gallons) of water every second, which is about 65 times that of all the rivers in the world. In this sense therefore, these "rivers of the ocean" are far more grand and majestic than their land counterparts.

## RINGS AND SPIRAL EDDIES

The fact that ocean currents are so strong often leads to another phenomenon, that of ocean eddies. When the speed and intensity of, for example, the Gulf Stream reaches a critical value the current itself becomes unstable, oscillates in direction, and spins off an eddy about 160 kilometers (100 miles) in diameter by pinching off this large oscillation or meander. Because of their near-circular shape these eddies are sometimes called rings. They often move in the opposite direction to the main current that generated them, and can retain their identity for up to three years. Perhaps this is nature's way of maintaining stability, by first limiting energy build-up and then redistributing that energy within the ocean.

Recently, through observations made from space, another type of eddy has been discovered. These have a spiral shape rather than the near-circular shape of rings, and are considerably smaller, being only about 16 kilometers (10 miles) in diameter. Little is yet known of these spiral eddies, other than that they all have a similar structure and are being discovered increasingly in many areas of the ocean. They appear to be made up of highly sheared currents: the strength of the current changes in steps as you move out from the center of the eddy. Once again, it is perhaps nature's way of maintaining equilibrium, by isolating shears and the energy associated with them when they build up in the ocean.

## VERTICAL MOVEMENT

The oceans also support vertical motion to a limited but vital extent. To understand the magnitude of this, one must first consider the disparity between the scales involved. Ocean basins are typically several thousand kilometers wide, but are only several kilometers deep even at their deepest points; hence the vertical motions should be reduced by at least this same factor.

Added to this, the oceans are horizontally stratified, which means that they are layered horizontally. As you go deeper and deeper in the ocean, the temperature progressively drops from around 25° C (77° F) near the surface at low latitudes to close to 0°C (32° F) toward the

# MAPPING THE SEABED

Swath mapping involves the remote sensing of the seabed on both sides of a ship's track by acoustic (or sound) energy. This produces a map of a swath of the seabed rather than a single line of soundings. The sound energy is pulsed out into the water column and the intensity of energy that is reflected or "backscattered" from the seabed is recorded by the instrument. The deeper the water, the wider the swath that can be mapped.

Swath mapping has allowed complete pictures to be made of submarine mountain chains, of the slumps and debris flows that are eroding the edge of the continental shelves, and of channel systems that meander across the deep-sea floor.

Several instruments have been developed for this purpose. The main differences between them lie in the frequencies they use and the width of the swaths they map. Higher frequency instruments provide greater detail but can map only relatively narrow swaths because more energy is lost at high frequencies. At present individual instruments can utilize only one frequency. However, instruments with multiple frequencies, which will allow multispectral imaging, are now being developed.

The instruments also differ in the way they are deployed—they can be hull-mounted, deep tow, or shallow tow. Hull-mounted systems, which include multibeam echosounders and some side-scan sonars, are easy to use, but need correction for ship motion. Shallow tow instruments, such as GLORIA, are towed on a conducting cable at depths of up to 50 meters (165 feet) below the surface. High resolution systems are generally used in shallow water, but can also be deployed in deep waters. Deep tow instruments, such as Seamarc I, are generally less buoyant than shallow tow ones, or have a depressor weight attached which forces them close to the seabed. The deep tow systems can provide more detailed images of the deep-sea floor, but there is a higher risk that instruments will be lost as a result of impact with the seabed. Some towed instruments can also produce contours of water depth. Swath mapping is used to maximum advantage by running adjacent ship tracks, which creates overlapping swaths that can be merged into a mosaic. As a result, maps of the seabed can be produced on the same scale as satellite images of the land.

**THREE KINDS OF SEABED MAPPING INSTRUMENTS**

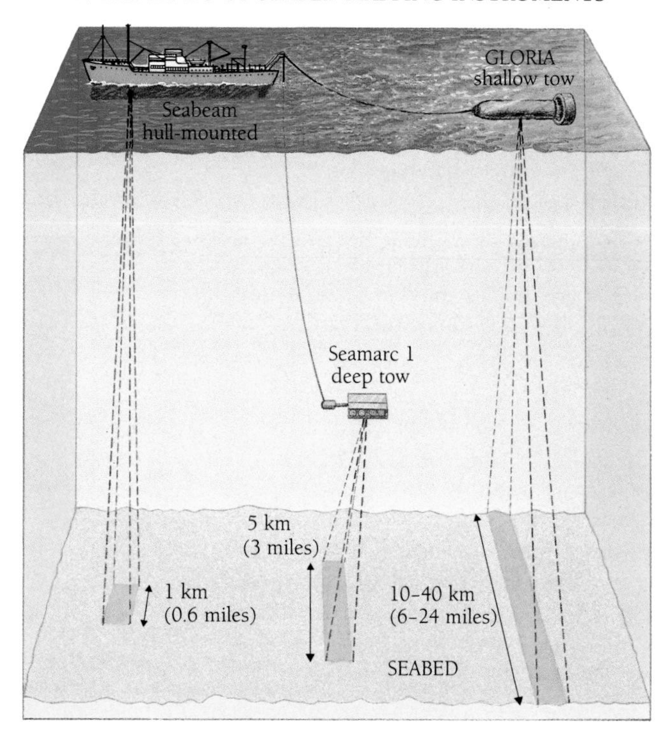


Two types of maps can be made. The first, known as bathymetry or water depth maps, uses swath echosounders, and are topographic in character. A swath echosounder produces a number (between 16 and 256) of depth measurements which are plotted as a series of contours on a map. Computer-generated perspectives of the seabed can also be made using this technique. The second type of map, known as acoustic character maps, uses side-scan sonars. The resulting images differentiate areas of contrasting backscatter and reflectivity of the acoustic energy. These variations depend primarily on the submarine topography, the texture, or surface roughness, and composition of the seabed.

DAVID JOHNSON

Marine Geophysical Laboratory, James Cook University

A 30-kilometer (19-mile) wide stretch of the seaward slope of the Great Barrier Reef, as measured by the side-scan sonar GLORIA.

bottom. Because the oceans are heated by the sun from the top down, they form horizontal layers with the warmer and hence less dense water near the surface, and colder and denser water further down. Since this is dynamically a very stable situation, there is very little tendency for this stability to be upset by vertical motion.

These combined effects severely limit the magnitude of vertical motions in the ocean. However, since the surface waters near the poles are themselves very cold, there does exist an extremely slow flow of this cold, dense water outward from the poles toward the equator, progressively sinking as it moves to its equilibrium depth. Over periods of hundreds of years this gradually replenishes the deep water throughout the oceans.

The most significant consequence of this vertical ordering in the oceans is that the horizontal currents, even the strong ones like the Gulf Stream and the Kuroshio, are significantly reduced with increasing depth, to such an extent that they are virtually nonexistent at depths of around 1,500 meters (4,900 feet). This is in strong contrast to what the situation would be in the absence of vertical structure, in which case the horizontal currents would extend from top to bottom of the ocean and remain undiminished in strength.

The only regions where this situation does not apply is where local winds blow the waters away from the coastline, and deeper waters must then flow vertically upward to fill the void. This is particularly evident along the eastern sides of ocean basins (the western sides of continents) where the local winds have a significant meridional, or north-south, component. Then, due to the effect of Coriolis, the water is transported away from the shore, and deeper, colder, and nutrient-laden water is brought up from below. These are known as upwelling regions, and are most important for the fishing industries in these areas since the planktonic food supply is thereby enriched.

Another, more subtle, effect of the vertical structure in the ocean is the existence of internal waves. This is brought about by the fact that the ocean temperature does not decrease uniformly with depth, but rather in three distinct steps. Near the surface, at depths down to several tens of meters, there is commonly a layer of warm water of constant temperature which has been uniformly mixed by the action of the winds. In the deep ocean the waters are cold with the

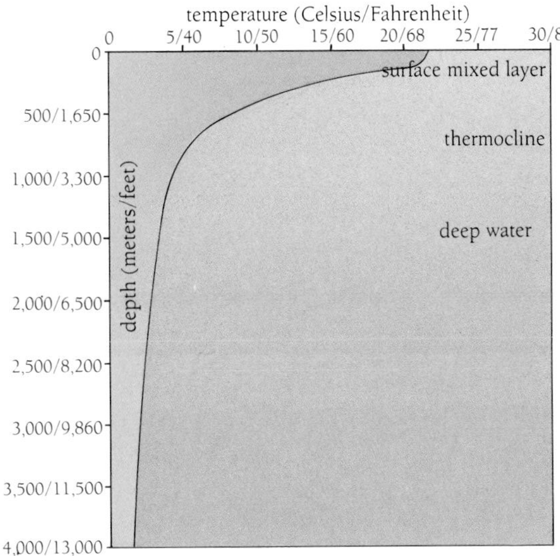


◂ A schematic view of ocean temperatures in tropical and subtropical waters. The thermocline, an area of rapid temperature change, separates the warm surface water from the cold water of the deep ocean below. Its precise depth varies according to latitude, season, and prevailing weather conditions. In polar latitudes the thermocline is absent, and there is little variation between surface and deep-water temperatures.

temperature decreasing only very slowly with depth. Between the warm upper mixed layer and the cold deep layer there is a third, transition, layer called the thermocline where the temperature, and hence density, changes rapidly with depth.

This rapid change in density produces conditions favorable to the propagation of waves, much like the change in density between the water and the air at the surface of the ocean which leads to surface waves. However, in this case it is not the winds that produce the waves but the "forcing function" of the tides. These internal waves are contained wholly within the thermocline and do not break the surface; but if conditions are right, they can have amplitudes much greater than surface waves. Hence, although of little consequence for ships, they can have marked effects on a submarine traveling at the depth of the internal wave.

## THE WAY AHEAD

The dynamics of the oceans present a rich tapestry of complex patterns which, given their diversity of size and persistence, make them difficult to monitor on a continuing basis.

Fortunately two modern technologies are now at the stage where they can combine to solve the problem: space remote sensing and ocean acoustic tomography, the latter being the equivalent of a CAT scan of the ocean using acoustic energy. On the one hand, space sensing allows the surface of the ocean to be measured regularly; while on the other acoustic tomography can tell us much about the interior of the ocean.

PAUL SCULLY-POWER

# 2 Ocean Wanderers

From minuscule plankton to mighty whales, creatures of many different kinds migrate seasonally through or over the world's oceans. Some are long-distance voyagers, traveling vast distances across the seas in response to biological impulses. Others migrate through smaller, more limited areas, but with a tenacity that continues to amaze observers. Seabirds cross the oceans high in the sky, following a course that is as predictable as the season. Turtles and seals move between feeding and breeding grounds. Communities of plankton, on the other hand, migrate vertically, moving up and down the water column in timeless natural rhythms.

Claude Carre/Jacana

▲ A microscopic view of zooplankton.

◄ Herring gull *Larus argentatus* in flight. This seabird has a worldwide distribution and is commonly seen winging its way above coastal waters.

## Plankton

Plankton is a body of small plants and animals that drifts in the ocean. Plankters (members of the plankton) range in size from minuscule microbes to jellyfish with a gelatinous "bell" up to 1 meter (3 feet) wide and tentacles extending over 3 meters (10 feet). The Greek word *plankton* means "wandering or roaming" and although plankton is considered to drift with currents and tides, most plankters are anything but passive. In fact, many animal and some plant plankters undergo daily vertical migrations over tens or sometimes even hundreds of meters.

### THE OCEANIC FOOD CHAIN

Phytoplankton ("plant plankton" in Greek) and zooplankton ("animal plankton") are important elements in all oceanic food chains. On land, through the process of photosynthesis, trees, grasses, shrubs, and flowers utilize sunlight, carbon dioxide, water, and nutrients to grow. The animals that eat them (herbivorous consumers) range in size from small insects to elephants. In the open ocean the plants (phytoplankton) are microscopic, but are often found in such high concentrations that the ocean can appear green. The main consumers of phytoplankton are microscopic animals rather than the large species found on land.

The most productive regions of the oceans are where upwelling water from the ocean depths brings nutrients toward the surface. Phytoplankton thrives under these conditions, and grazing zooplankton multiplies with the

▼ The oceanic food chain enables the productivity of phytoplankton to be transferred to zooplankton, and then to increasingly powerful carnivorous predators. Feces and animal remains fall to the bottom to provide nutrients for bottom-dwelling organisms. The amount and direction of transfer vary with latitude, time of year, and climatic conditions.

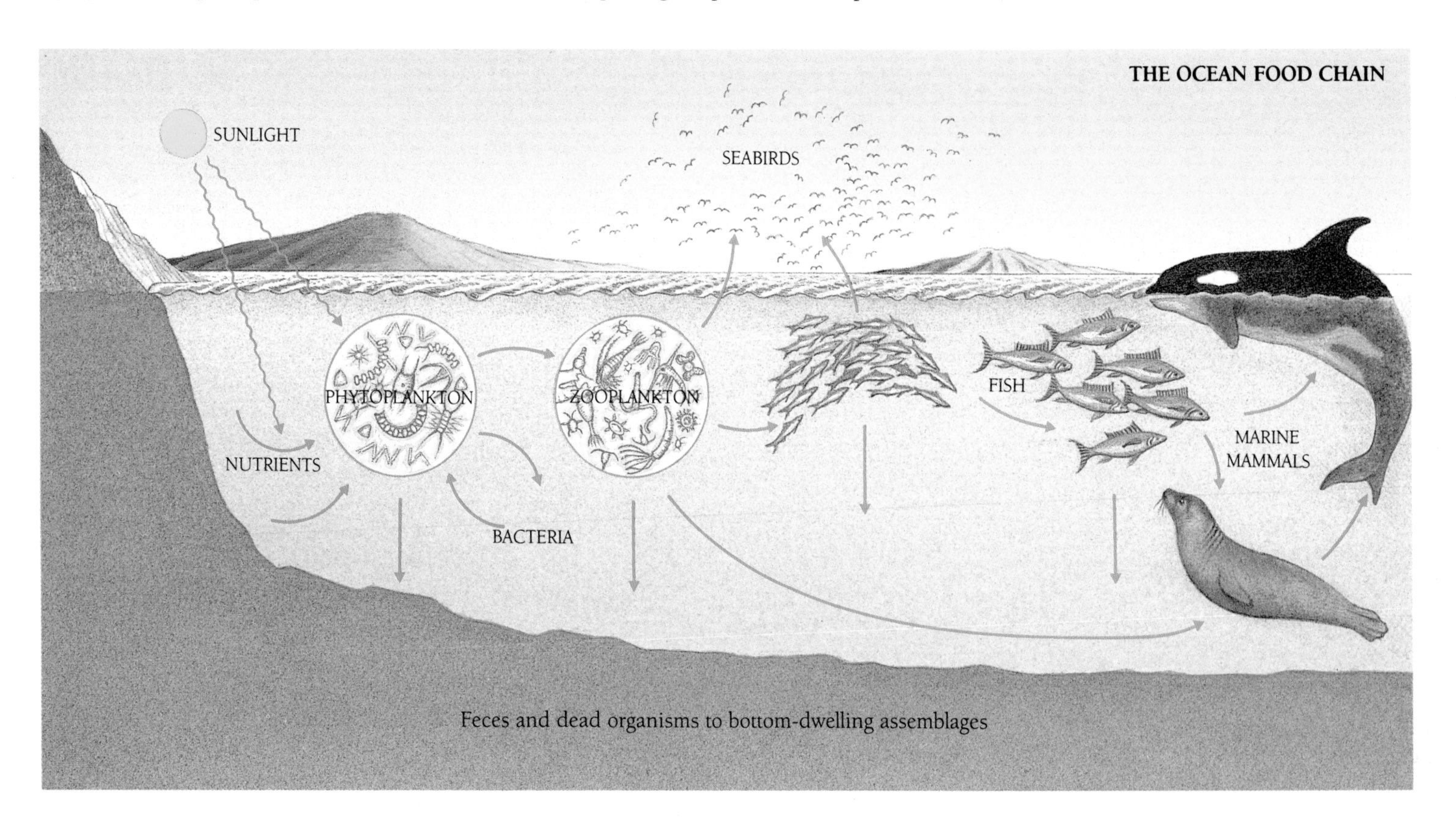

abundance of food. Herring, sardines, and anchovy feed on the abundant plankton and are in turn preyed upon by large fish and birds. Some of the world's richest fisheries are found in areas of upwelling.

If upwelling does not occur, the effects on the animals that depend on plankton resources can be disastrous, and fisheries generally suffer as a consequence. The so-called El Nino phenomenon along the coast of Peru has been shown to result in a disruption to normal upwelling. In the longer term, scientists are concerned that changes in weather patterns caused by an increased concentration of greenhouse gases in the atmosphere will alter cycles of upwelling. The Earth's supply of atmospheric oxygen depends in part on phytoplankton, which, like other plants, releases oxygen as a byproduct of photosynthesis.

Peter David/Planet Earth Pictures

▶ Appearances can be deceptive: gelatinous zooplankters like this siphonophore are voracious predators of other zooplankton.

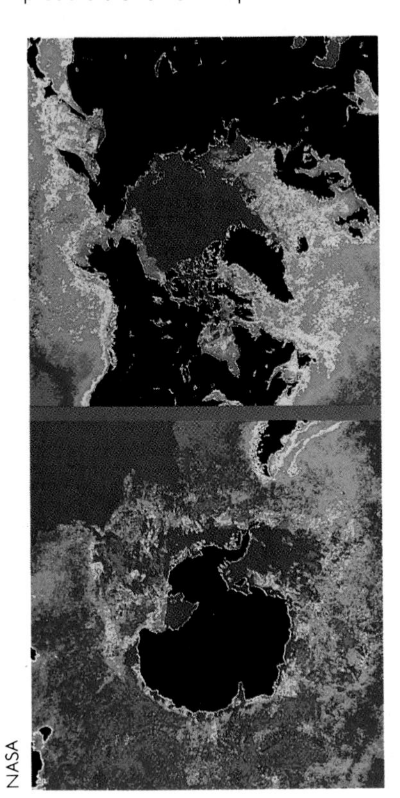
NASA

▲ Highly seasonal phytoplankton blooms characterize the polar waters, where photosynthesis is strongly influenced by the availability of light. These satellite photographs show summer blooms in the Arctic (*top*) and Antarctic (*above*).

### SEASONAL PATTERNS AT DIFFERENT LATITUDES

Cycles of plankton abundance vary in different parts of the world. In polar regions, plankton populations crash during the winter when there is constant darkness and an extended field of ice. In summer, with almost total light, plankton reaches peak abundance. In temperate regions there is usually a spring bloom of phytoplankton and a small bloom in autumn. Although concentrations of phytoplankton are generally low in tropical waters, high rates of reproduction by phytoplankton and high rates of grazing by zooplankton both contribute to a rapid turnover of plankton.

**THE ANNUAL CYCLE OF PLANKTON PRODUCTION**

▶ The abundance of zooplankton closely follows the seasonal changes in phytoplankton production. In high polar latitudes, phytoplankton blooms in a short summer burst; zooplankton productivity peaks shortly afterward. In tropical latitudes, concentrations of both phytoplankton and zooplankton are low but fairly regular. In temperate latitudes, the abundance of zooplankton reflects the spring and autumn blooms of phytoplankton.

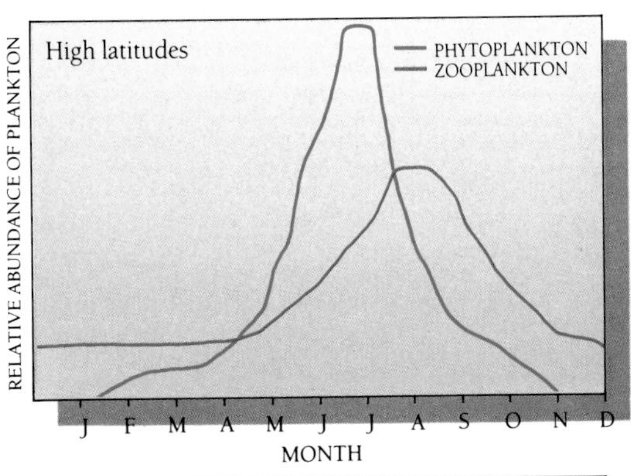


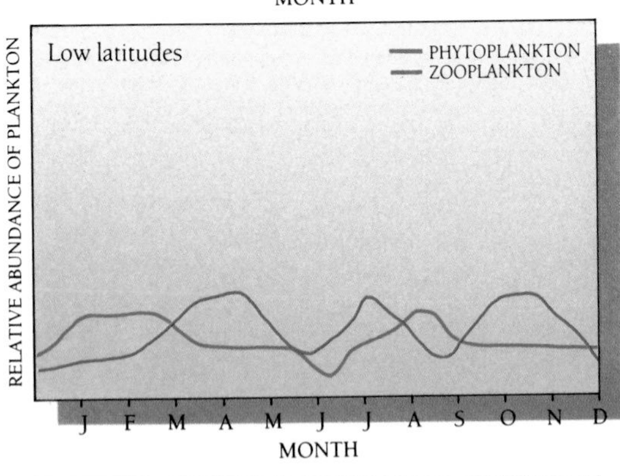


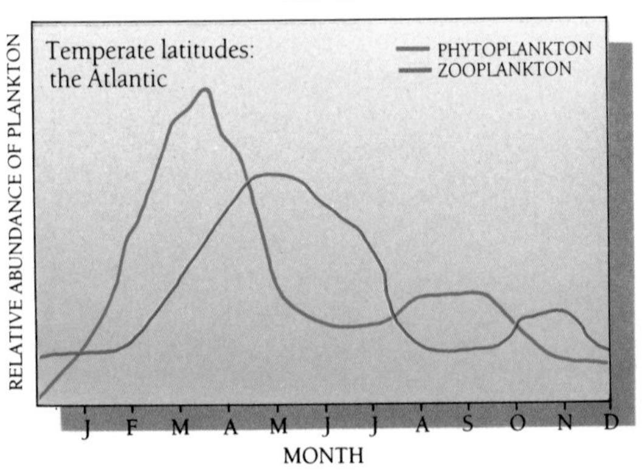


### KINDS OF PLANKTON

The range of oceanic plankton is best described on the basis of size. The smallest kinds of plankton are bacteria with a minimum size of about 0.001 millimeter. Photosynthetic phytoplankters range in size from 0.002 millimeter to over 1 millimeter (0.04 inch). Only recently has it been recognized that plankters below 0.02 millimeter are responsible for a large proportion of oceanic primary production, especially in areas where nutrient levels are relatively low. Tiny bacteria gain some of their nutrition by clustering around phytoplankters that are actively photosynthesizing and utilizing organic material released by the phytoplankton.

Among the single-celled organisms that belong to the phylum Protista (in Greek, "first of all"), the diatoms and dinoflagellates are extremely important elements in the plankton. Diatoms are plant cells that range in size from a few thousandths of a millimeter to 1 millimeter (0.04 inch). Individual cells often form chains a few centimeters long, while dinoflagellates photosynthesize as plants and ingest solid food particles. Tiny flagellate cells consume bacteria and the smallest of phytoplankton. Other groups of protists include foraminifera, radiolarians, acantharians, and ciliates. Ciliates are an important source of food for recently hatched fish in the wild and are used in aquaculture.

Zooplankers larger than 0.05 mm (0.002 inch) are grazers of phytoplankton, and some are predators. They are divided into two general categories: meroplankton, which spend only part of their life as plankton, usually as larvae; and holoplankton, which spend their entire life as plankton. Many animals that are familiar to us in their adult form, such as crabs, barnacles, lobsters, and sea urchins, spend time as plankton, and these larval forms bear no resemblance to their adult appearance.

Important grazers and consumers of microbes include copepods, cladocera, larvaceans, and salps. Many holoplanktonic animals prey on other plankton: predators include jelly plankton and arrow-worms (chaetognaths). Predators can have a marked effect on some groups of plankton. For example, it has been argued that because jellyfish consume large numbers of larval fish, very high concentrations of jellyfish may reduce the size of fish populations. Some zooplankters are omnivorous: euphausids, or krill, for example, usually feed on phytoplankton, but will happily consume zooplankton if it is available as easily targeted packages.

Light rarely penetrates sea water below depths of more than 150 meters (490 feet). Above this is the photic zone in which the plant life of the ocean photosynthesizes. A wide variety of organisms that live on or near the bottom below the photic zone depend for survival on a "fall-out" of material from the upper layers. Plant cells, fecal pellets, microbes, live and dead plankton become bound together as mucus-like blobs called marine snow, which facilitates the passage of material from the upper layers of the ocean to the bottom, where it is eaten by the bottom-dwelling animals.

## VERTICAL MIGRATION

Despite their tiny size, most planktonic organisms migrate regularly through their environment. The best-known migrations of plankton—both phytoplankton and zooplankton—are vertical, where organisms move up and down in the water column. Movements of phytoplankton are generally achieved by controlling the amount of gas, oil, or salt within the organism. The production of gas or oil, or the removal of salt, will cause the organism to rise, while their release or absorption will cause the organism to sink. Some phytoplankters use mobile whip-like hairs called flagellae to help them move through the water.

Three patterns of vertical migration have been recognized: nocturnal migrations, where plankton moves toward the surface at night and descends to deeper water during the day; reversed migrations, where plankton moves toward the surface during the day and to deeper water at night; and twilight migrations, in which organisms migrate at dawn and dusk. The most common pattern is nocturnal migration. The distance zooplankters migrate varies considerably and in the open ocean scientists have recognized a "ladder" of migrations. For example, some species may migrate only through the upper 100 meters (330 feet) of the water column, while others migrate through 1,000 meters (3,300 feet) of water each night.

It is believed that many zooplankters move toward the surface at night to prey on food that is abundant in the photic zone; moving back to deeper water before dawn may reduce the chance of being eaten by predators. Moreover, some zooplankters reproduce while they are in surface waters. The timing and duration of migrations may be regulated by the light-dark cycle and the presence or absence of prey, predators, or mates in surface waters.

Horizontal movements of plankton are generally limited. However, by migrating vertically, plankters sometimes move considerable horizontal distances because currents often move in different directions at different depths.

Claude Carre/Jacana

▲ This strange planktonic animal is the phyllosoma stage of a *Macrura reptantia*, and is scarcely recognizable as the larval form of a lobster. Phyllosoma larvae spend 11 to 13 months in the plankton, during which they undergo several molts.

Peter David/Planet Earth Pictures

◄ Krill (euphausid prawns) are found worldwide, reaching their largest size in the cold water around Antarctica. They are an important component of many marine food chains, and huge numbers are devoured by baleen whales during the Antarctic summer.

Charles Arneson

▲ Copepods have a complex system of mating and reproducing. After mating is completed, and dozens of eggs have been fertilized, the eggs are stored in a sac-like structure that bulges from both sides of the female's abdomen.

REPRODUCTION OF COPEPODS

Copepods are grazing zooplankters related to more familiar crustaceans such as prawns and rock lobsters. Although they are small, usually under 3 millimeters (0.12 inch) long, they have a complex mating behavior and interesting reproductive biology. Males detect females from some distance by chemical cues. When they sense the presence of a female, their swimming behavior changes from swimming in straight lines to performing loops and twists in an attempt to locate her. Once the male finds the female he will grasp hold of her with his hinged antennae, and swing around to grab the female's abdomen with a highly modified fifth leg. When the pair is locked in position, the male passes a packet containing sperm, called a spermatophore, to the female. The spermatophore adheres to the female and has a small tube leading into the genital opening. Eggs are fertilized as they move down the oviduct, past the spermatophore tube and into two sac-like structures on each side of the female's abdomen.

ANCIENT PLANKTON

The skeletons of some plankters, such as the single-celled foraminifera and radiolarians, are well suited for preservation in sediments, and some of the best records of animals and plants that lived in ancient seas come from sediments and sedimentary rocks that contain plankton. Micropaleontologists examine sediments or thin sections of rocks to describe the microfossils they contain. Fossilized plankton has provided important information about plate tectonics. Sediments with relatively young fossils are found close to areas with active sea-floor spreading. The age of fossil plankton increases with distance from ridges where new sea floor is being produced through volcanic activity. Because we know the age of many fossil zooplankters, they are sometimes used to date rarer fossils found in the same sedimentary layers.

The presence of some plankters in ancient seas is conspicuous to us even now. For example, white chalk cliffs in Europe, 65–100 million years old, are made up primarily of coccoliths—small planktonic organisms with calcareous skeletons. The breakdown products of ancient marine plants and animals have also contributed to reserves of petroleum trapped in geological formations.

M. J. KINGSFORD

# FISH

Like the Flying Dutchman, some fishes are bound to spend their entire lives on the open sea. Others temporarily leave fresh water or the coastal margins in exchange for oceania, and do so for a variety of reasons: to avoid competition with their own by dispersal from their birthplace; to take advantage of a great feast; or to avoid the seasonal variations of a temperate climate in favor of a warmer, tropical environment.

Evolutionary biologists who analyze life history strategies suggest that anything a species can do to improve its reproductive success is desirable, and that those individuals that live in a predictable and benign habitat will thereby be more successful. So, if one's habitat is variable, often harsh and unpredictable, then trade it for a better one—and that is just what many migratory and oceanic fishes have done. Such fish are typically highly fecund species, early to mature, and often short-lived. Some migratory species have extended this to the extraordinary reproductive strategy of semelparity (suicidal reproduction). Coastal species that exist in more variable and unpredictable habitats often live longer and mature later. They are less fecund, and reproduce repetitively, a strategy scientifically known as iteroparity.

The absence of strong geographical and topographical barriers to oceanic fish movement has limited their opportunities for speciation, and for that reason there are considerably fewer oceanic species than those that occupy the nearshore and fresh water. However those that have adapted to oceanic life have done so remarkably well.

FROM FRESH WATER TO THE SEA

The phenomenon of migration from fresh water to the ocean, or vice versa, is called *diadromy*. Fish that spend most of their life in the sea and return to fresh water to breed—such as shads, sturgeons, and Pacific salmon—are said to be *anadromous*; diadromous fish that spend most of their lives in fresh water and migrate to the sea to breed, such as the freshwater eel, are called *catadromous*. Although diadromous fish make up only 1 percent of the world's species total, their desirability and the enormity of their numbers during migration make them ideal candidates for exploitation by humans.

The best-known example of anadromy is that of the five species of Pacific salmon of the genus *Oncorhynchus*. These fish leave the coastal waters of the north Pacific and cross the ocean, often as far as Japan, returning several years later to their precise home stream to spawn and die. Biologists have discovered that they probably navigate by means of an internal compass, aided by geomagnetic clues. Once within their home river system, they complete their journey by following chemical cues and odors which are unique to their parental spawning site and are recalled from imprinting at the time they were fingerlings. Anadromous behavior is more common in temperate waters, and is presumably related to the ratio of available food from both oceanic and freshwater sources.

Catadromy, on the other hand, is a rarer and tropical event, and reflects the inversion of that ratio. The Atlantic freshwater eels, *Anguilla anguilla* and *A. rostrata*, undergo an incredible and still poorly understood migration, whereby the adults, before spawning, presumably leave the Mediterranean and the Americas and congregate in the deep water of the Sargasso Sea. There, they breed and die, spawning the eel larvae which will ultimately return to the American and European rivers. Diadromous fish, particularly the anadromous species, are not numerous in species, but are so in biomass, and for that reason their effect upon the food chain of the oceans is quite significant.

Tom McHugh/Photo Researchers

▲ The eel *Anguilla anguilla* spends most of its life in fresh water, but then undergoes a most unusual migration to the Sargasso Sea to breed and die. Eel larvae eventually find their way back to American and European rivers.

## THE TUNAS: ADAPTATION AT ITS BEST

The fish that perhaps best represent adaptation to the oceanic environment are the tunas and their relatives. These 58 species are divided among three families within the suborder Scombroidei, which to a non-ichthyologist means the tunas, bonitos, seerfish, ceros, sierras, wahoos, mackerels, sailfishes, marlins, and swordfish. They live in the surface waters of all tropical and temperate oceans and seas, generally in the mixed layer and above the thermocline (the top 150 meters/490 feet or so). They are well camouflaged for such an existence, with a dark blue-green dorsal surface, silvery sides, and a bright underbelly. Such a livery makes them difficult to

Jeff Foot/Survival Anglia

◄ After several years roaming the Pacific Ocean, a journey of thousands of kilometers, a pair of sockeye salmon *Oncorhynchus nerka* have returned to spawn in their natal river in British Columbia.

▲ The efficient hydrodynamic bodies of these bluefin tuna *Thunnus thynnus* enable the fish to cover great distances very quickly.

see from above or below by presumptive predators or their own potential prey species.

The tunas, billfishes, and their relatives, have attained the ultimate specializations for swimming in the open sea. Their bullet-shaped hydrodynamic bodies are so streamlined that they swim through the water without visible effort. In fact, tuna cannot stop swimming, for should they do so they would be unable to breathe. They have achieved this streamlined elegance via a Faustian bargain with their respiratory system: the muscle action required to pump water across their gills has been discarded in favor of swimming with their mouth open like a ram-jet ventilator. Their oxygen and energy demand is very high, needing a food intake of as much as one-quarter of their body weight each day. Their torpedo-like bodies are covered with small scales and their median fins fit into grooves while swimming so as to reduce drag. Such a high hydrodynamic efficiency, coupled with lunate caudal fin of low torque and high stall speed design, allow the wahoo *Acanthocybium solandri* to reach a measured velocity of 76 kilometers per hour (47 miles per hour) and the sailfish *Istiophorus platypterus* to attain speeds of 109 kilometers per hour (68 miles per hour).

These oceanic tunas have a muscle composition and physiology which is also designed for sustained swimming at high speed. The unusually high proportion of red to white muscle mass allows the northern bluefin tuna *Thunnus thynnus* to cross the Atlantic in 119 days, a distance of at least 7,770 kilometers (4,800 miles). In the Pacific, skipjack *Katsuwonus pelamis* have been tagged and rediscovered nearly 10,000 kilometres (6,100 miles) away, and albacore

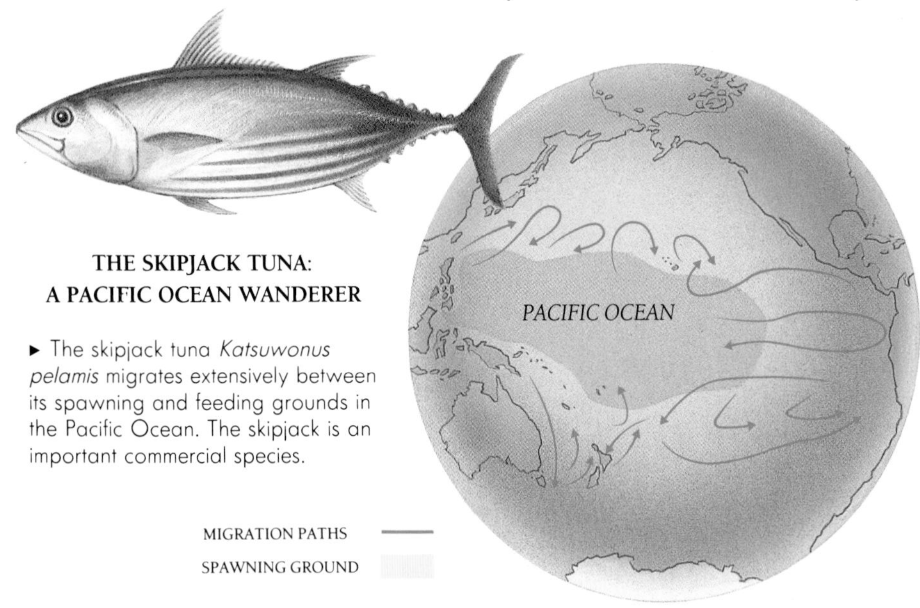


**THE SKIPJACK TUNA: A PACIFIC OCEAN WANDERER**

► The skipjack tuna *Katsuwonus pelamis* migrates extensively between its spawning and feeding grounds in the Pacific Ocean. The skipjack is an important commercial species.

*Thunnus alalunga* are known to travel from Japan to Californian coastal waters each year. Migration is simplified for most tunas through anatomical specializations which result in elevated body temperatures. The heat generated by the red muscle mass is captured (rather than dissipated to the cooler surrounding water) by special arterial and venous shunts which other fish lack, and thereby elevates the temperature of the blood entering the gills and surrounding the viscera. This warm-bodiedness, rather than warm-bloodedness as higher vertebrates have achieved, allows for greater muscle efficiency and strength as well as the ability to inhabit a broad range of environmental temperatures.

The adaptive strategies of oceanic and diadromous fish are a marvel of the process of natural selection, but are unprepared for the development of high-seas fisheries. The absence of geopolitical boundaries and the increasing demand for protein increase the difficulty of stock assessment and management. International cooperation will be required if oceanic and migratory species are to be conserved.

JOHN E. McCOSKER

## SHARKS

It is commonly presumed that the open ocean is rife with sharks. Few would disbelieve Herman Melville when, in *Moby Dick*, he described the abundance of sharks around the *Pequod* at anchor: "Any man accustomed to such sights, to have looked over her side that night, would have almost thought the whole round sea was one huge cheese, and those sharks the maggots in it."

Yet such is no longer the case. The limited food sources available to sharks in the open ocean, a reproductive behavior which limits their abundance and capacity to respond to population pressures, and the recent onslaught of high-seas shark fishing have reduced the worldwide population of sharks.

Norbert Wu/Planet Earth Pictures

▲ The graceful blue shark *Prionace glauca* is abundant in temperate and tropical oceanic waters, riding the currents on its migration paths. It is easily recognized by its long snout and pectoral fins.

There are about 350 living species of sharks, and of them, only five or six species are truly oceanic. Most are coastal species and a few inhabit the deep-sea bed. Even if one includes the 425 living species of skates, rays, and sawfish (considered by modern ichthyologists to be merely flat sharks), only one or two additional oceanic wanderers may be added.

### LIMITS TO DISTRIBUTION

Both the reproductive behavior of sharks and rays, and the physiological limitations imposed by seawater temperatures, play a role in limiting the geographic distribution of elasmobranchs (sharks and their relatives). Elasmobranchs either give live birth to a very few young that at birth appear like miniature adults, or lay a few very large eggs (unlike the millions of minute eggs produced by a bony fish). Elasmobranch eggs or young are therefore unable to drift or swim in the plankton for months at a time, and their range reflects this limitation. A few pantropical species exist, such as the tiger shark *Galeocerdo cuvier* or the blacktip shark *Carcharhinus limbatus*, which is probably due to the casual pace of their evolution. There are also several pantemperate species, such as the great white shark *Carcharodon carcharias* or the plankton-feeding basking shark *Cetorhinus maximus*. Their distribution may be explained by their ease of navigating the Cape of Good Hope or Cape Horn as well as their ability to use islands as stepping stones. Yet all the above-mentioned species are limited by the prey they consume to oceanic islands or continental coastlines.

Truly oceanic sharks range in size from the poorly known and puny crocodile shark *Pseudocarcharias kamoharai*, which grows to 110 centimeters (43 inches), to the whale shark

Mark Conlin/Planet Earth Pictures

◀ The blue shark is a voracious predator, consuming anything it can fit in its mouth, including this mackerel offered for a photo opportunity.

Marty Snyderman/Planet Earth Pictures

▲ Growing over 12 meters (40 feet) long, whale sharks *Rhincodon typus* are the largest in the world. These rare and beautiful creatures are completely harmless to humans, feeding entirely on plankton.

*Rhincodon typus*, by far the world's largest fishlike creature and possibly reaching 18 meters (59 feet) in length. The only other sharks that primarily occupy the high seas are the blue shark *Prionace glauca*, the shortfin mako *Isurus oxyrinchus*, the oceanic whitetip shark *Carcharhinus longimanus*, and the silvertip shark *Carcharhinus albimarginatus*.

THE WHALE SHARK

Like the great whales, the whale shark is so large that it is incapable of capturing almost any sizeable prey. So it too has turned to feeding on abundant and microscopic phytoplankton and zooplankton. In so doing, each shark probably consumes a considerable biomass. However the whale shark is not common, and is rarely seen in a group. Upon a small coastal lagoon or atoll its effect may be profound, but upon the ocean it is insignificant, except for its magnificence. The whale shark is tolerant of human company, and it is harmless.

MAKO, SILVERTIP, AND WHITETIP

The mako shark, however, is a fearsome creature, and were it is not for its offshore habitat, it would often be in the headlines for having consumed humans. It is fast and powerful, has large, sharp, and bladelike teeth, and grows to nearly 4 meters (13 feet). It is a common offshore species in tropical and warm temperate waters, and can appear almost anywhere if the water is warmer than 16° C (61° F).

Other dangerous oceanic shark species include the requiem sharks of the genus *Carcharhinus*. The silvertip and the oceanic whitetip can also be found in most tropical seas, but fortunately live offshore. Due to the rarity of resources on the high seas, they are catholic in diet, and consume bony fish and squids, birds and turtles, and carrion of almost any size or kind. With their long, broad, paddlelike pectoral fins, the whitetips are most impressive. They swim languidly at or near the surface and have been observed to enter calmly a school of small, frenetically feeding tuna, open their capacious maw, and wait for a tuna to swim recklessly in. The strength of their jaws and the sharpness of their teeth were well known to whalers who often saw their floating catch dismembered by whitetips or silvertips, attracted by the feast.

James D. Watt/Planet Earth Pictures

► Oceanic whitetips *Carcharhinus longimanus* traverse most tropical oceans and feed indiscriminately on a wide variety of marine animals. They are characterized by their lobed fins and the white splotch on the dorsal fin.

THE BLUE SHARK

And finally, the blue shark, probably the most abundant and adaptable of living sharks. It was originally named *Squalus glaucus* by Linnaeus in 1758; its type locality "Habitat in Oceano Europaeo". This graceful, dark blue shark reaches about 4 meters (13 feet) in length, and is circumglobal in all temperate and tropical seas. It is often found in large aggregations and occasionally near to shore. Tagging studies of blue sharks demonstrate that they ride the clockwise currents in the north Atlantic, crossing from America to Europe with the Gulf Stream and returning to the Caribbean by the westward-flowing North Equatorial Current. In the Pacific they seem to move south in the northern winter and northward in the summer. They too will eat almost any protein they confront, but are limited by the smaller size of their mouth and jaws.

Blue sharks are quite fecund, as elasmobranchs go, and upon reaching five years of age females can give birth to as many as 135 young. But their abundance is declining, worldwide, as a result of the heavy fishing pressure wrought by humans. Tens of thousands die annually in oceanic barrier nets as an incidental by-catch of more valuable fish. The increased need for protein, as well as the Epicurean desire for sharkfin soup, has now directed fisheries in search of sharks, rather than killing and discarding them as pests. Population data for oceanic sharks are rare and difficult to obtain, but necessary if fisheries are to continue without major and irreversible depletion of the stocks. International cooperation is necessary but will be difficult to achieve for creatures of the sea that so repulse and frighten us and, as Melville wrote, "are the maggots in it".

JOHN E. McCOSKER

# WHALES AND DOLPHINS

The distribution and migration of animals are governed largely by their food supply. In the oceans, food for whales and dolphins is quite patchy; in some areas fish, cephalopods, crustaceans, and other organisms abound, while large tracts of ocean are virtually barren. For the most part, tropical seas are relatively poor in nutrients. Polar seas contain up to one hundred times the concentration of organisms found in most tropical oceans.

A few areas in the tropics are, however, very fertile, and support teeming masses of plankton, fish, and squids—for example, the sea off the Galapagos Islands, the Caribbean Sea, and the Arabian Sea. Upwelling occurs in some tropical regions, bringing nutrients to the surface to support the network of organisms in those regions. Very large schools of dolphins, numbering in the thousands at times, are found there, some remaining for most if not all the year, together with such predominantly tropical whales as Bryde's whales *Balaenoptera edeni* and female sperm whales *Physeter catodon*.

Many species of whales and dolphins are local, living over or near the continental shelf of landmasses, or very close inshore, and in sheltered bays and inlets. They live singly or in small groups, and make up a relatively small total biomass which their prey (mainly fish, cephalopods, crustaceans, and other small invertebrate species) can sustain.

THE GREAT WHALES

Before this century, when pelagic whaling using

Al Giddings/Survival Anglia

▲ Despite their enormous bodies, humpback whales *Megaptera novaeangliae* are strong swimmers, covering thousands of kilometers in their annual migration from tropical breeding grounds to polar feeding grounds. They have no teeth, but use a curtain of baleen plates to filter out planktonic food such as krill and small fish. This pair in Arctic waters is gulping food at the surface.

▼ Spotted dolphins are often found in small social groups, although very large schools are seen in the oceans.

Howard Hall/Planet Earth Pictures

Francois Gohier/Photo Researchers

▲ Largest of all whales is the female blue whale *Balaenoptera musculus,* which reaches well over 30 meters (100 feet).

factory ships was introduced, the great whales, including blue, fin, sei, and humpback whales, were in very large numbers and comprised a huge biomass. These are tropical whales that still mate and give birth in the tropics, usually in areas of the sea in which there is little or no food. They make regular annual migrations from tropical breeding areas to feeding areas in high latitudes. The evolution of this migratory pattern is uncertain, but it must be related to their having ranged from their breeding areas and found sources of food in colder waters, probably beginning in the Tertiary period about 30 million years ago. Climatic and other changes possibly led to an increased geographical separation between breeding and foraging areas.

The cold-water oceans that are so rich in small organisms can sustain a massive biomass of

▶ A sperm whale *Physeter catodon* about to sound. Sperm whales are toothed whales which feed predominantly on giant squid, and are capable of diving to depths far in excess of 1,000 meters (3,330 feet) in search of their prey.

Barbara Todd/Hedgehog House New Zealand

predators, which include the great whales, seals, seabirds, and large fish. The abundance of food in these polar seas enabled the whales to increase in individual size and total numbers. Many of them have become highly specialized, particularly with regard to their feeding apparatus. The great whales lost their teeth and developed baleen plates, the horny downgrowths from the palate, frayed at their tips, that together form a "curtain" to filter out the small organisms that make up the whales' diet.

The annual migrations of the great whales involve some risk, because in returning each year from the polar seas to breed in their tropical "home", they leave a feast to migrate through vast tracts of ocean in which there is very little food, to reach breeding areas where there is virtually none. Adult females are particularly vulnerable during this migration, because while fasting they support a rapidly growing fetus on the way to the tropics and must produce milk for a suckling calf as they venture out to feeding areas in polar seas again. They lose a vast amount of weight, and the stress is undoubtedly too much for some of them, which die during the long journey.

EXPLAINING THE MIGRATIONS

Two questions about these phenomenal migrations beg an explanation: why do these whales, that can withstand cold so well, return to tropical waters to breed, rather than stay in high latitudes where food is abundant? And why is food so abundant in polar seas?

The answer to the first question is probably that in cold waters the food supply is reduced in the winter months when the sea freezes over in the high latitudes, and the small organisms that make up the zooplankton descend to greater depths and virtually "hibernate". In order to maintain their body temperature, whales probably have to move about to some extent, and energy loss is considerably less in the tropics than in the freezing polar seas. Moreover, because the newborn whale calves have a relatively large skin surface area and virtually no blubber to insulate them against the cold, they may die from body heat loss in freezing sea. However, some whale species, including white whales and narwhals, do remain in the cold Arctic sea year-round, so they have obviously managed to find ways of conquering this problem.

▼ By far the greatest proportion of fertile areas of the ocean are in the high latitudes. Only a few relatively circumscribed areas of tropical seas are particularly fertile.

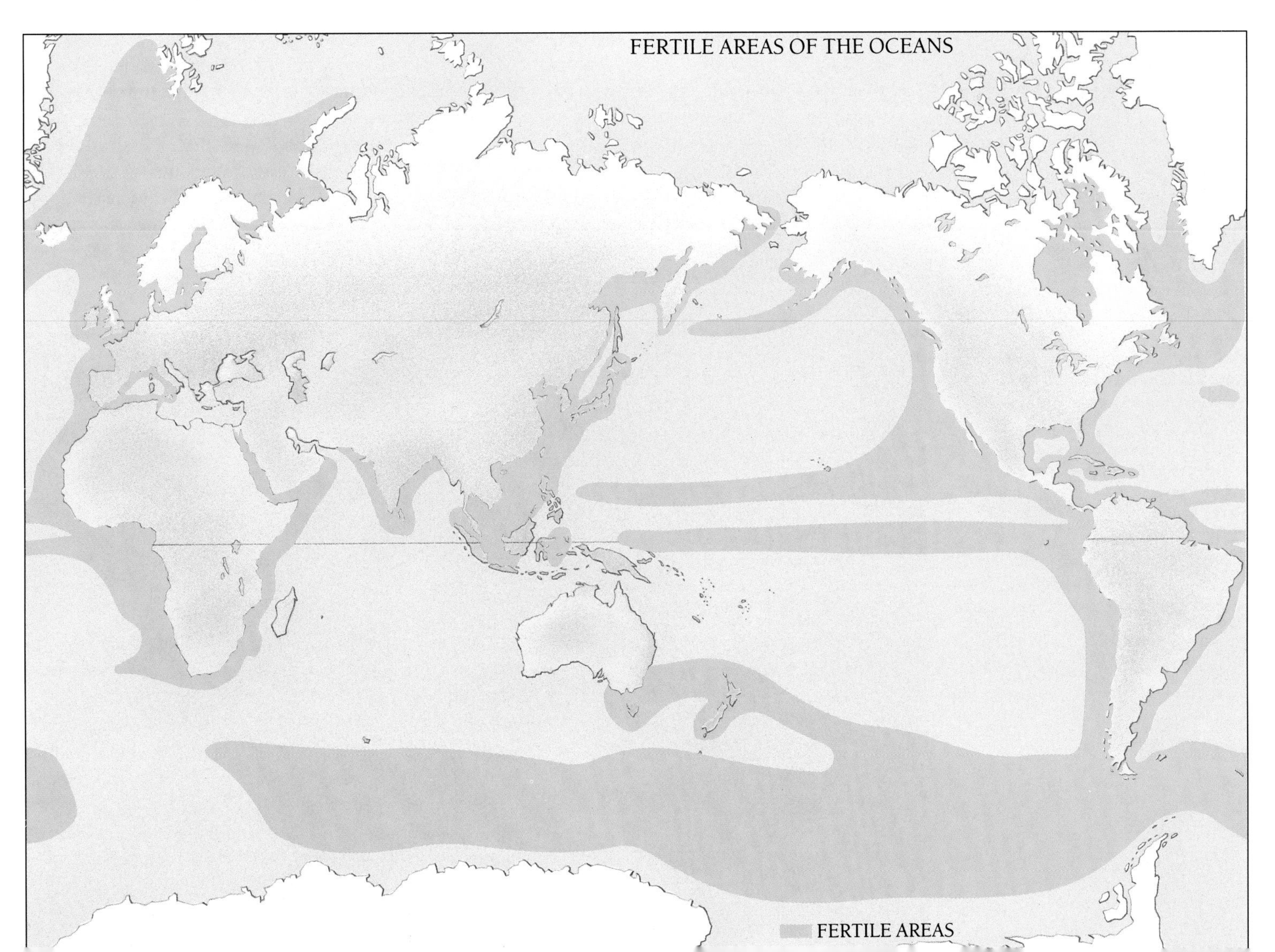

Annie Price/Survival Anglia

▲ Newly hatched green turtles *Chelonia mydas* run for their lives toward the sea from their breeding ground on Ascension Island. Their soft bodies are tasty morsels for predators such as seabirds.

The answer to the second question is more straightforward. The enormous living resources of the polar seas result from several oceanographic factors. There is a drift of nutrients along the sea floor toward the poles, where it rises to the surface and moves toward warmer water. So the surface layers of ocean near the poles are extremely rich in the nutrients required by phytoplankton, the tiny plant organisms in the sea. With the help of sunlight, these synthesize organic matter from carbonic acid and water. Carbonic acid is more soluble and oxygen is more plentiful in cold than in warm water. Conversely, destructive bacteria thrive in warmer temperatures. Consequently, the ingredients are present in polar waters to permit mass blooming of phytoplankton in the summer months, when sunlight is available for photosynthesis almost 24 hours a day. The phytoplankton is the foundation of an enormous standing crop of zooplankton on which the whales and other large predators feed. But in winter, when there is little sunlight, the phytoplankton stops growing, the surface water freezes to form sea ice, the zooplankton descends in the water column and virtually stops developing. Thus it is unavailable to predator species, including the great whales.

MICHAEL BRYDEN

# SALMON, TURTLES, AND SEALS

Some of the most notable migrations among the marine vertebrates, other than marine eels, tuna, and the great whales, are those of salmon and certain species of turtles and seals.

## SALMON

Because of its economic significance, the life history of the Pacific salmon *Oncorhynchus* has been studied more intensively and is better understood than that of its European relative of the *Salmo* genus.

Salmon hatch and spend the first part of their lives in oxygen-rich freshwater streams or lakes, and remain there for one to five years depending on location. The hatchlings grow through various stages named "fry", "parr", and "smelts". They then swim down to the river mouth, where they

Colin Monteath/Hedgehog House New Zealand

▶ Seals spend most of their lives at sea, returning to land only to breed and molt. These Weddell seals *Leptonychotes weddellii* inhabit the Antartic region and are members of the family Phocidae. Phocids have limited mobility in their hindquarters, which considerably restricts their movement on land.

become adapted both to the salinity of sea water and a new diet.

Young Pacific salmon emerging from streams and rivers of the American Pacific coast seem to head northwest or west (those from the west coast of Alaska go south), to lead a pelagic life where growth is rapid, in contrast to their relatively slow growth in fresh water. Like all carnivorous marine fish, they travel long distances, remaining close to the surface and feeding on small fish and species of crustaceans.

At about two or more years of age, and occasionally after only one year, salmon return to rivers—a very high proportion to the rivers of their birth—to spawn. They do this in the period from June to November. They fast during migration and spawning, drawing on their considerable fat reserve for energy. The pelagic stage of the Pacific salmon's migration cycle may cover distances of a few thousand kilometers, and the freshwater portion may cover more than 2,000 kilometers (1,240 miles).

Spawning occurs in clear freshwater streams. Between 8,000 and 26,000 eggs are laid at a time, in small hollows excavated in the gravel of the riverbed. Body reserves become greatly depleted and most animals die after spawning. The few survivors either spend the winter months in deep-water patches of rivers, or return at once to the sea where they rapidly build up their body reserves again. Less than 10 percent return to spawn a second time after one or two years. Survival during the migrations is related to the distance traveled.

TURTLES

Although sea turtles are the most spectacular migrants among the reptiles, information about the distribution and migration of various species is not obtained easily, and the only reasonably detailed knowledge of migration refers to three nesting populations of the green turtle *Chelonia mydas*. The general pattern of return migration, at least of those groups of green turtles that breed along the coasts of Costa Rica and Ascension Island, is known quite well, despite gaps in knowledge of some aspects.

Eggs are buried as they are laid, above the high-tide line on sandy shores of tropical and subtropical mainland and island coasts. Some six to eight weeks later the young hatch and immediately migrate across the sand to the sea, into which they effectively disappear. Nothing is known about where they go or what they feed on during their first year of life. It is assumed that, as they leave the beaches of their birth, they drift with the current for a period. This period may be short, perhaps only a few weeks, during which the young turtles arrive in nursery areas where they may live among the floating rafts of *Sargassum* weed. A more likely explanation, however, is that they live a pelagic existence for a year, feeding on zooplankton as they go. Even when they adopt an adult-type herbivorous diet, young green turtles do not remain long in one place, and their migrations could well be exploratory while they perform spatial learning and establish a familiar area.

Adult green turtles feed predominantly on turtle grass, an underwater flowering plant, off continental coasts. At maturity they select an appropriate beach for resting, and the second phase of the lifetime migration circuit begins. Grazing and breeding areas may be 2,000 kilometers (1,240 miles) apart. Every two or three years females leave the feeding areas, migrate to the breeding areas, lay from three to seven batches of 100 or so eggs at 12-day intervals, and then return, migrating with the surface current, to the grazing areas.

The migratory pattern of males is less certain, but undoubtedly occurs, because males are also seen around the breeding areas. Copulation takes place offshore, when the eggs that will be laid two or three years later are fertilized.

SEALS

All seals, sea lions, and walruses, known collectively as pinnipeds, are amphibious, alternating between periods in the water and periods on land, on mainland coasts, islands, sandbanks, ice fields, or ice floes. For each species, or among breeding groups within species, births are concentrated into a relatively short time of the year, a period that is followed fairly quickly by the main mating season. An exception is the Australian sea lion *Neophoca cinerea*, which has a protracted gestation and breeding season, and is believed to have an 18-month breeding cycle.

Dispersal in pinnipeds is associated with

Colin Monteath/Hedgehog House New Zealand

▲ The most abundant seals in the world are crabeater seals *Lobodon carcinophagus*. However, despite their name they do not eat crabs; instead they use their teeth to sieve krill from the sea.

Annie Price/Survival Anglia

▲ These sleek and elegantly colored king penguins *Aptenodytes patagonicus* breed in colonies around Antarctica. Breeding pairs take it in turns to incubate an egg in a fold of skin between their feet.

feeding; convergence is associated with birthing, copulation, molting, and resting. Pinniped species such as the northern fur seal *Callorhinus ursinus* and the harp seal *Phoca groenlandica*, that are distributed at well-marked foci during the breeding season, disperse after breeding as individuals migrate different distances in different directions. Species like the southern sea lion *Otaria byronia*, which breed along continental coasts, and the crabeater seal *Lobodon carcinophagus* and leopard seal *Hydrurga leptonyx*, which breed among or at the edge of the circumpolar band of Antarctic pack ice, disperse when individuals move different distances rather than in different directions. Some species, such as the leopard seal and the Ross seal *Ommatophoca rossii*, are more or less solitary throughout their lives. Other species, such as the northern fur seal and the elephant seals *Mirounga* are intensely gregarious during the breeding season, but more or less solitary while at sea, though some convergence occurs when food is abundant.

There is a suggestion that species that feed predominantly on fish and cephalopods may disperse more than other species, and they may also be inclined to have individual home ranges. On the other hand, species that feed on invertebrates such as krill or bottom-dwelling molluscs and crustaceans may show less dispersal. Those species that breed on oceanic islands seem to show greater convergence during the migration to the breeding grounds whereas coastal breeders and ice breeders seem to show progressively less convergence during this phase of the migration.

Until recently, most of what is known about the migration of seals has been learned from resightings of marked animals, and there were, therefore, large gaps in our knowledge. In the past few years the fitting of time-depth recorders and, more recently, satellite recorders, to individual animals, has given us much greater insight into the distribution, range, and migration of some seal species.

MICHAEL BRYDEN

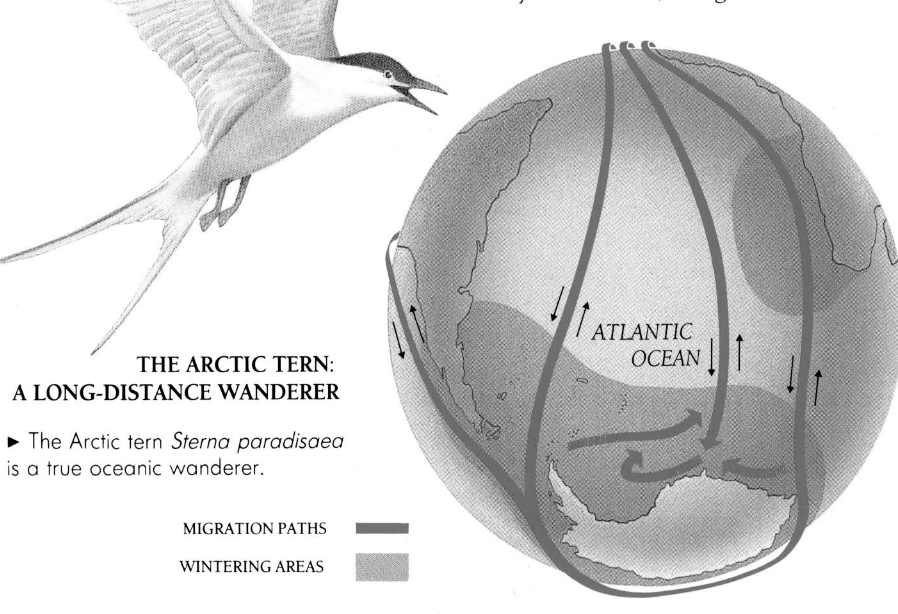


**THE ARCTIC TERN: A LONG-DISTANCE WANDERER**

► The Arctic tern *Sterna paradisaea* is a true oceanic wanderer.

# OCEANIC BIRDS

The first birds evolved on land about 160 million years ago. Even as these feathered reptiles were learning to flap from tree to tree, toothed versions of today's loons were learning to swim. Birds have been associated with the sea ever since. Their use of the marine environment is constrained by two requirements: air to breathe, and land on which they can nest.

## NESTING AND BREEDING

Characteristically, seabirds nest on islands where they are safe from land-based predators. At the height of breeding there can be few more spectacular sights in the natural world than an island of nesting birds, an exhilarating mix of noise, smell, and sheer fecundity. Some birds nest in burrows, some in dense colonies on the surface, and others on cliff ledges, laying eggs much narrower at one end than the other to prevent them from rolling off. These aggregations of birds have a profound effect on the islands on which they breed, concentrating nutrients from the surrounding seas into mountains of excreta or guano, which is mined as a source of fertilizer. Particularly famous are the islands off Peru where mountains of guano 90 meters (295 feet) thick had accumulated over thousands of years before they were carried away to Europe and America during the nineteenth century. Today, in both Peru and southern Africa, the removal of guano is regulated.

Because such productive islands are home to tens of millions of breeding birds, a prolific source of food is needed nearby. This is especially

Richard Matthews/Planet Earth Pictures

▲ Gannets nest in large colonies, mostly on small offshore islands. These birds are long lived, usually breeding in their fifth year or later. One egg is laid and parents take turns to incubate the egg on a nest of seaweed and guano.

true for colonies of penguins—birds that have become so well adapted to the sea that they have paddles for wings. Because they cannot travel as far as flying birds, penguins need to catch nearly all the food for their young within 20 kilometers (12 miles) of their breeding colonies. Their only compensation is that they can dive deeper than other birds. The largest species, the emperor penguin *Aptenodytes forsteri*, can feed at depths of more than 250 meters (820 feet). All other seabirds are confined to the top 20 meters (65 feet) of ocean.

Flying seabirds are more mobile than penguins. Some leave their nests for days or weeks at a time while they tour the ocean in search of food. In 1989 a wandering albatross *Diomedea exulans* was tracked by satellite over a distance of 12,000 kilometers (7,400 miles), flying from the Crozet Islands in the southern Indian Ocean to the Antarctic continent and back before relieving his mate at the nest. Other species may well travel as far across the ocean.

It seems remarkable to us that seabirds can find their often tiny nesting islands in such a vast expanse of ocean. Birds, however, can probably pinpoint their position from the pattern of waves, winds, and currents just as we might read a map. Seabirds also have a strong sense of smell, which they use to find both food and their nesting sites when, as many do, they return after dark.

This homing ability is all the more extraordinary considering the distances traveled by seabirds on their migration paths. The Arctic tern *Sterna paradisaea*, for instance, breeds north of the Arctic Circle between May and August, then travels the length of the globe to spend from November to March in the pack ice of Antarctica. As a reward it lives in almost constant daylight. Many other seabirds, such as shearwaters and storm petrels, migrate similar distances but breed in the Southern Hemisphere then pour north in tens of millions as the days shorten. Others move laterally, some albatrosses and petrels in the Southern Hemisphere simply following the wind. Unimpeded by continents, they probably circumnavigate Antarctica several times between breeding attempts.

## DIVERSITY AND ADAPTATION

All seabirds are carnivores, with the majority taking fish, squid, or krill. To find their prey, seabirds must be particularly sensitive to water temperatures and currents. Some species favor warm water, others cool. Those of southern seas sometimes follow the Humboldt or Benguela currents formed of cold, upwelled water rich in

**HOW SEABIRDS CATCH THEIR PREY**

▲ Seabirds utilize a variety of feeding methods to exploit the resources of the ocean: most, in fact, are able to use more than one method. Some birds, such as noddies, skuas, and frigatebirds, feed in flight, snatching food from the ocean surface or just underneath. The storm petrel "patters" across the surface layer. Some, including scavengers such as giant petrels, swim across the surface layer, or dive under it. Some, including penguins, cormorants, and diving petrels, dive into the water and pursue their prey for some distance. Others, including pelicans, tropicbirds, terns, and gannets, plunge through the surface of the ocean to capture mobile prey at some depth.

nutrients which run northward along the western coasts of South America and southern Africa respectively. Often, productivity is greatest where waters of different temperature meet, particularly where cold, nutrient-rich water wells to the surface at the edges of continental shelves or in eddies, tide rips, and other turbulence.

Especially rich are the cold waters of higher latitudes, and these support the greatest diversity of seabirds. Of the four major groups of birds adapted to life at sea, two are concentrated in the Southern Ocean, one around the Arctic, and only one in the much larger area of tropical sea. In the south live all but one of the 17 penguin species and most of the so-called tubenoses, the petrels and albatrosses. In the Northern Hemisphere there is a great diversity of larids, the order of birds containing gulls, terns, puffins, murres, auklets, and guillemots. The group with greatest variety in the tropics is the pelecaniforms, which includes the pelicans, boobies, tropicbirds, frigatebirds, and shags.

These groups have adapted to their marine environment in different ways, though they have in common webbed feet for swimming and salt glands for removing excess salt from their blood. The most aquatic are the flightless penguins, though some of the auklets and murres in the north and diving petrels in the south flap their tiny wings underwater just as they do above it. By contrast, the petrels and albatrosses have the most efficient gliding flight of any bird. Their long, narrow wings, as much as 3 meters (10 feet) from tip to tip in the wandering albatross, obtain lift from the small breezes generated by waves moving up and down. In the endlessly windy

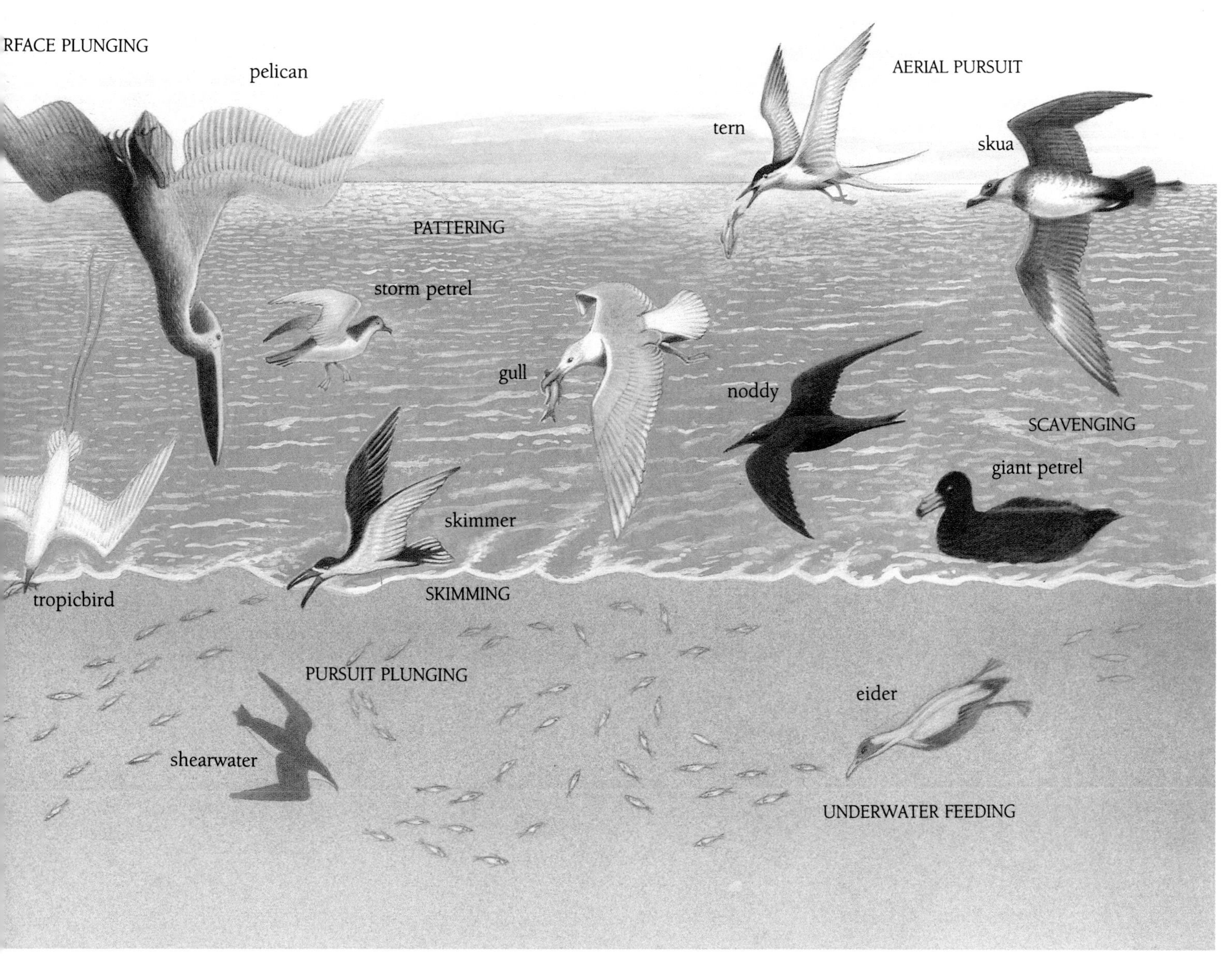


southern oceans they have only to spread their wings to be effortlessly airborne. The boobies and tropicbirds are streamlined for plunging into the ocean, sometimes closing their wings and falling vertically through the air for 30 meters (100 feet) or more to capture fish. They, like most other seabirds, have a waterproof plumage, but their close relatives, the frigatebirds, have apparently sacrificed this capacity to save weight and energy. As a result they can stay airborne for weeks at a time, snatching food from the ocean's surface or stealing it from other seabirds.

THREATS AND PROBLEMS

Despite the expanse of their ocean environment, many seabirds are threatened. Coming to land only on remote islands, most are absurdly tame, and several species have been exterminated by human exploitation for food or feathers. Introduced predators, such as rats and cats, have resulted in the extinction of many others. Now there are also threats at sea. Thousands of birds die annually in gill and drift nets on the open ocean, and albatross populations have been declining steadily for the last 20 years as a result of longline fishing. Others are poisoned by pollution. Oil removes the waterproofing of the feathers, so birds freeze to death. Even Antarctic penguins show traces of pesticides, and nearly all seabirds contain plastic in one form or another that has been pecked from the ocean surface. Birds are such an important part of the marine ecosystem that it would be tragic if they were decimated by what is essentially human carelessness and folly.

STEPHEN GARNETT

# 3 LIFE IN THE OCEANS

Despite many similarities and the mixing of waters along their boundaries, each ocean has a distinct identity. The Pacific, Atlantic, Indian, and Southern oceans are typified by their physical nature—temperature and salinity, for example—and biological characteristics. Similarities between the oceans, such as the transfer of energy through food webs which encompass microscopic plankton through to large marine mammals, ensure that life is maintained. Differences—often resulting from the varying geological history of each ocean—make life in the oceans particularly fascinating.

Andrew Mounter/Planet Earth Pictures

▲ The siphonophore *Physalia physalia*, also known as Portuguese man-o'-war or bluebottle, is a colony of tiny animals, each modified for a specific purpose. Some feed, some protect or attack, and others are reproductive.

## THE PACIFIC OCEAN

The diversity and abundance of life in the ocean depend ultimately on the basic fertilizing nutrients available in its waters. Productive surface waters occur in the northern and southern extremities of the Pacific, where the near-freezing water welling up from the ocean floor carries prodigious quantities of fertilizing minerals. In spring and summer the waters bloom with millions of tonnes of microscopic plankton upon which even large creatures such as baleen whales or basking sharks can directly feed. Apart from the microscopic organisms which photosynthesize carbohydrates, plankton also consists of the eggs, larvae, or young of marine creatures.

Ocean current systems sweep the rich cold waters toward the Americas and toward equatorial regions. Thus, in terms of fishery economics, the northern, southern, and eastern regions of the Pacific are the most productive. However, this is not to deny the economic importance of the western regions or the biological interest of the myriad forms of marine life that inhabit them, particularly around the islands, coral reefs, atolls, and cays that are so characteristic of the western Pacific Ocean.

The waters of the northern and southern halves of the Pacific tend to remain separate and many of the creatures inhabit only one of the ocean's two main systems. Some salmon, for instance, including the prized sockeye *Oncorhynchus*, occur naturally only in the north Pacific. On the other hand, widely roaming oceanic wanderers such as the giant whale shark, the Pacific marlin, as well as some dolphins and whales, freely cross from one hemisphere to the other.

◄ Immune to the deadly stinging tentacles, a clownfish hovers above the anemone it calls home. This intriguing relationship is a feature of all coral reefs: the clownfish gains immunity by slowly covering itself in a layer of mucous from the anemone, preventing an attack by tricking the nematocysts (stinging cells) into recognizing the fish as "self".

Jane Burton/Bruce Coleman Limited

◄ The brightly colored Pacific lobster *Enoplometopus occidentalis* has five pairs of legs, one pair of which is modified as chelae (pincers). True lobsters have claws on their first three pairs of legs, which they use for cleaning (as seen here) and manipulating food particles.

Kev Deacon/Dive 2000

▲ Brilliant red and yellow hues of a soft coral against the ocean blue. Soft corals have eight tentacles per polyp, which places them taxonomically in the coelenterate group Alcyonaria. Although related to hard corals, they do not secrete a limy skeleton.

CORALS AND REEFS

The Pacific immediately conjures up visions of coral islands and reefs. These coral ramparts, built of immense quantities of limestone, are produced by innumerable diminutive relatives of the jellyfish and sea-anemones, the so-called coral polyps. Each coral organism has the capacity to secrete some form of protective tube about its body. As it grows, it buds off younger individuals which remain attached to it, thus rapidly forming a colony of individuals joined with a common flesh. In some species—the soft corals—the protective coating is a tough, fleshy material with the result that the whole colony may be quite flexible and, at first sight, be mistaken for kelp.

In general, reef-building corals occur in the warmer waters of the ocean with the active part of the reef never extending more than about 37 meters (120 feet) below the surface. Solitary corals, which resemble large anemones but with a limy exoskeleton, are generally found in deeper, cooler Pacific waters.

Corals and all the members of the zoological group to which they belong—including jellyfish, box-jellyfish, Portuguese men-o'-war, and sea-anemones—have tentacles which are equipped with microscopic stinging cells capable of ejecting a poison barb into the flesh of any potential prey that touches them. Any surfer who has been stung by the long trailing tentacles of the Portuguese man-o'-war *Physalia* (also known as the bluebottle) can attest to a painful experience. In the more tropical waters of the Pacific various box-jellyfish, or sea-wasps, *Chironex* have venomous stings which have sometimes killed humans within minutes.

A coral reef harbors an amazing variety of fish, crustacea, sea-urchins, starfish, brittlestars, molluscs, and worms. One of the most famous of the worms is the palolo worm *Leodice* of the waters around the coral-fringed islands of the central Pacific. As with many of its relatives, the eggs and sperm develop in the rear half of the creature's body. At the appropriate time, the rear section detaches from the parent body and swims up to the surface. The remarkable thing about the palolo's swarming is the precision of its timing. It occurs on the day of the last quarter of the October-November moon. The swarming is so dense that it is easily harvested to provide the islanders yearly with a delectable feast.

MOLLUSCS AND CRUSTACEA

Innumerable species of molluscs inhabit the Pacific, many of them unique to specific regions. The rock oyster *Crassostrea* of the eastern

Australian coast is famed for its superior taste and for the fact that, unlike many other oysters, its offspring hatch and develop away from the parent. Three other molluscs unique to the ocean are all members of the squid/octopus group. The blue-ringed octopus *Octopus maculosus* of the southwestern region, although comparatively small, has a highly venomous bite. The Humboldt squid in the waters off South America is the largest of the Pacific forms, weighing in at some 50 kilograms (110 pounds).

Perhaps the most interesting of the three is the pearly nautilus *Nautilus pompilius*, which inhabits the waters of the coral reefs and islands of the south Pacific. While resembling the squid or octopus in having a head surrounded by tentacles, it also possesses a stout, spirally coiled, lustrous shell into which the animal can completely retract itself. With this large chambered shell, the creature resembles the long-extinct ammonite.

Crustacea, particularly prawns and lobsters, are important human food resources. Pacific lobsters lack the claws of their northern European counterparts but have a pair of stout, spiny antennae which, when lashed about, are effective at deterring an enemy. While not unique to the Pacific, various species of prawns are certainly its most characteristic and widespread crustacea. These too differ from the prawns and shrimps of European waters in several ways. Apart from anatomical points and the fact that they grow much larger, the principal difference is that, when they spawn in the deep sea, their fertilized eggs are set free and not stuck to the body of the female, as is the case with other types of prawns and, indeed, crustacea in general.

## SHARKS AND RAYS

The Pacific harbors a number of archaic creatures such as the Port Jackson shark *Heterodontus* and the frill-gilled shark *Chlamydoselachus*. The several species of the former live inshore of the coasts of eastern Australia and neighboring islands while the latter inhabits the waters of the northwestern Pacific. *Heterodontus* is a harmless, mollusc-eating creature with powerfully muscled jaws for crushing the shells of its principal food. With its elongated sinuous body, *Chlamydoselachus* resembles an eel more than a shark. Both creatures have a long geologic history, being almost unchanged in form since Jurassic times some 150 million years ago.

Sharks evoke less than romantic images of the Pacific. While some deserve their bad reputation, most seem to be blameless. Closely related to sharks are the rays. The largest of these unique to the Pacific is Captain Cook's stingaree *Bathytoshia*, which reaches lengths of over 4 meters (13 feet) and can weigh more than 200 kilograms (440 pounds). Larger still is the more widely distributed manta ray, often referred to as the devil fish despite being a non-aggressive feeder on plankton.

## MARINE MAMMALS

Some 20 species of whales, dolphins, and porpoises are exclusive to the Pacific. They do, however, share the ocean with some of the other more widely distributed species. Of the 20 endemic species, there is one baleen whale, the Pacific right whale *Eubalaena*. The remaining 19 are all toothed creatures which actively pursue their food. It is a mixed group consisting of five species of beaked whales, 11 of dolphins and two of porpoises. The nineteenth is the Pacific pilot whale *Globicephala*, the first of the large whales to be kept and trained in captivity.

In the Southern Hemisphere, seals are generally regarded as inhabiting the Southern and Antarctic oceans. Nevertheless, the leopard seal *Hydrurga* occasionally appears along the east Australian coast and fur seals *Arctocephalus* are a common sight on the shores of New Zealand and South America. The north Pacific, however, harbors quite a variety, the best known of which is the California sea lion *Zalophus*, an eared seal that has made a name for itself as a performing seal. Other noteworthy examples are the monk seal *Monarchus* in the region of the Hawaiian Islands and, further north in the Pacific, the northern elephant seal *Mirounga*.

A marine mammal not related to either the seal or whale groups is the slow-moving browser of seaweeds, the dugong or sea-cow *Dugong dugong* which occurs in the tropical waters of eastern and northern Australia. Until the nineteenth century, the dugong had a close relative in the north Pacific, Steller's sea-cow, which became extinct from over-hunting.

Another Pacific mammal is the sea-otter, which

▼ The sea-otter *Enhydra lutris* inhabits the kelp beds of the north Pacific, and is renowned for its extraordinary feeding habits. A stone brought from the sea floor acts as a "table", on which the otter's food is placed and opened. After intensive hunting by fur traders attracted by their thick pelts, these animals were endangered by the beginning of this century. Their numbers over their entire range from Alaska to California are now increasing, thanks to a 1911 treaty prohibiting the trade in otter pelts and national marine mammal protection legislation passed in 1973.

Jeff Foott/Bruce Coleman Limited

occurs in the colder waters of the American shores of the north Pacific. The largest of the otters, over 1 meter (3 feet) long, it rarely leaves the water and, to rest or sleep, beds itself in surface kelp or seaweed. Another species, not quite as fully marine, but having the same general behavior, lives far to the south in Tierra del Fuego. Sea-otters have the remarkable habit of fetching a flattened stone from the depths and, while floating on their back at the surface, placing the stone on their chest. They then proceed to open such food as sea-urchins, molluscs, and crabs by cracking them against the stone.

J. R. SIMONS

# THE ATLANTIC OCEAN

The Atlantic, the second largest ocean, contains about 25 percent of all the water in the world ocean, and is about one-half the volume of the Pacific. Oceanography had its birth in the Atlantic, and concepts developed in the North Atlantic have been applied to other oceans with varying degrees of success. Generalizations have not always stood the test of time, since the North Atlantic differs in fundamental ways from other parts of the world ocean.

## HYDROGRAPHIC STRUCTURE

The Atlantic may be visualized as an immense fiord, opening at its southern end into the Indian, Southern, and Pacific oceans, and essentially closed in the north. Its northern end is connected with the North Polar Sea, often called the Arctic Ocean, although its volume is far too small for it to be considered as a true ocean. Exchanges between the North Polar Sea and the North Atlantic are minuscule compared with the exchanges that occur in the south.

Because of the fiord-like nature of the Atlantic, its northern and southern components are very different. The North Atlantic is the warmest and most saline of all the world's oceans, while the South Atlantic is more like the Indian and Pacific oceans, although it is slightly fresher and more saline than the oceanic average. The saltiness of the North Atlantic results from the outflow of the Mediterranean Sea, which discharges prodigious amounts of extremely saline water. This Mediterranean outflow dominates the eastern North Atlantic just as the Gulf Stream dominates the western North Atlantic.

Because the North Atlantic is not open at the north, its circulation is much more closed than that of the South Atlantic, and its currents are much stronger. In the South Atlantic, the warm Brazil Current is a weak analogue of the powerful Gulf Stream, and the Benguela Current is a warmer analogue of the Canary Current in the North Atlantic, transporting cool water and pelagic organisms.

▼ Puffins *Fratercula arctica* are commonly known as sea parrots because of their large and colorful bills. Most of their lives are spent at sea but they come ashore annually to breed, here at Runde Island in Norway. They nest in large colonies on bare ground or in burrows.

Uwe Walz/Bruce Coleman Limited

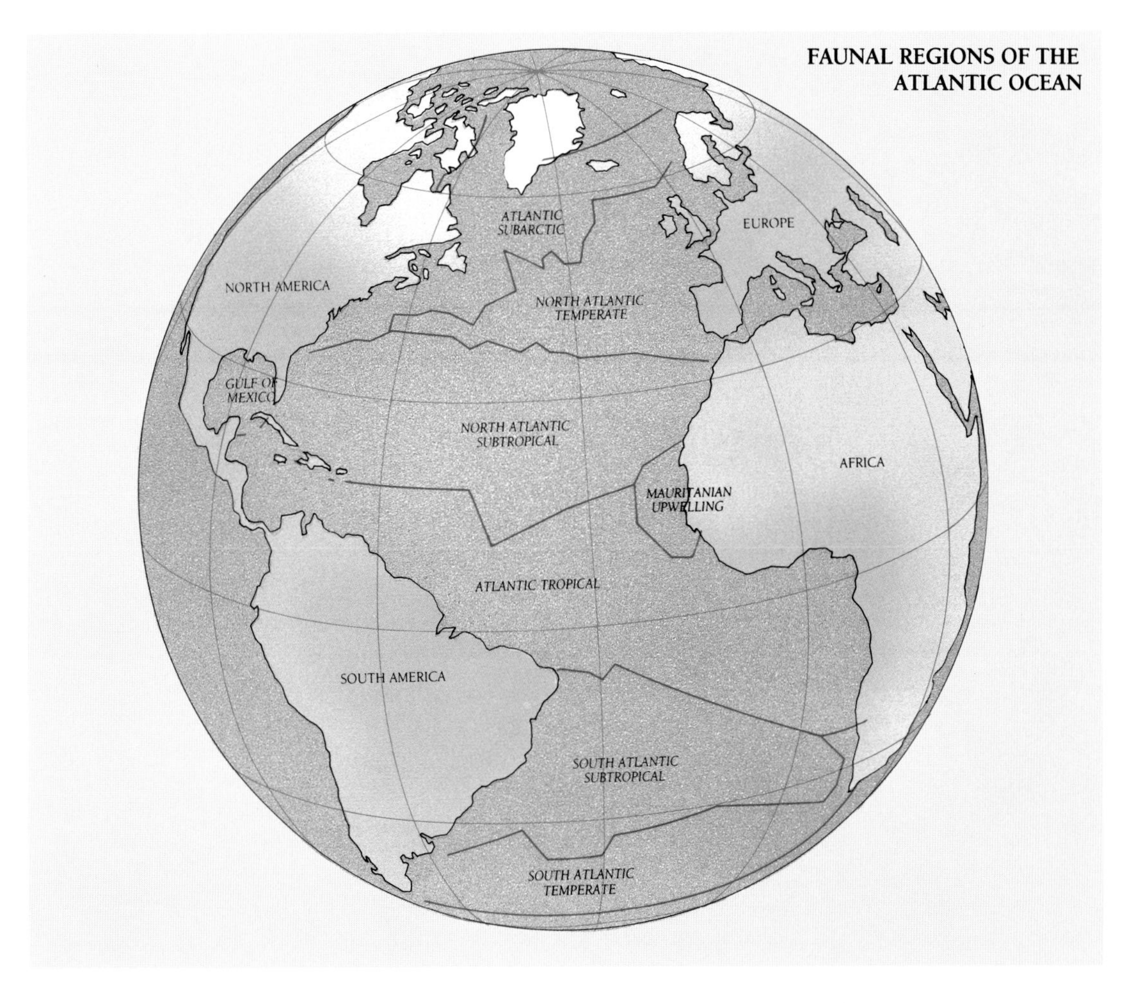


◂ The division of the Atlantic Ocean into faunal regions has provided a framework for the study of midwater fishes and other marine animals.

FAUNAL REGIONS

The Atlantic Ocean has recently been divided into eight immense faunal regions by R. H. Backus and co-workers of the Woods Hole Oceanographic Institution, based on collections of midwater fishes. They tried to find physical oceanographic boundaries that corresponded with the limits of midwater fish distribution, thereby linking faunal patterns with specific water masses. From north to south, these regions are: the Atlantic Subarctic, the North Atlantic Temperate, the North Atlantic Subtropical, the Gulf of Mexico, the Mauritanian Upwelling, the Atlantic Tropical, the South Atlantic Subtropical, and the South Atlantic Temperate. These faunal regions serve well to describe the distribution of a wide variety of pelagic vertebrates and invertebrates.

The Atlantic Subarctic is a transition zone between the frigid waters of the North Polar Sea and the North Atlantic Temperate Region. It is characterized by extremely low diversity, with only a single species, *Benthosema glaciale*, making up the majority of the midwater fish biomass. Puffins and other seabirds nest along its coasts and on its islands; pilot and fin whales traverse its breadth on their oceanic journeys.

As one moves south toward the equatorial regions, biological diversity increases. The North Atlantic Temperate Region extends from the shores of Europe to the northeastern part of the United States, and is highly productive, with marked seasonal changes in water temperature. The Grand Banks and North Sea, historically very important centers of fisheries activity, are located in this region, their harvests including pelagic herring, sardine, and anchovy, and demersal cod, flounder, and Atlantic perch.

The South Atlantic Temperate Region, extending to the south to the Antarctic Convergence, is open to the Indian and Southern oceans, and is bounded only on its western side by land (the east coast of Argentina). Since it is open to the Southern Ocean, species such as whales, penguins, and seals that occur close to the Antarctic continent are also found here.

The North and South Atlantic Subtropical Regions are areas of low primary production, but high diversity. Warm water extends to great depths, and the surface waters are very clear and very low in the nutrients necessary for the growth of plants. This is because of the deep permanent thermocline, which prevents nutrient-rich deeper waters from reaching the surface. In the North Atlantic, the western portion of the Subtropical Region forms the Sargasso Sea. In the South Atlantic, a discrete "sea" does not form, because it is open to exchanges with the Indian and Southern oceans.

The Atlantic Tropical Region is centered on the thermal, rather than the geographical equator. The

# THE SARGASSO SEA: AN OCEAN DESERT

Larry Lipsky/Tom Stack & Associates

A well-camouflaged sargassum fish.

Christopher Columbus encouraged his officers and men to continue their westering voyage, bound for the new world, with these words from his log: "The admiral says that on that day, and ever afterwards, they met with very temperate breezes, so that there was great pleasure in enjoying the mornings, nothing being wanted but the song of nightingales. He says that the weather was like April in Andalusia. Here they began to see many tufts of grass which were very green, and appeared to have been quite recently torn from the land. From this they judged that they were near some island, but not the main land, according to the Admiral, 'because,' as he says, 'I make the main land to be more distant.' "

Portuguese seamen called this "weed" *sargazo*, because the floats on the *Sargassum* look a bit like little grapes. They're yellow, but at least they are the right shape. The *Sargassum* is a brown alga which looks very much like an attached marine plant, but which exists entirely in the pelagic, open-ocean environment. Like many plants, the *Sargassum* can reproduce in an entirely asexual manner—the plant grows on one end and dies away on the other.

A SEA BOUND BY OCEAN

The *Sargassum* is only one of hundreds of life-forms unique to the Sargasso Sea, the only sea in the world bordered not by land but by the Atlantic Ocean. The great north Atlantic currents—the Gulf Stream, the North Atlantic Current, the Canary Current, and the North Equatorial Current—are a single closed-circulation cell called the North Atlantic Gyre. This circulation cell, and those in all the other major oceans of the world, are the primary transporters of water, and thus of heat, on the surface of our planet. They move hot water from the equatorial regions to the higher latitude (and therefore "temperate") regions, and cold water from high latitudes to the equatorial regions where it can be warmed once again.

The clockwise north Atlantic circulation has a number of special consequences. The Coriolis effect, a result of the Earth's spin, causes water in motion to move to the right in the Northern Hemisphere (it moves to the left in the Southern Hemisphere). This causes anything that floats to become concentrated in the center of the gyre. Thus flotsam and jetsam in the north Atlantic can circuit the ocean for years, but eventually collect and concentrate in the Sargasso Sea. The gyre is lens-shaped, varying in thickness from about 200 meters (650 feet) at the edges to perhaps 100 meters (330 feet) under the Gulf Stream and in the center, under the Sargasso Sea.

Life in the ocean, as on land, depends upon plants, which are the base of the food chain. Plants need sunlight and nutrients in order to grow. In the ocean, nutrients are concentrated in the deep water (where there is no light), but come to the surface in the coastal zone, and at the equator, where upwelling and wind-mixing stir the water and transport nutrient-rich deep water to the surface. By contrast, the central gyres, like the Sargasso Sea, have no local nutrient supply. The water is thus biologically very poor, nearly devoid of life. It is the bluest, most translucent water in the ocean. Because there is nothing to reflect the light, little life, and no sediment, sailing through the Sargasso one can see many tens of meters down into the water. Life is sparse, but jellyfish, salps, and other large animals are visible.

In the last twenty years, the nutrient transport system into the Sargasso Sea has become much better understood. As

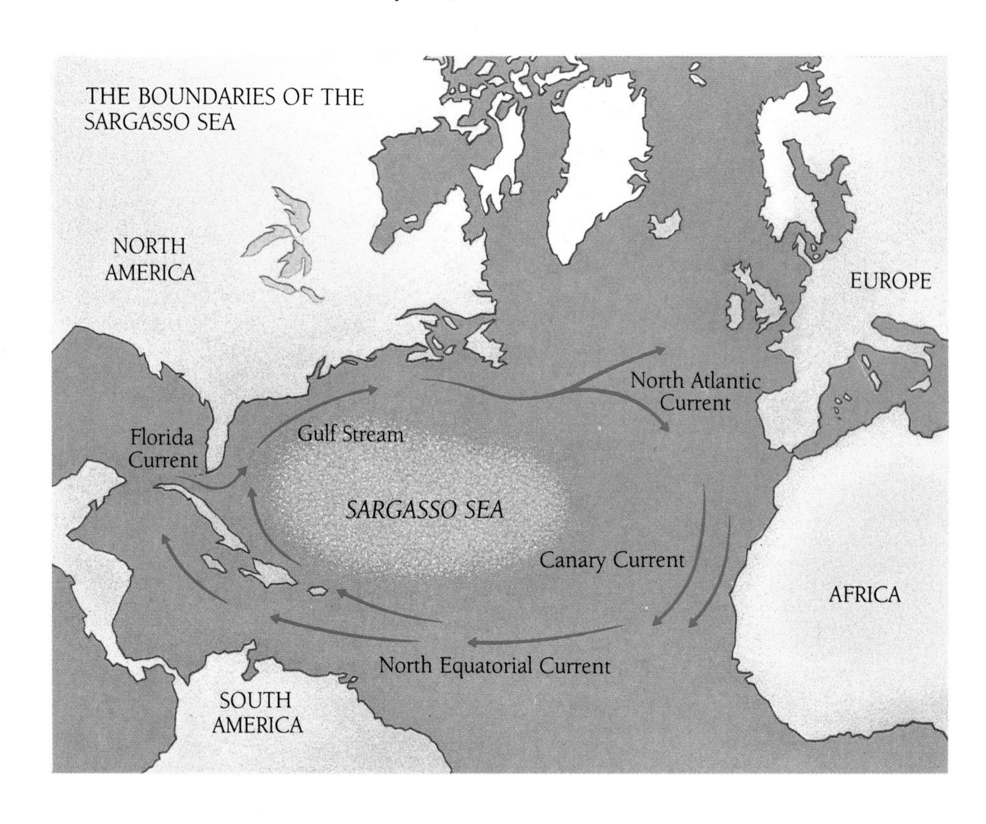

Robert F. Sisson

The *Sargassum* supports a number of life-forms, living within its swirling mass.

the Gulf Stream moves north along the east coast of the United States, the stream meanders. Some of the meanders break off from the main current and become closed-circulation cells, called rings. Some of these cells entrap nutrient-rich coastal water and transport it offshore, into the center of the Sargasso Sea. These rings have been tracked for several weeks by ships and now satellite data show that they may persist for months.

### LIFE IN THE *SARGASSUM*

It is within the *Sargassum* itself that the real drama unfolds. A number of life-forms have co-evolved with the *Sargassum* and have taken on the yellow-brown coloration of the alga. They live their entire lives in its camouflage. Perhaps the most exotic is the sargassum fish *Histrio histrio* which crawls through the *Sargassum* with nearly prehensile fins. There are two common shrimps, and a small crab, as well as a nudibranch. A tiny pipefish, a relative of the seahorse, mimics the thin strands of the *Sargassum* perfectly. Several more cryptic animals, hydroids, bryozoans, corals, and other small marine invertebrates also live only in the *Sargassum*.

The most famous inhabitants of the Sargasso Sea are not permanent residents. The eel *Anguilla,* a delicacy that graces European and American tables, spawns in the Sargasso Sea. The larvae, called *leptocephali* after their small heads, spend from one to three years in the gyre. They then migrate out of the oceanic waters into rivers in North America, western Europe, and the Mediterranean. By the time they reach the rivers they have metamorphosed into a more eel-like form and are harvested as elvers. Eventually those not taken as elvers grow to full size—perhaps 50 centimeters (20 inches)—in the rivers. They continue to be used as food, often as smoked eel. When they are fully mature, the eels return to the sea and migrate back to the center of the Sargasso to mate and spawn. This behavior which is described as catadromous (ocean-fresh water-ocean), is the opposite of the anadromous (fresh water-ocean-fresh water) lifestyle of fish like the salmon. Speculatiŏn about the life history of the eels dates to Aristotle and Pliny the Elder, who suggested that they had no gender and were spawned by rubbing their slime off on rocks.

The Sargasso Sea is one of the most fabled parts of the ocean. The subject of many literary myths, it is really known only to professional seafarers who cross it on ship tracks from Europe to the Caribbean. It is the graveyard for hundreds of ships set adrift in the North Atlantic Gyre. More importantly, perhaps, it is among the most peaceful, least visited, and most beautiful parts of the world ocean.

JAMES C. KELLEY

Ron Gilmer

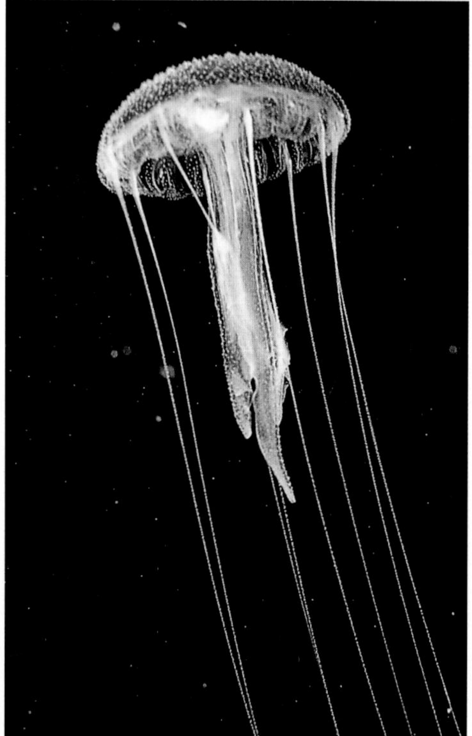

Ron Gilmer

Ron Gilmer

▲ The exquisite siphonophore *Physophora hydrostatica (above)* inhabits the Atlantic Subarctic and Atlantic Temperate regions. Another delicate beauty is the luminescent jellyfish *Pelagia noctiluca (above center)*, which occurs throughout most of the Atlantic. The upper surface (bell) and the oral arms of the jellyfish are dotted with pale bundles of stinging cells, and eight slender tentacles trail below as it moves through the water. The ctenophore (comb jelly) *Bolinopsis infundibulum (above right)* is another inhabitant of the Subarctic and Temperate regions of the Atlantic. Comb jellies move through the water by rapidly beating rows of comb-like paddles.

► *Callianira antarctica* is a ctenophore (comb jelly) inhabiting the temperate regions of the South Atlantic and the Southern Ocean. Although gelatinous and resembling jellyfish, comb jellies belong in a separate, exclusively marine, phylum, the Ctenophora.

thermal equator is the region with the warmest surface waters, and runs diagonally in a wide swath from the Caribbean Sea to the coast of Angola. Its currents are very complex, and change direction seasonally. It is a region of extremely high faunal diversity.

In addition to these major Atlantic regions, Backus and his colleagues described two smaller regions that did not fit into the overall pattern. The Mauritanian Upwelling, off the coast of northwestern Africa, is extremely productive, and one species of midwater fish, *Lamadena pontifex*, is unique to it. On the other hand, while the Gulf of Mexico has no unique pelagic species, it has a fauna that is a mixture of temperate, subtropical, and tropical species.

## THE SPECIAL NATURE OF THE ATLANTIC

The Atlantic Ocean started forming about 200 million years ago when the continents of North and South America began to separate from the continents of Europe and Africa, along a massive rift valley. Because the Atlantic is so young, the bottom-living (benthic) animals that inhabit it are

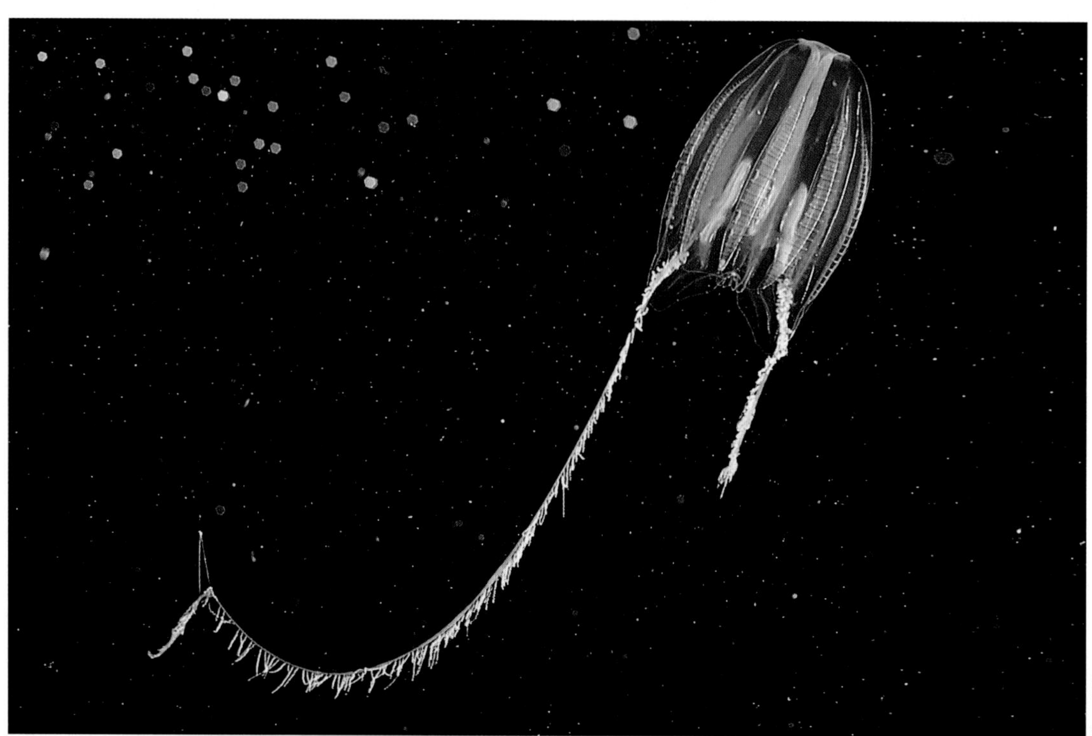

Ron Gilmer

Pam & Willy Kemp/Oxford Scientific Films

▲ Large schools of lutjanids are common in coral reef habitats; because they are easily caught on handlines they are frequently known as "snappers". These golden snappers *Lutjanus lineolatus* inhabit the deeper coral reefs of the Seychelles and the Indo-West Pacific.

descended from colonists from the other oceans and seas. Therefore, the diversity of the benthic fauna is not as great in the Atlantic as it is in the Indo-Pacific. For example, many genera of reef-building corals of the Pacific are absent altogether in the Caribbean.

Because the waters of the North Atlantic wash the shores of Europe and North America, it is not surprising that oceanography had its beginnings in the North Atlantic. Early studies on the speed and direction of the Gulf Stream were prompted by economic forces. Those who could use the extra impetus provided by this "river in the ocean" could cut their transit times across the Atlantic, and thus increase their profits. Several nineteenth-century oceanographic expeditions had their focus in the Atlantic, and, as a result, many of our present ideas about the nature of the world ocean are largely based on studies made in the Atlantic.

However, as we learn more about other parts of the world ocean, the atypical nature of the North Atlantic has become apparent. Processes such as the "spring bloom" (a rapid and overwhelming growth of phytoplankton as days become longer) are entirely lacking in other parts of the world ocean, such as the north Pacific. It is clear that a greater understanding of the world ocean as a whole is necessary before we can generalize with confidence.

RICHARD HARBISON

## THE INDIAN OCEAN

The Indian Ocean, the third largest in the world, is less well studied than the other oceans. Life in the Indian Ocean is often considered as part of a wider region—the Indo-West Pacific—implying that there are considerable similarities between the Indian Ocean and the western Pacific. Many marine organisms have pelagic eggs or larvae which are dispersed by tides and currents. The passage of sea water across the north of Australia allows these organisms to pass between the two oceans either as adults or as larvae. The Indian Ocean covers a similar range of southern latitudes to the Pacific, and many species therefore occur in both oceans.

The Indo-West Pacific region has the richest marine fauna in the world. The central area (roughly the Malay Archipelago, the Philippines, and New Guinea) has been suggested as the principal evolutionary center from which the entire Indo-Pacific has been populated. Diversities are higher there for corals, sponges, medusae, crustaceans, echinoderms, and fish. The fauna becomes progressively impoverished with distance from this center, but there are also many species that occur only in the Indian Ocean.

Open-ocean areas of the Indian Ocean are characterized by relatively low primary productivity. Phytoplankton and associated

► A magnificent coral cod *Cephalopholis* spp. swimming away from its crevice on a reef in the Red Sea. Coral cod are members of the Serranid family which are generally hermaphroditic, typically changing sex from female to male.

▼ The abundant seagrass meadows of the Indian Ocean are a productive coastal resource, supporting extensive food webs, acting as a nursery ground for some species, and helping to stabilize areas of coastline.

zooplankton densities are generally very low, and it is only adjacent to coasts that productivity is stimulated. Where plankton densities are more concentrated, either at particular depths or places, large filter-feeding organisms such as whale sharks and baleen whales aggregate. The smaller filter feeders and predators attract larger predators including sharks, dolphins, and whales.

FISH

There are between 3,000 and 4,000 species of tropical shore fish in the Indian Ocean, often forming bright colorful "clouds" around corals, but with many different roles to play in reef ecology. Many species have restricted distributions within the ocean, particularly in the Red Sea and the Arabian Gulf. Both these areas have been subjected to isolation because of sea-level changes and have thus been provided with opportunities for evolution of new species.

There are fewer species of fish on the continental slope and in deeper water, but they too exhibit a high degree of endemism: 58 percent of the species are found only in the Indian Ocean. Pelagic fishes are also abundant, including needlefish which may be a meter (3 feet) in length. Truly oceanic fish include sharks, flying fishes, tunas, marlins, and ocean sunfishes. Most of these are widely distributed throughout the tropics and migrate between feeding and breeding grounds. Tuna fisheries are well developed in the Indian Ocean, and the species caught include albacore, yellowfin, southern bluefin, and billfishes. These are very mobile and are distributed across the ocean, with some migration around the southern pole.

CORAL REEFS

Where the water temperature never falls below 18–20° C (64–68° F) in winter (between 30° latitude north and south of the equator), coral reefs are formed on shorelines throughout the Indian Ocean. However, this pattern is modified where there is freshwater runoff, particularly with a heavy sediment load off the land as occurs off the coast of India. There, mudflats and mangrove swamps form less picturesque, but no less important, marine habitats.

Coral reefs in the Indian Ocean may be fringing reefs, patch reefs, or atolls. Deep-water atolls occur throughout the ocean. The longest fringing reef occurs in the Red Sea, with a total length of 4,500 kilometers (2,800 miles), but there are also extensive fringing reefs off eastern Africa and along most suitable coasts in the Indian Ocean. Patch reefs are extended areas of comparatively shallow seabed with simple reefs. These are common in the Red Sea but also occur elsewhere in the Indian Ocean.

SEAGRASS BEDS

This region has some of the most diverse and abundant seagrass beds in the world, particularly on the Western Australian coastline. Seagrasses are descended from terrestrial plants and, unlike algae or seaweeds, have true roots and flower

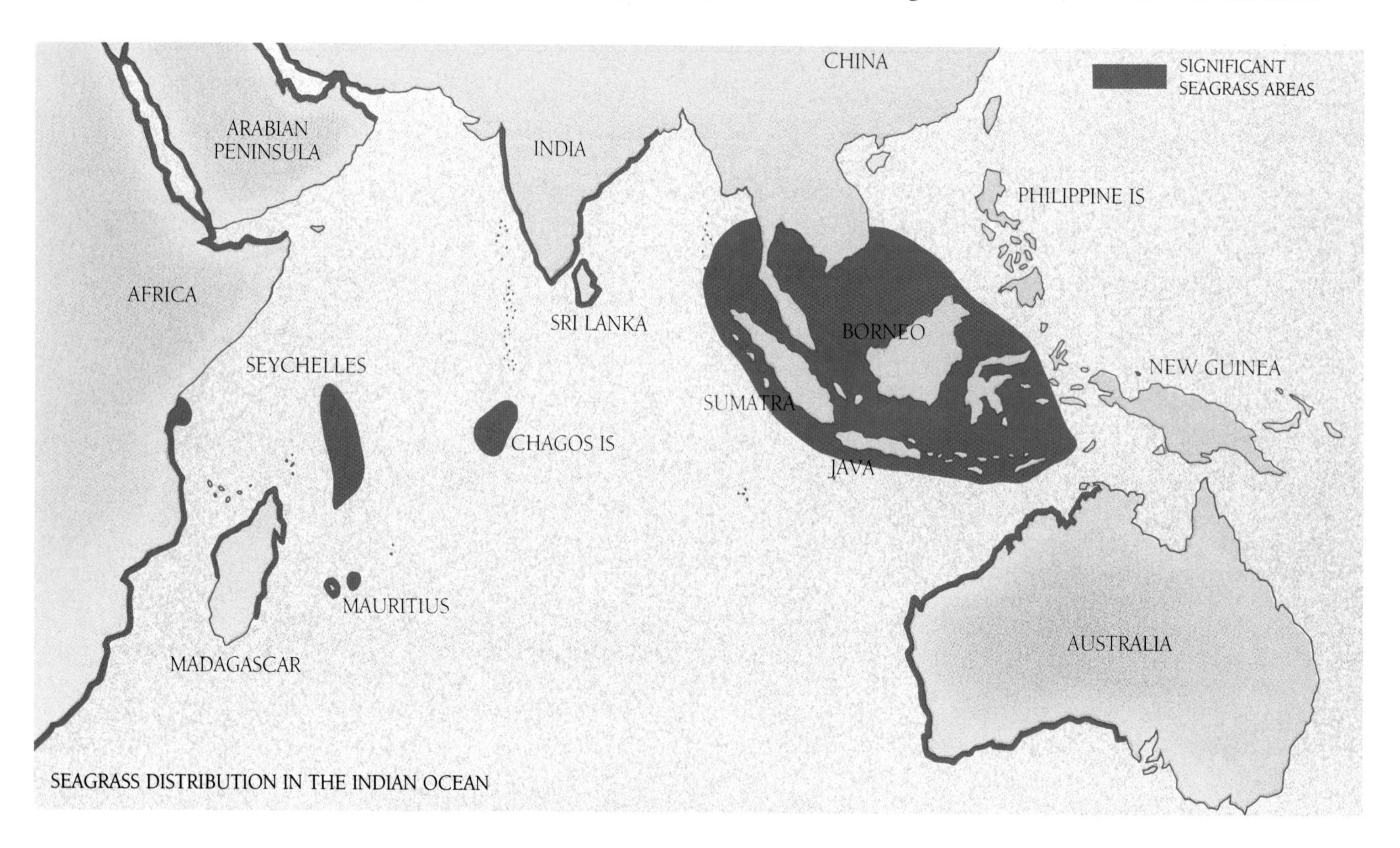


SEAGRASS DISTRIBUTION IN THE INDIAN OCEAN

Peter Scoones/Planet Earth Pictures

Leo Meier/Weldon Trannies

▲ Extensive meadows of seagrass are a feature of the coastline around Shark Bay on the west coast of Australia. Seagrass, which relies on sunlight for photosynthesis, thrives in the shallow waters of the bay, and supports a range of marine animals including seasnakes, turtles, and a large dugong population.

under water. They grow mainly on sand in areas that have some protection from open-ocean swell, in coastal lagoons, and behind reefs. Seagrasses form a food supply for turtles and dugongs, and some of the world's largest remaining dugong herds occur in the Indian Ocean, especially in Shark Bay, Western Australia. There are still dugongs elsewhere in the Indian Ocean, for example in the Red Sea and the Arabian Gulf, but these populations have been greatly reduced by human impact.

Seagrass beds also act as a nursery for juvenile stages of fish and prawns, providing a refuge from predators and thus allowing these species to grow to adult size. In some areas, such as the northern and western coasts of Australia, lobsters also use

▶ On the borders between sea and land, mangroves are among the earliest colonizers. These pioneer plants are often the first stage in a succession of plant and animal communities that transform the intertidal habitat. Much of the Indian Ocean coastline is fringed by mangrove forests.

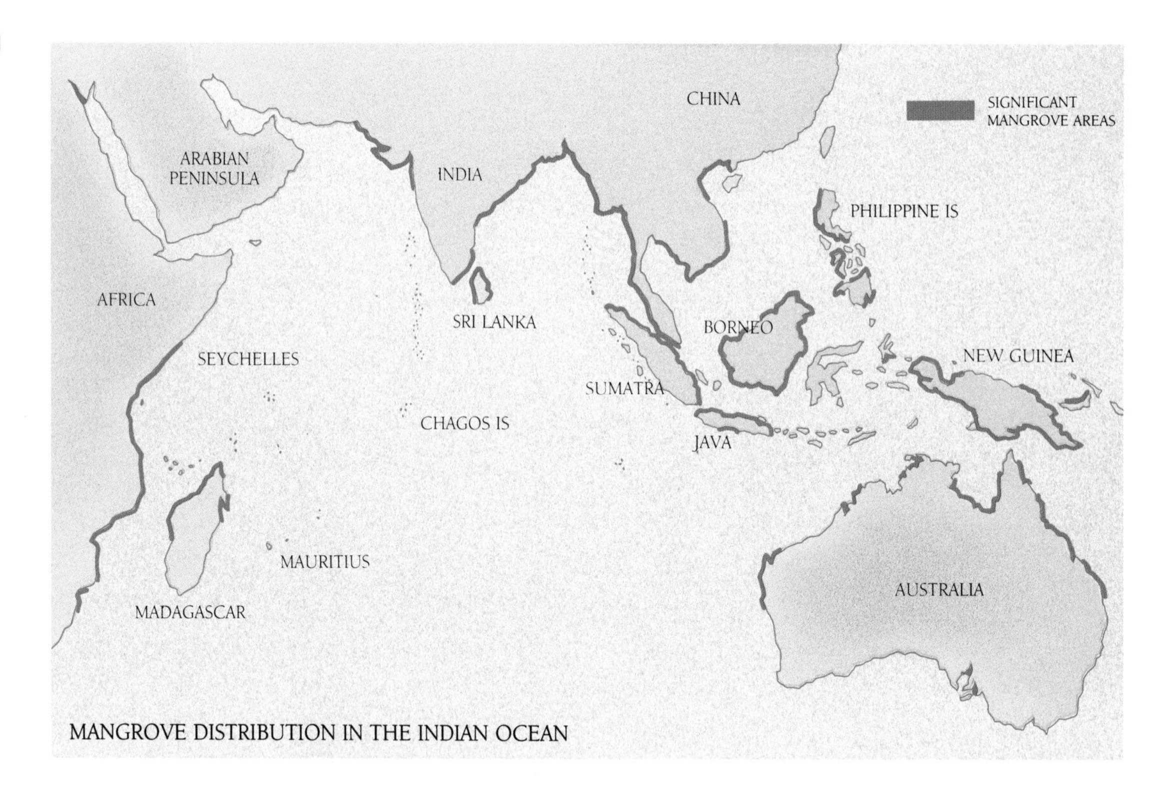

Kathie Atkinson

▲ Grey mangroves *Avicennia marina* encircle an estuary on Burrup Peninsula in Western Australia. The woody community of a mangrove forest is highly productive, and is an important habitat for many species of animals. By stabilizing sediments, mangroves also provide considerable protection to coastal areas.

seagrasses as their larders, foraging for small animals and debris that accumulate in seagrass meadows. Seagrasses slow the rate of water flow over them, helping to prevent erosion and sometimes building up extensive sandbanks. Over geological time these structures can accumulate more sand than even a coral reef. This sand comes partly from the water passing over the seagrass, but also from the calcareous animals and algae associated with the seagrass leaves. A good example of this is in Shark Bay, where sand structures over 10 meters (33 feet) thick and many kilometers in length have accumulated in about 5,000 years.

### MANGROVES

Mangroves dominate the intertidal mudflat zones in tropical and some subtropical coasts of the Indian Ocean, and are associated with estuaries. There are many different types of mangroves, the most widespread being *Avicennia*, the red mangrove, and *Rhizophora*, the black mangrove, which are circumtropical. *Sonneratia* is common in the Indo-Pacific as the most seaward species, grading into *Avicennia* and then inland into *Rhizophora*. Three species of *Rhizophora* are found in the Indian Ocean: *R. mucronata*, which is the commonest from east Africa to New Guinea; *R. stylosa*, which occurs from Malaysia to Australia; and *R. apiculata*, which has a similar distribution to *R. stylosa*, but extends to India. Mangroves are important sources of productivity, but are very vulnerable to human impact. Reclamation work in heavily populated areas of the Indian Ocean has lead to almost complete removal of mangroves from some areas, such as around Singapore and elsewhere in developed Southeast Asia.

DIANA WALKER

## THE SOUTHERN OCEAN

The Southern Ocean completely encircles the Antarctic continent, its currents flowing in a predominantly west to east direction. Its northern boundary is the Antarctic Convergence at about 45–55°S where the ocean meets and flows under the warmer waters of the north. This is a region of strong winds, turbulent seas, and an almost continuous cloud cover. Southward from the convergence the temperature of the ocean becomes progressively cooler and ranges from 8 to 11° C (46–52° F) in the north to minus 1.8° C (29° F) at the coast of Antarctica. In winter, sea ice forms first around the coast of Antarctica and extends northward to cover more than 50 percent of the Southern Ocean.

### A PRODUCTIVE ENVIRONMENT

The seemingly inhospitable nature of the Southern Ocean belies its productivity, which is probably as high on an annual basis as the oceans in more temperate climates. Plants and animals are tolerant of the cold and have features that fit

M.P. Kahl/Bruce Coleman Limited

▲ A crabeater seal *Lobodon carcinophagus* basks on drift ice near the Antarctic Peninsula. Crabeaters are planktivorous, using their teeth to sieve mainly krill from the water. The high seasonal productivity of the Southern Ocean ensures an abundant supply of food.

them to their environment. The blanket of ice calms the surface of the ocean. Its undersurface provides a habitat for zooplankton and allows sea-ice algae to develop; and the upper surface supports breeding seals and emperor penguins.

At the base of the food web are primary-producing organisms—the microscopic algae of the phytoplankton and the sea-ice microbial communities. In the presence of light, these organisms produce food in the form of carbohydrates from carbon dioxide and water. Because the amount of daylight varies considerably, and at the coast of Antarctica ranges from almost continuous darkness in winter to continuous daylight in mid-summer, primary production is markedly seasonal. It is highest in the coastal zone where upwelling brings nutrients to the surface. In spring the biomass of phytoplankton increases rapidly, to be consumed by the microzooplankton (mostly protozoans and zooplankton larvae) and the zooplankton.

There are three major groups of herbivorous zooplankton: the euphausids, the salps, and the amphipods, all of which usually form dense aggregations. Of these, the euphausid *Euphausia superba* or krill is the most important as it provides the major link in the transfer of energy from the primary producers to the larger carnivorous organisms including the baleen whales, seals, seabirds, fish, and squids. Humans are also predators of krill and harvest up to 500,000 tonnes a year.

## FISH AND SQUID

The fish of the Southern Ocean are predominantly bottom-dwelling species inhabiting the coastal shelves of Antarctica and the subantarctic islands. Of the world's 20,000 species of fish, only 120 occur in the Southern Ocean—but most of these are found nowhere else. About 75 percent of the fish species belong to the group of Antarctic cods, and many live in an environment of minus 1.8° C (29° F). Their body fluids contain anti-freeze substances. One group of "icefish" (Chaenechthyds) have no hemoglobin, the substance that carries oxygen in the blood. Instead they absorb oxygen directly into their tissues, principally through the gills. Many of these species have been harvested commercially and some have been substantially overfished, particularly around the islands of South Georgia.

Squid are important in the food web of the Southern Ocean as they appear in the diet of many species of albatross and penguins and in particular the sperm whale. However, as they are difficult to catch, little is known of their biology.

## SEALS AND SEABIRDS

Four species of seal breed on the Antarctic pack ice. The Weddell seal *Leptonychotes weddelli* prefers the stronger ice nearest the Antarctic coast and maintains ice holes that allow it to move between the air it must breathe and the sea where it feeds, principally on fish. With a population of 2 million or more, the crabeater seal *Lobodon carcinophagus* is the most numerous of all the world's seals. It inhabits the inner pack-ice zone in the vicinity of its major food source, the Antarctic krill. Leopard seals *Hydrurga leptonyx* and Ross seals *Ommatophoca rossii* breed in the outer pack-ice zone on ice that is broken into floes, allowing easier access to the ocean. Both feed on krill; the leopard seal also feeds on penguins and has been observed to attack the pups of the crabeater seal. The Antarctic fur seal *Arctocephalus gazella*, which breeds on the subantarctic islands of the Indian and Atlantic oceans, is now recovering from heavy sealing and is expanding its range from South Georgia down to the Antarctic Peninsula where it also feeds on Antarctic krill.

Some 35 species of seabirds breed on the Antarctic continent and subantarctic islands. Typical are the penguins, so specialized for pursuing their prey under water in a cold marine environment, and the albatrosses, equally specialized marine species able to ride the Antarctic gales with seemingly effortless ease. Other species include the petrels, prions and storm petrels, and members of the gull family (skuas, a gull, and a tern). The Adelie penguin is the most important of the seabirds, both numerically and in its consumption of krill, which in total approaches that consumed by the crabeater seals and about half that of the baleen whales. It inhabits the entire coastline of Antarctica, often nesting in large colonies.

E. Mickleburgh/Ardea London

▲ An epitaph to the whaling years: a scene of ecological devastation at the abandoned whaling station at Prince Olaf Harbor on South Georgia Island. Although whaling ceased in 1965, much of the debris from the station remains. More importantly, the population levels of the whale species hunted in this region have never recovered.

## WHALES AND THEIR FUTURE

Seven species of baleen whales, which feed predominantly on krill, and eight species of toothed whales, which feed on squid and fish, and some on penguins, seals, and other whales, flourish in the Southern Ocean. The whaling industry focused on the largest of the baleen whales—blue, sei, fin, and humpback—and reduced their populations to very low levels. Even with the continued prohibition on whaling of these species, recovery of the populations to former levels is doubtful, and the very survival of the blue whale is at risk. The toothed sperm whale has also been reduced to low levels and is now a protected species.

While they are in Antarctic waters during the summer months, the present stock of baleen whales consumes some 40 million tonnes of krill, compared with about 150 million tonnes before whaling. The difference, termed the "krill surplus", may not necessarily be available to the large whales to rebuild their numbers. The penguins, seals, and smaller species of whales, which have a shorter generation time, are able to build up their populations more rapidly than the great whales, and thus become more powerful competitors. Currently, there is in fact an increase in numbers of penguins and Antarctic fur seals.

## CONSERVATION OF LIVING RESOURCES

An international agreement, the Convention for the Conservation of Antarctic Marine Living Resources, has been in force since 1982. Its prime purpose is to ensure rational use of the living resources of the Southern Ocean and it is attempting to manage the fisheries in an ecosystem context. Though progress is slow, the fundamental issues are set in a conservation standard which requires that all dependent and related species, and the recovery of depleted stocks (such as whales), are taken into account in the management of the fisheries. This convention applies to the whole Southern Ocean and is thus unique in the world.

KNOWLES KERRY

# 4 The Ocean Depths

Because of its inaccessibility and extreme conditions, the deep sea has always been regarded as a mysterious and forbidding place. This sense of awe has been accentuated by the names given to its greatest depths, the "abyss" and the "hadal" zones, conveying images of bottomless pits and hell. It has at once been considered as home to fantastic monsters and as lifeless as space. Over the last century technological developments have allowed scientists to probe deep below the waves and into the abyss. Surprisingly, far from being an uninhabited wasteland, the deep sea has proven to be home to a diverse collection of fish and invertebrates that are adapted to living within this environment of extremes.

Norbert Wu

▲ The deep-sea anglerfish *Melanocetus johnsoni* uses a modified dorsal spine as a built-in "fishing pole". The spine is topped with bioluminescent tissue.

◄ These squid *Pyroteuthis margaritifera* inhabit the pelagic ocean depths. They swim via jet propulsion and, like all cephalopods, have a well-developed sensory system.

## The Deep Sea

The deep sea is that area of the ocean below 200 meters (650 feet). This is the depth where, in the clearest oceanic waters, so much light is absorbed that photosynthesis does not support phytoplankton. Why is this environment so different from the one we inhabit? The most important factors, for defining the deep sea and shaping the fauna within it, are pressure, temperature, and darkness.

### PRESSURE

On land we are accustomed to living under one atmosphere of pressure, the weight of the entire air column above us. Water is much heavier than air; in water, for each 10 meters (33 feet) of depth, the pressure increases by another atmosphere. Therefore, in the deepest parts of the ocean the pressure is over 1,000 atmospheres.

However, while this pressure certainly has an effect on the physiology of living organisms, it does not preclude life. Under high pressure water is virtually incompressible, but gas-filled spaces can be crushed. Most deep-sea animals have no air spaces, such as lungs and swim bladders, in their bodies. Those deep-living fish that have a gaseous swim bladder are able to control its internal pressure, unless they change depth rapidly. What appears forbidding to humans need not be a hindrance to other life-forms.

### TEMPERATURE

Unlike the ocean surface, where tropical water temperatures can be above 30°C (90°F) while polar temperatures dip below freezing, the deep sea is cold throughout, ranging from minus 1 to 2°C (30–34°F). The variety of surface-water temperatures creates many habitats, to which different animals have adapted. Thus, shallow-water communities differ from the polar seas to the tropics. In contrast, the constancy of deep-sea latitudinal zones: the deep sea does not get warmer from the poles to the equator, but is rather a single, cold bath. As a result, deep-sea animals are free to live throughout the abyss without having to adapt to different climates.

This might seem surprising in view of the vastness of the environment. The deep sea extends over 341 million square kilometers (132 million square miles)—67 percent of the globe—at an average depth of 3,800 meters (12,465 feet). Yet there are only three basic habitats: the water column, or pelagic zone; the soft bottoms of accumulated terrestrial particles and planktonic remains; and the rocky bottoms of undersea mountain ranges, volcanoes, and seamounts. Sampling of these habitats reveals little relationship between latitude and the distribution of organisms. Populations of closely related animals are found from one end of the deep ocean to the other. Species vary from place to place, but the genera they belong to are found throughout. This evidence indicates that, despite

▼ For centuries, marine explorers believed the oceans were inhabited by fearsome creatures. Myths of their dreadful deeds generated illustrations of their bizarre appearance. This colored woodcut, dated 1550, depicts the sea monsters supposedly found in the north Atlantic.

Granger Collection

LIFE IN THE DEEP SEA
1
2
3
4
5
6
7
8
9
10
11
12
13
14
15
16
17
18
19
20
21
22
23
24
25
26
27
28
29
30
31

its vastness, the deep sea has relatively homogeneous conditions throughout. Thus, the deep sea represents the largest continuous environment on Earth.

### THE ABSENCE OF LIGHT

The factor of greatest biological importance in this environment is the absence of light. Light is fundamental to all life. Sunlight is the energy source that powers the food web upon which all animals, including humans, depend. Plants—in the ocean the microscopic phytoplankton and seaweeds—are at the center of this web, using solar energy to convert water and carbon dioxide into carbohydrates through the process called photosynthesis. Carbohydrates are the "bricks" of nature, the macromolecules that organisms need to grow, reproduce, and maintain themselves. The plants, in turn, are eaten by herbivorous animals, which are themselves eaten by carnivorous animals, and thus the entire food web owes its existence to solar energy. Clearly, the absence of light is a critical feature of the deep-sea environment. Without light there can be no photosynthesis and thus no plants, no local production of food. Therefore, all the creatures that live in this vast expanse are dependent upon food which is ultimately produced in the thin layer of sunlit surface water. But only about 1 percent of the food produced at the surface reaches the bottom of the deep sea! A crude analogy might be one in which only the inhabitants of the top floor of a large apartment building produce food; whatever they do not eat themselves trickles down to the floors below. Residents of each successive floor eat their portion of whatever they can catch.

ROBERT R. HESSLER, SCOTT C. FRANCE, AND MICHEL A. BOUDRIAS

**DIVISIONS OF THE MARINE ENVIRONMENT**

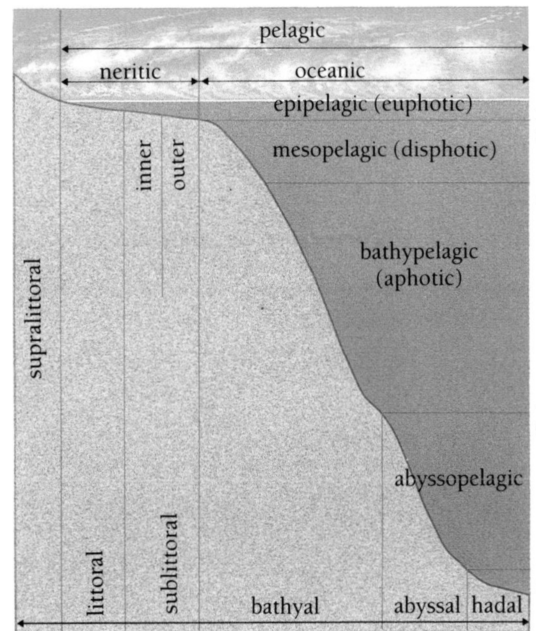


◂ Like the terrestrial world, the deep-sea environment is divided into a number of areas, each of which supports a diversity of organisms.

## LIFE IN THE DARK

Sunlight penetrates 1,000 meters (3,300 feet) into the ocean, though it supports photosynthesis only in the upper 200 meters (650 feet). Between 200 and 1,000 meters is a twilight zone, where only blue light remains. By far the greatest volume of the deep sea is below that, enveloped in a world of perpetual darkness where the only light perceived is biologically produced.

### BIOLUMINESCENCE

Many midwater organisms use bioluminescence in a variety of ways. Some animals flash unique patterns to identify and attract mates in the dark.

◂ Below 1,000 meters (3,300 feet), the deep sea is dark and, to most species, forbidding. But some creatures have adapted to this habitat, and have developed ways of surviving in this environment.

1 cod
2 green turtle
3 mackerel
4 flying fish
5 Portuguese man-o'-war
6 dolphin
7 bluefin tuna
8 prawn
9 sperm whale
10 squid
11 dragonfish
12 shark
13 hatchetfish
14 swordfish
15 lanternfish
16 viperfish
17 demersal octopus
18 gulper
19 anglerfish
20 deep-sea eel
21 bivalve
22 anglerfish
23 whalefish
24 whalefish
25 rattail
26 brittlestar
27 crinoid
28 short-armed starfish
29 glass sponge
30 tripod fish
31 lamp shell

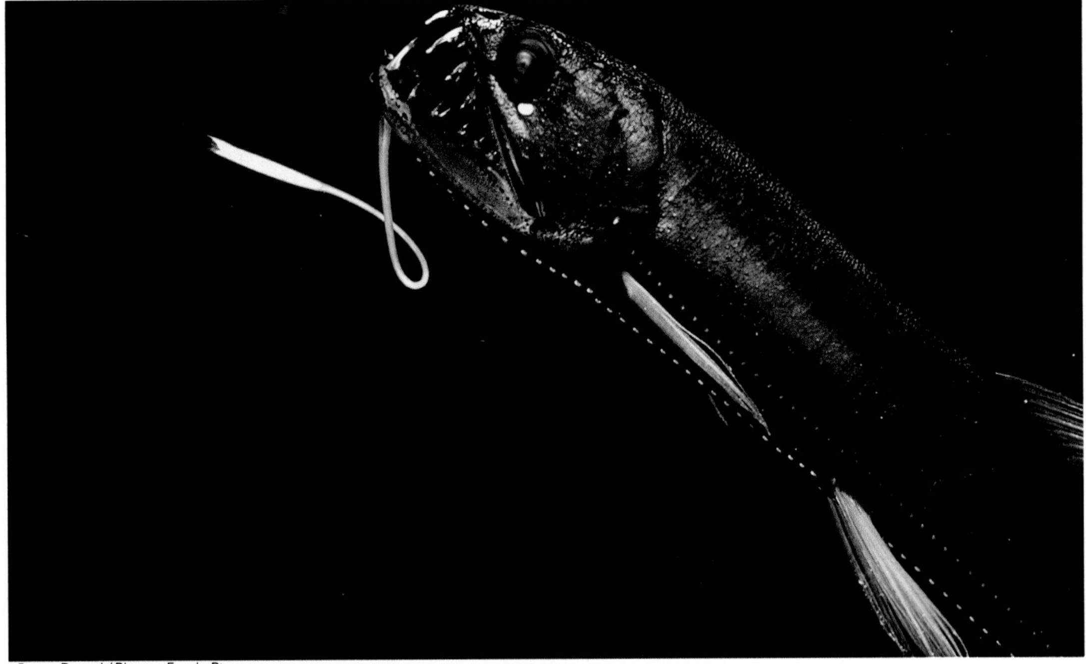

Peter David/Planet Earth Pictures

◂ The deep-sea snaggletooth *Astronesthes gemmifer* has an impressive array of teeth, distinctive ventral light organs, and a luminous barbel beneath its chin. Surprisingly, this strange creature is related to salmon and trout.

Peter David/Planet Earth Pictures

▲ The deep-sea fish *Eutaeniophorous* spp inhabits depths to 1,600 meters (5,280 feet), in an environment where "boom and bust" food supplies are a fact of life. This individual provides a rather extreme example: its stomach is grossly distended from swallowing a large number of copepods.

In the twilight zone, several kinds of fish and shrimp possess bioluminescent organs on the lower half of their bodies which act as a camouflage mechanism. These animals mimic the wave-length and intensity of sunlight filtering down from the surface, thus vanishing from the view of upward-looking predators that are searching for the silhouette of prey. In the deeper zones of the pelagic deep sea, large gelatinous organisms (ctenophores, salps, medusae) use bioluminescence as a startling mechanism. When touched, they produce a burst of light that may temporarily blind or distract their predators. Some copepods expel bioluminescent clouds to confuse their attackers as they escape.

Bioluminescence may also be a hunter's tool. Many deep-sea predators expend little energy searching for prey, but instead find ways to attract them. The anglerfish, the most notorious of this genre, waves a bioluminescent lure at the top of its head to entice smaller fish, then captures them with a rapid gulp of its large, armored mouth. It has been suggested that krill emit light from the eye, like a flashlight, to illuminate prey which have struck its antennae.

## DETECTING PREY

Body pigmentation may also be used as camouflage against predators. Below 500 meters (1,640 feet) most deep-sea fish are black, while crustaceans are deep red. In limited blue light, this renders them effectively invisible. How, then, do predators detect their "invisible" prey? Many have evolved alternate sensory modes. Numerous fish species detect sound waves generated by motion. Sensory hairs in internal canals or on the body surface respond to small changes in water pressure around the animal. Some shrimp have antennae, several times as long as their body, which spread out in the water waiting to be touched. Many of these animals are also capable of detecting low concentrations of odor. They can smell food, or even potential mates, from several meters away, and then follow the odor trail wafting from its source. These strategies are also employed by bottom dwellers.

Peter David/Planet Earth Pictures

► *Caulophryne jordani* is a deep-sea anglerfish with more than just a bioluminescent "fishing pole" (modified dorsal spine). The fine hairs covering its body are part of the fish's nervous system and assist it in detecting prey.

### CONSERVING ENERGY

Life in the pelagic deep sea is not only dictated by darkness itself, but is strongly affected by the resulting low food levels and the need to utilize energy effectively. For example, in order to reduce the amount of energy spent maintaining position in the water column, animals have found ways of floating passively: by replacing heavy bone with lighter material such as cartilage; by incorporating fat, which is lighter than water, in body tissues; or by increasing the body's water content. On the whole, life in the bathypelagos proceeds at a languid pace with most animals casually swimming or drifting, simply waiting for food to come to them.

ROBERT R. HESSLER, SCOTT C. FRANCE, AND MICHEL A. BOUDRIAS

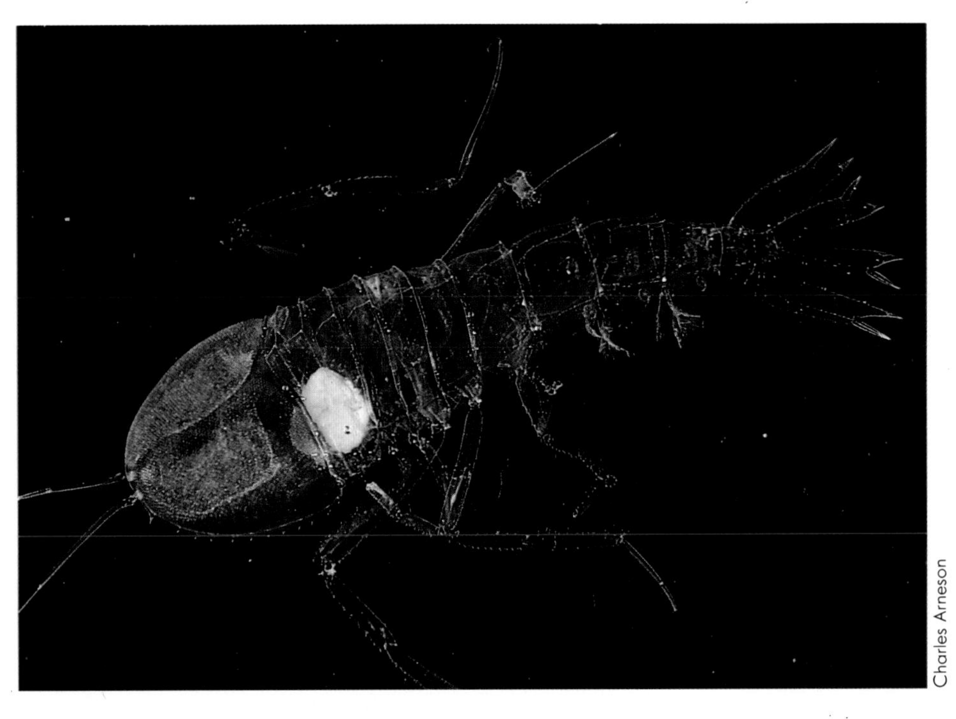

Charles Arneson

▲ One way of avoiding predators in the mesopelagic environment is to become transparent. Apart from eye pigmentation and stomach contents, the body of this deep-sea amphipod *Cystisonia* is as clear as glass.

# BOTTOM DWELLERS

Bottom-living animals differ from water column inhabitants in many ways. For unknown reasons, bottom dwellers do not use bioluminescence. Hence vision is unimportant, and most benthic creatures either have no eyes or are otherwise blind. Bottom dwellers have little or no pigmentation. The occasional splash of color—orange foraminifera, reddish brittlestars, yellow sea-lilies—may result from ingested food or be a byproduct of a physiological process.

The animals that live on the bottom of the deep sea generally belong to the same taxa (groups) one finds in shallow water: for example, annelids (bristle worms), molluscs (snails and bivalves), and crustaceans (such as shrimp, amphipods, and isopods). But at the lower levels of taxonomic classification—species, genus, and more rarely family—the animals are usually different. A species that can live in shallow water could not survive in the deep sea, and vice versa. Interestingly, most deep-sea animals are small: a typical deep-sea isopod, for example, is smaller than one from shallow water. As well, taxa composed of small animals make up a greater proportion of the community in the deep sea.

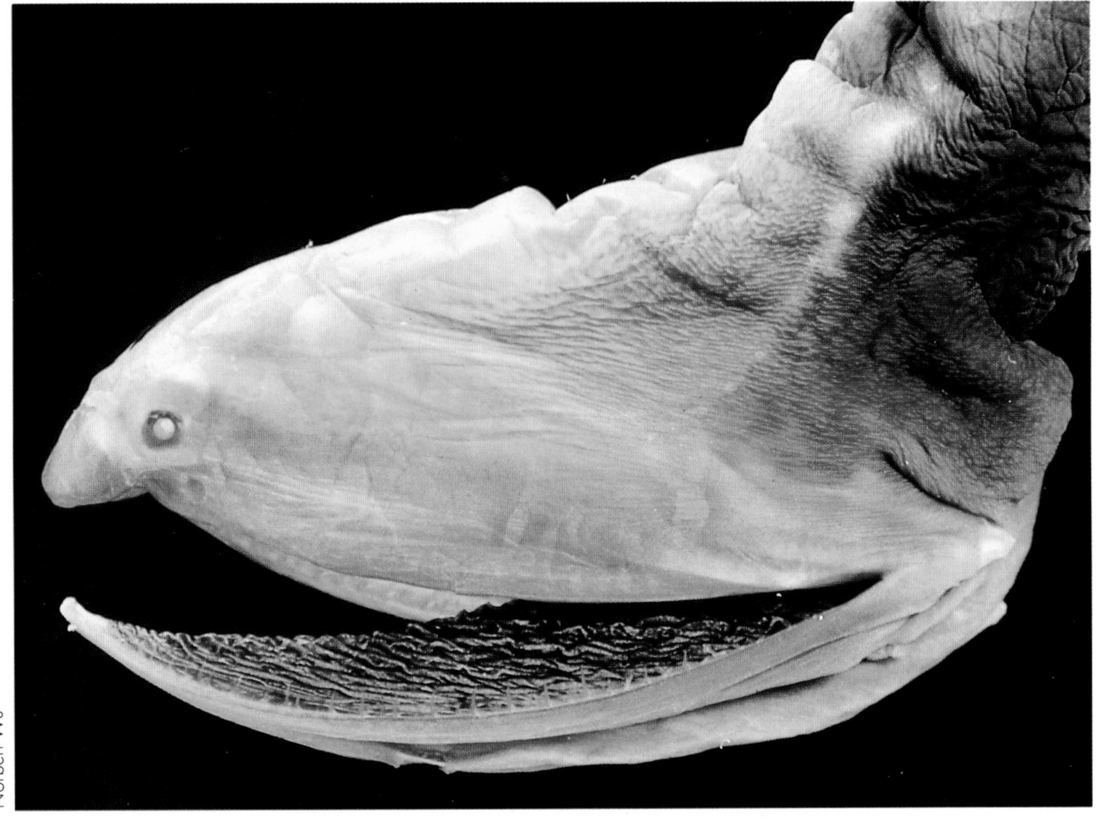

Norbert Wu

◄ Strangely disproportionate to the size of its body, the gigantic pelican-like jaws of the gulper eel *Eurypharynx* spp enable this fish to swallow large prey. As with most bathypelagic vertebrates, the gulper's teeth and skeleton are flimsy.

Peter David/Planet Earth Pictures

▲ The bottom-dwelling deep-sea prawn *Nephropsis atlantica* is a deposit feeder. The well-developed setae on its front limbs assist it to detect food particles on the sediment surface.

▶ The hard-bottom substrate of the mid-ocean ridges is dominated by suspension-feeding organisms such as anemones and stalked glass-sponges.

IFREMER

OBTAINING FOOD

The single most important factor that molds the way deep-sea animals live is paucity of food. In the depths of the sea, nutrition comes as a weak rain of particles from the surface, ranging in size from small zooplankton and fecal pellets to plant debris and large fish or marine mammal carcasses. Bottom-living animals feed on this food in three basic ways: they pick up deposited detritus from the bottom; filter suspended food out of the water; or eat other animals. All three feeding types live in the deep sea, but in different proportions, depending on the currents and whether the bottom is soft sediment or rocky.

Deposit feeders are most common. These animals either live a sedentary life, often inhabiting tubes or burrows, or roam over the mud leaving tracks in their wake. Some, such as holothurians (sea cucumbers), ingest the surface layer of mud and digest what little food it contains. Others, including most of the crustaceans, carefully select only the nutritious particles. The most diverse group of bottom dwellers, the polychaetes or bristle worms, include some with long tentacles which extend from burrows to sweep the bottom for food particles.

Suspension feeders obtain food either by filtering large volumes of water or by spreading a net of arms or tentacles to capture their meal. This way of life is difficult in the deep sea, where few particles are in suspension and currents carrying particles toward animals are typically slow, moving only 2 centimeters (1 inch) per second. Many suspension feeders also need hard surfaces for attachment. For these reasons, suspension feeders are less common than deposit feeders in the deep sea, particularly on soft bottoms.

Because suspension feeders depend on current flow to bring their food, they benefit by being exposed to faster moving water. As current speed is greater slightly above the bottom, many suspension feeders have evolved ways to gain additional height, thus placing themselves within the best zone of suspended food. For example, glass sponges grow on long, delicate stalks; deep-sea corals can be 2-3 meters (6-10 feet) high, with complex branching patterns reminiscent of small trees; and many echinoderms, such as the brittlestars and brisingid starfish, climb the stalks of other animals.

On hard bottoms, topographic relief may intensify the speed of passing currents. It is here that suspension feeders are most likely to be concentrated. Even on small scales, such as the top surface of manganese nodules, only centimeters above the bottom, small suspension feeders can dominate, taking advantage of slightly enhanced water movement.

Strict carnivores are relatively uncommon on the deep-sea bottom, perhaps because there is too little for them to hunt. However, many fish and crustaceans that swim above the bottom are specialized for consuming carrion. Much like vultures, they are able to detect infrequent food falls and descend on them within hours of their arrival on the bottom. Since they must often go for long periods without food, they are effective in gorging themselves when the opportunity arises. Thousands of amphipods have been recorded at bait, even in trenches at depths up to 10,000 meters (33,000 feet). They attack fish carcasses like maggots and within hours devour them.

As a result of the lack of food, the abundance of life in the deep sea is low, hundreds of times lower than in shallow water. For decades scientists studying soft bottoms thought that because of this low abundance, the diversity of animals must also be low. We know now that this

P. Briand/IFREMER

▲ Dead creatures falling to the ocean depths are rapidly devoured by mobile scavengers such as *Alicella gigantea*, the largest of the amphipods, which can grow to a length of 34 centimeters (13 inches).

gence Nature/NHPA

◄ This extraordinary creature, *Colossendeis colossea*, is a deep-water species of marine spider or pycnogonid found at depths of 5,000 meters ( 16,500 feet). Deep-water spiders have no eyes and feed by using their proboscis to extract juices from worms and other soft-bodied invertebrates.

is untrue. The variety of life in deep-sea soft bottom communities is extraordinary, almost as high as it is in tropical shallow waters. The reason for this is still unknown, but may well be related to the unusual stability and predictability of the deep-sea environment.

ROBERT R. HESSLER, SCOTT C. FRANCE, AND MICHEL A. BOUDRIAS

# THE SULFUR VENTS

In 1977, geologists studying the Galapagos spreading center in the equatorial eastern Pacific made a wondrous discovery. There, in the ice-water 2,600 meters (8,500 feet) below the surface, warm water spewed forth from cracks in the ridge basalts. Clustered around these vents were strange and beautiful creatures new to science. Luxuriant thickets of worms, 1 meter (3 feet) long, with blood-red, tentaculate plumes emerging from white tubes, gave the impression of a giant rose garden. Enormous 30-centimeter (12-inch) clams and piles of large mussels lay among the worm thickets. Crabs and shrimps by the dozen clambered over the sedentary fauna in search of food; tiny bristle worms projected their tentacles from tubes on the rock; small anemones carpeted the hard sea floor; and strange spaghetti-like acorn worms and gelatinous "dandelions" (siphonophores) sat delicately around the edges of the vent field.

The discovery of this deep-sea hydrothermal vent community opened the door to an exciting period of research that has revealed vents on spreading center ridges throughout the ocean.

Al Giddings/Ocean Images

▸ The hydrogen sulfide environment of deep-water vents has produced some anomalous animals. These giant tubeworms can reach lengths of 1 meter (3 feet) and are fed by their symbiotic sulfur-oxidizing bacteria.

IFREMER

▾ The tripod fish *Benthosaurus* sp. exemplifies the "sit and wait" feeding strategy of many deep-sea creatures. Modified fins allow this fish to sit above the bottom in the faster-moving currents where it is better able to detect odors from food.

## OASES IN THE OCEAN

The vent ecosystem is an anomaly in the deep sea: an "oasis" where a well-fed, teeming community thrives in an otherwise food-poor habitat. How can these animals grow so large and in such abundance in the deep-sea "desert"? The answer lies with the chemical content of the emerging hot vent water and the bacteria that take advantage of it. At these undersea vents, water seeps down through cracks in the rock and is superheated to temperatures greater than 600° C (1,100° F) as it nears lava-filled magma chambers. The composition of dissolved constituents in the hot water is altered via reactions with the rock. Biologically, the most important change is the addition of hydrogen sulfide, which sulfur-oxidizing bacteria use as an energy source to manufacture carbohydrates through chemosynthesis. This process is similar to photosynthesis, but differs significantly in that it does not rely on sunlight, the fundamental source of energy on Earth. At vent ecosystems, sunlight is supplanted by geothermal energy.

Scientists studying the vent water found abundant chemosynthetic bacteria living in the hot water and growing on the rocks around the vent openings. More interestingly, when biologists closely examined the vent animals they found that several species had a mutually beneficial relationship, or symbiosis, with chemosynthetic bacteria living within their tissues.

For example, the giant vent tubeworm has no mouth or intestinal tract. The trunk of its body contains a large organ packed with sulfur-oxidizing bacteria. The worm extends its plume into the warm water to absorb hydrogen sulfide and other necessary inorganic chemicals which the blood carries to the bacteria. The bacteria benefit by being provided with an ideal incubation chamber; the worms benefit by receiving nutrition from the bacteria. Symbiotic relationships with chemosynthetic bacteria have also been noted in vent bivalves and snails, which harbor the bacteria in their gills.

## ADAPTING TO THE VENT ENVIRONMENT

Living in a community based on hydrogen sulfide is not without problems. Hydrogen sulfide is toxic to one of the basic metabolic processes of all animals, cellular respiration, and poisons oxygen transport in the blood by pre-empting the oxygen-binding site of hemoglobin. Thus we have a paradox: a compound which is a beneficial energy source is at the same time poisonous to the organisms that rely on it!

But vent organisms have evolved ways of coping with this toxicity. The tubeworm binds the sulfide to a second, special site on the hemoglobin molecule. Thus one molecule can transport both oxygen and sulfide in the blood. The vent clam has evolved an entirely new molecule whose sole function is to bind and transport sulfide, thus leaving hemoglobin with the sole task of oxygen transport. In the crabs, which have no symbiotic bacteria, sulfide is converted to a less toxic compound in the liver. The fact that alternative mechanisms have evolved to overcome sulfide toxicity is a good indication that the vent habitat, with all its food, is a preferred place to live and that vents have been around long enough to permit this evolution.

## LOCATING NEW HABITATS

Even though the hydrothermal vent habitat has been in existence for millions of years, individual sites are much more ephemeral. In the process of heating sea water, the magma chamber slowly cools off, with the result that circulation slows and the chemical changes diminish. Alternatively, the plumbing in the rock can be closed off by earthquakes or chemical precipitation. In either case, the vent dies, and the community dies with it. To avoid extinction, it is imperative that vent species be able to disperse to new vent habitats. For most species, the adults are either attached to the rock or too clumsy to be able to walk very far. They must rely on dispersal of their larvae. Scientists have yet to determine how offspring locate new vent sites. But, since food is plentiful, organisms can produce many young. These may drift in the water until, with luck, they find the needle in the haystack, another hydrothermal vent. Most will not, and they will die.

Much remains to be studied in this fantastic, newly discovered habitat. How does the community change through time? What are the similarities and differences between vent communities in different oceans? What new kinds of animals are waiting to be discovered? The hydrothermal vents hold secrets which will keep deep-sea scientists busy for many years.

ROBERT R. HESSLER, SCOTT C. FRANCE, AND MICHEL A. BOUDRIAS

Robert Hessler

◂ A clump of mussels and clams cluster at a fissure at 2,500 meters ( 8,250 feet) on the "Rose Garden" deep-water vent site in the Galapagos rift zone. Anemones carpet the adjacent bottom.

# FISHES OF THE DEEP SEA

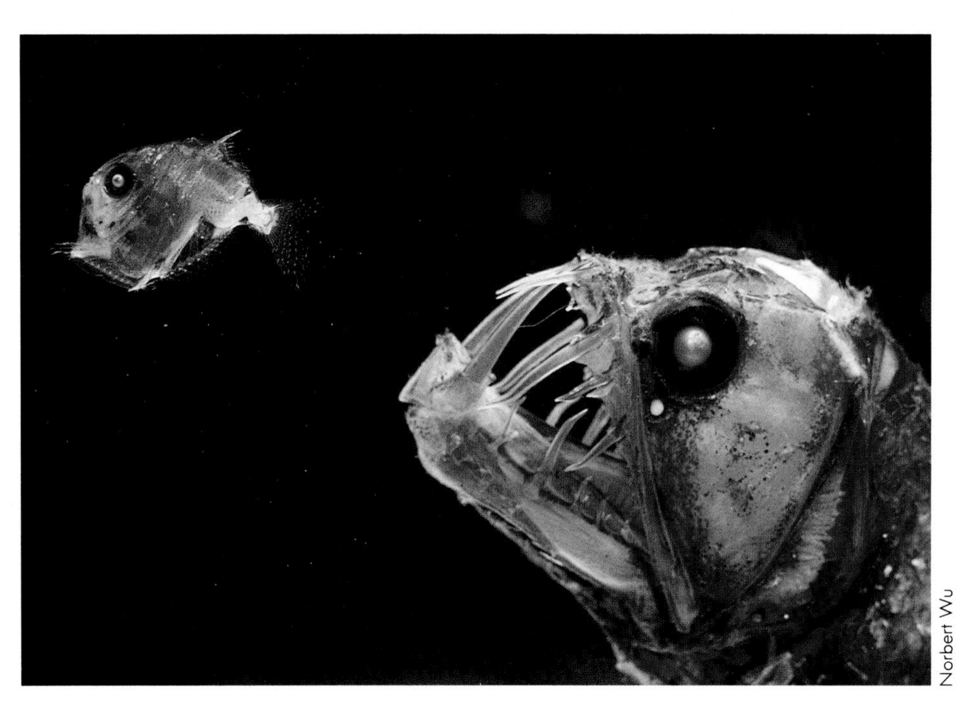

Norbert Wu

A deep-sea viperfish and its prey, a hatchetfish.

There are two basic kinds of deep-sea fishes: the pelagic fishes of the midwaters and the bottom-dwelling benthic fishes. Midwater fishes between 200 and 1,000 meters (650–3,300 feet) live in the twilight or mesopelagic zone where the remaining light is slowly absorbed by sea water. Those midwater fishes below 1,000 meters (3,300 feet) live in the total darkness of the true deep sea, the bathypelagic and abyssopelagic zones.

Although the deep sea is the world's largest habitat, only some 2,500 species—about 10 percent of all fishes—live there. Of these, about 1,500 species are bottom dwelling, 850 pelagic species live in the twilight zone, and 300 midwater species live below 1,000 meters (3,300 feet). Some species move between the zones. Rattails are the dominant family of bottom fishes, while lanternfishes, lightfishes, and dragonfishes predominate in the twilight zone. Anglerfishes and whalefishes are the most common bathypelagic fishes below 1,000 meters (3,300 feet).

Most deep-sea fishes belong to primitive groups, like sharks, eels, and the less advanced bony fish families. Their environment is harsh, with low levels of food, light, and temperature. Presumably because of competition from the more advanced groups of bony fishes, deep-sea fishes have successfully adapted to this habitat. The lack of light and low level of food have been the most important evolutionary forces molding their striking adaptations.

## BIOLUMINESCENCE

Living light, or bioluminescence, is characteristic of most of the fishes of the twilight zone, some of the benthic fishes, and the bathypelagic anglerfishes. Twilight zone fishes, such as the aptly named lanternfishes and lightfishes, have light organs or photophores in their skin, usually on the lower half of the body, and can control the amount of light they produce. Some of the bottom-dwelling rattails and other species have internal light organs, and the light shines through scaleless windows on the belly or through the body wall. Here light is produced by a colony of luminous bacteria living in the light organ. Similarly, the light in the lures of deep-sea anglerfishes comes from a colony of bacteria within the lure.

This living light has several functions. In twilight zone fishes, the silhouette will be obliterated to predators hunting from below if the light produced matches the intensity of the downcoming sunlight. Most lanternfish species have distinct patterns of light organs that could be recognized for schooling or mating (much as color is used by shallow-water fishes). Large light organs near the eyes presumably help the fish to see food. The elaborate luminous filaments on the heads of anglerfishes and chins of dragonfishes must lure prey toward the fishes' enormous mouths. Finally, some fishes are able to flash or squirt a luminous cloud to startle or divert a predator.

## VERTICAL MIGRATION

Many twilight zone fishes migrate from the daytime depths of 500 to 1,000 meters (1,640–3,300 feet) to the upper 200 meters (650 feet) at night, to feed in the rich surface waters. During daylight hours these fishes would be easy prey for the surface predators in the shallows. But at night their adaptations to the dark deep sea make them equally at home in shallower water. A number of bottom-dwelling fishes also rise from the bottom, particularly at night, but the distances they travel are usually much less. There is no evidence that the midwater fishes living below 1,000 meters (3,300 feet) undertake a vertical migration. With no light penetrating into their domain, there would be no daily cues for such movement.

## FOOD AND FEEDING

As all food is produced in the surface waters, the amount of food decreases with increasing depth. A striking feature of most deep-sea fishes is a very large mouth, often with large or specialized teeth. Anglerfishes, for instance, have long, slender teeth that bend into the mouth as prey enter, but lock upright if prey attempt to go backward out of the mouth. Other fishes and crustaceans are the food of most pelagic deep-sea fishes, and a number of species, like anglerfishes and swallowers, eat fishes longer than their own body. With sparse food and slow digestive rates because of low temperatures, meals may be infrequent.

### SENSE ORGANS

The eyes of deep-sea fishes are variable, depending on the zone they inhabit. Twilight zone fishes typically have large eyes, and some focus upward to take advantage of the shadows of prey caused by the downcoming light. In the black bathypelagic zone, the eyes are usually tiny. Most whalefishes, for instance, have eyes with a maximum diameter of about 2 millimeters (0.08 inch); there is no lens, so no image can be formed.

In this lightless environment the lateral line system, which picks up pressure waves created by swimming animals, is the most important sense organ and may be very large. In some bathypelagic fishes the nasal organs are most highly developed in males, indicating that mate selection may involve female pheromones. The ear stones or otoliths of most lanternfishes are large and sculptured, suggesting that sound reception may be important in this group of twilight zone fishes. However, ear stones of most bathypelagic species are small and featureless.

### REPRODUCTION

The male and female sexes of twilight zone fishes are usually separate individuals. They breed seasonally and produce a moderate to large number of small eggs; their larvae inhabit the food-rich upper waters. In the harsher bathypelagic zone, striking modifications have occurred. All male anglerfishes are dwarfs without lures. Some become parasitic on the females; after the male bites into the skin of a female, a placenta-like connection forms around the male mouth and all nutrients are received from the female. Males and females are thus guaranteed to be together in the breeding season. The few male whalefishes known are also dwarfs, less than 5 centimeters (2 inches) long, so they do not compete for resources with mature females, which may be up to 40 centimeters (16 inches) long.

A number of pelagic deep-sea fishes, particularly among eel families, breed only once and die. Some bottom-dwelling fishes are hermaphroditic, with both sexes present in one individual. These rare species do not have to find the opposite sex for a successful mating, just another of their own species; and self-fertilization of their own eggs may be possible.

There is still much to learn about the biology of deep-sea fishes. Great advances should be made when we can keep some of these extraordinary animals alive in aquaria.

JOHN R. PAXTON

Norbert Wu

The dwarf male deep-sea anglerfish attaches itself parasitically in front of the female's tail fin.

# 5 THE FRINGES OF THE SEA

Mark Mattock/Planet Earth Pictures

▲ Periwinkles washed up onto the beach. Soon these dead shells will be crushed or eroded by wave action, the pieces contributing to the shifting sediments from which beaches are constructed.

◀ Low tide exposes the magnificent white sand swirls carved by the ever-changing channels in this estuary in Queensland, Australia. Protected from the fresh water of the estuary, a small fringing reef nestles in the lee of the peninsula.

The shallow waters fringing the continental coastlines support concentrated and diverse populations of plants and animals. Below most coastal waters are continental shelves—gradually sloping underwater extensions of the land, sometimes carved with deep submarine canyons. The meeting of land and sea produces a range of environments where plants and animals thrive on sandy beaches and rocky shores, in river estuaries, dense mangrove forests, saltmarshes, and seagrass meadows. The habitat of the plants and animals of this zone is sometimes harsh and always unstable, often buffeted by waves, and subject to the constant ebbing and flowing of tides. The organisms that inhabit the fringes of the sea have adapted to their unpredictable environment in a remarkable variety of ways.

## THE CONTINENTAL MARGIN

Around all the continents and the major islands are shelves that extend from the shore to depths of 100 meters (330 feet) or so. These continental shelves grade into continental slopes that meet, in most places, a continental rise at depths between 3,000 and 4,000 meters (10,000–13,000 feet). Together, the shelf, slope, and rise make up the continental margin.

The topography of these submarine features is the result of erosion and deposition during the rise and fall of sea level through the most recent ice ages. When the sea level was at its lowest, some 20 million years ago, the areas that now make up the continental shelves were laid bare of water. Waves cut at a much lower level than today, rains and winds carved the shelf surfaces, and rivers flowed across them, building deltas to form the continental slope. As the glaciers melted and the sea rose and gradually covered the shelves, beaches, barrier islands, and sand flats were formed. Although the beaches became somewhat masked as the sea continued to rise, many are still identifiable, such as those containing mastodon bones on the continental shelf off New Jersey.

As depicted on charts, the shelves seem to be flat, with average depths of 60 meters (200 feet) and terminating at depths around 130 meters (430 feet). There are, however, hills, longitudinal and transverse gullies, sand ridges, and sand waves scattered over all the surfaces. The average slope across all shelves is seven minutes of a

David George/Planet Earth Pictures

◀ Jagged promontories and caves exposed by the erosion of lava tubes on Lanzarote Island, Canary Archipelago. The outside of a lava flow cools first, while lava in the middle continues to flow for some time, creating a "tube". When the inner flow finally cools, long hollows and caves often remain in the basalt.

Covell Publications/Horizon

▲ Sunrise gilds the majestic limestone stacks of the "Twelve Apostles", along the coastline of the Port Campbell National Park, Victoria, Australia. Sculptured by wind and waves, these rock pillars are graphic testimony to the strength of the forces acting on the interface between ocean and coast.

degree: a slope less than that allowed on the best of billiard tables!

Data from shipborne seismic profilers provide the basis for classifying the structure of the world's continental margins. Though all have been modified by the action related to changing sea levels, wave erosion, changes from currents, and slumping of sediments down the slopes, the underlying structural framework has remained undisturbed.

## TYPES OF SHELVES

Earthquake faults in the Earth's crust are common around the continental slopes. Faults are to be expected at the edge of continents, for this is where many earthquakes originate. The shelf inside the faulted slopes may have several origins. Older rocks, folded and faulted, may have been eroded into a rather flat shelf as the sea level rose. The material from the eroded rocks would have slumped onto the faulted slope, or if the scarps were steep, slid off into deep water.

Other shelves have been built along coasts where prodigious supplies of sediments poured out from the adjacent land, forcing the underlying crust to sink continuously from the weight of the deposits. This type of shelf and slope, formed under conditions of subsidence, has a structure similar to those of great river deltas.

The most intriguing shelves are those with a geologic "ridge" at their outer edges. Whether the ridge is part of the basement rock of the continent, a coral reef, salt domes, or a volcano, the projection acts as a dam to catch the sediments from the land as the sea level rises. Many shelves have been formed in this way. Coral-reef dams occur off the east coast of Florida and in the Java Sea. On the wide shelf off Texas, salt domes make up the dam, and potentially hold major pools of oil. For the most part, volcanic dams—those off the eastern Asiatic coast and in the Arctic—are responsible for the widest shelves. The fill in these huge shelves has been provided by the great rivers entering the sea across the respective coasts.

## SHELF WIDTH AND DEPTH

The average width of continental shelves is 75 kilometers (46 miles). Off the coast of the United States, the width varies from zero at Cape Canaveral, Florida, to just over 200 kilometers

▶ The island of Newfoundland lies above the continental shelf off the eastern coast of North America. Just as the coastal fringes have been shaped by erosion and weathering, so the continental margin is the result of weathering and deposition during the rise and fall of sea levels.

J. L. Manaud/Odyssey

**KINDS OF CONTINENTAL MARGIN**

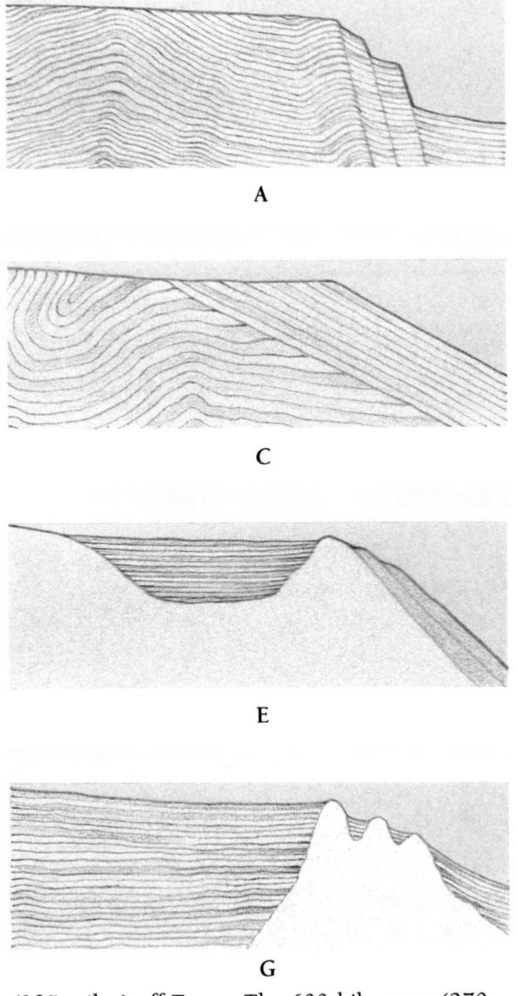


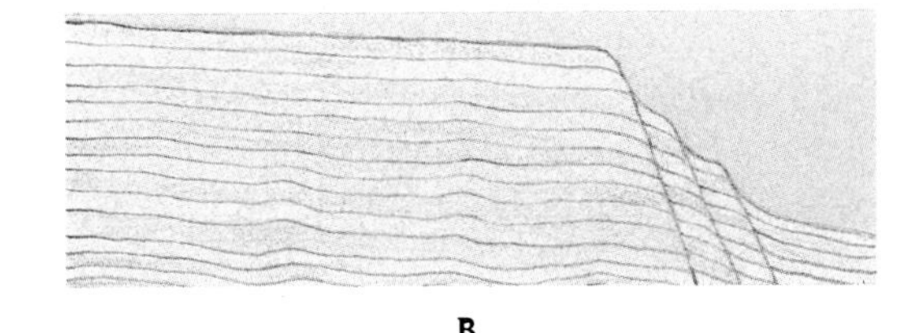


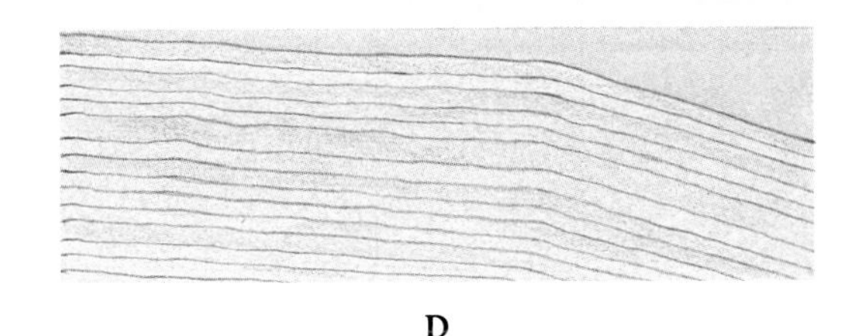


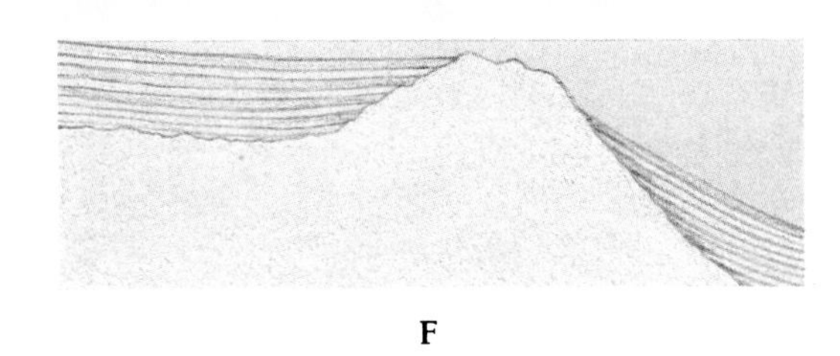


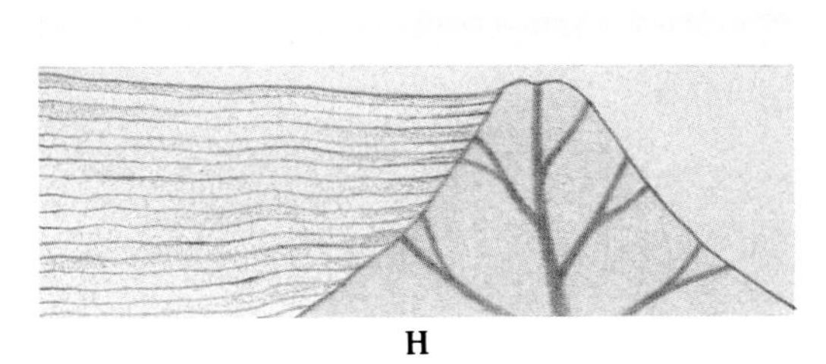


**A** Wave-beveled and faulted margin
**B** Prograded faulted margin
**C** Fold-beveled inner margin, prograded at outer edge
**D** Prograding margin
**E** Basement rock forms outer sediment dam
**F** Coral reefs form outer sediment dam
**G** Salt domes form outer sediment dam
**H** Volcanic action forms outer sediment dam

(125 miles) off Texas. The 600-kilometer (370-mile) wide shelf of the Bering Sea is the widest off any American coast. Cape Canaveral is probably the only place in the world where there is no shelf: it is unique because of millions of years of erosion by the Gulf Stream. Even where the shelf is narrow, just a few kilometers wide, there is always a slope diving off into thousands of meters. This occurs off the coasts of California, southern Saudi Arabia, Peru, Chile, New South Wales, Mozambique, and the east coast of India.

Wide shelves have many interesting features. Take, for instance, the shelf off northern Australia. At one end it terminates with New Guinea whereas north of Darwin it suddenly drops off into the Indonesian Trench. No other shelf has such an island-trench relationship. Crossing over that huge, deep trench that grades into the Java Trench to the west, Sumatra, Java, Borneo, and the Malaysian peninsula lie atop the shelf and are the most vigorous volcanic shelf islands in the world.

The shelf off western Europe contains the entire United Kingdom, plus the North Sea, the Skagerrak, and the English Channel. Other shelves on which islands lie include those that hold Wrangel and the New Siberian islands north of the Soviet Union and the huge island of Novaya Zemlya that separates the Barents from the Kara seas. Across the Arctic Ocean, that great archipelago of islands north of Canada that includes Hudson Bay, and Victoria and Baffin islands, lie on Canada's Arctic shelf.

## SUBMARINE CANYONS

The existence of canyons cutting deep into the continental margins has been known for more than a century, although their origins and significance in marine processes have been a subject for debate.

Submarine canyons have steep walls and sinuous valleys with V-shaped cross-sections. Their axes slope outward continuously, and many have reliefs comparable to those of the largest land canyons. Some, such as La Jolla off southern California, have heads that are nearly at the shore. Others, like those off the west coast of Corsica, appear to be recently submerged continuations of land canyons or, perhaps, the inner portion of submarine canyons elevated to become land canyons. At the mouth of the Congo River, a large submarine canyon extends 25 kilometers (15 miles) into an estuary and is 450 meters (1,475 feet) deep at the bay mouth.

As logical as the association between modern rivers and canyons may seem, there are some canyons that appear to be related to ancient rivers flowing during times of lowered sea level. The Hudson Canyon is a good example. It is fairly easy to project the present mouth of the river at New York across the shelf to the head of the canyon. The same is true of the huge Monterey Canyon, off the coast of central California. This canyon has depths and a cross-section that compare with the Grand Canyon of Arizona. Although no river enters the ocean today at the canyon's head near Monterey, geological data indicate that the Salinas River entered the sea there during the ice ages, modifying and exhuming this ancient canyon several times.

Some canyons have no river associations, either past or present. Two such are the Bering Canyon, the longest in the world, and the Great Bahama Canyon, with the highest walls of any we know. The Bering Canyon lies on the north side of the Aleutian Islands. It is more than 1,100 kilometers (680 miles) long, has a large number of tributaries on the south side from the eastern end of the Aleutian Island chain, and its head is some 500 kilometers (300 miles) from the mainland of Alaska. Perhaps the Yukon and Kuskokwim rivers flowed into the Bering Canyon during times of lower sea level, but there is no evidence to prove this hypothesis.

The Great Bahama Canyon is indeed a puzzle. It lies between the low Great Abaco and Eleuthera islands yet has wall heights approaching 4,350 meters (14,270 feet): greater than any land canyon! It has a total length of 280 kilometers (175 miles) and two V-shaped branches, one of which, the tongue branch, heads into the broad floor of the Tongue of the Ocean which itself can be traced southward for 100 kilometers (60 miles) at depths to 1,460 meters (4,790 feet). Fossils from the canyon walls indicate that this great valley existed during the 20 million years or so that the Bahama Banks have been growing upward apace with the sinking of the plateau.

## THE ORIGIN OF THE CANYONS

The origin of submarine canyons has been a hotly debated topic for many years. Erosion during a lowered sea level appears to have been a major contributor only in the Mediterranean Sea. There is good evidence that at times in the geologic past the sill at Gibraltar closed and the waters of the Mediterranean evaporated to a level much lower

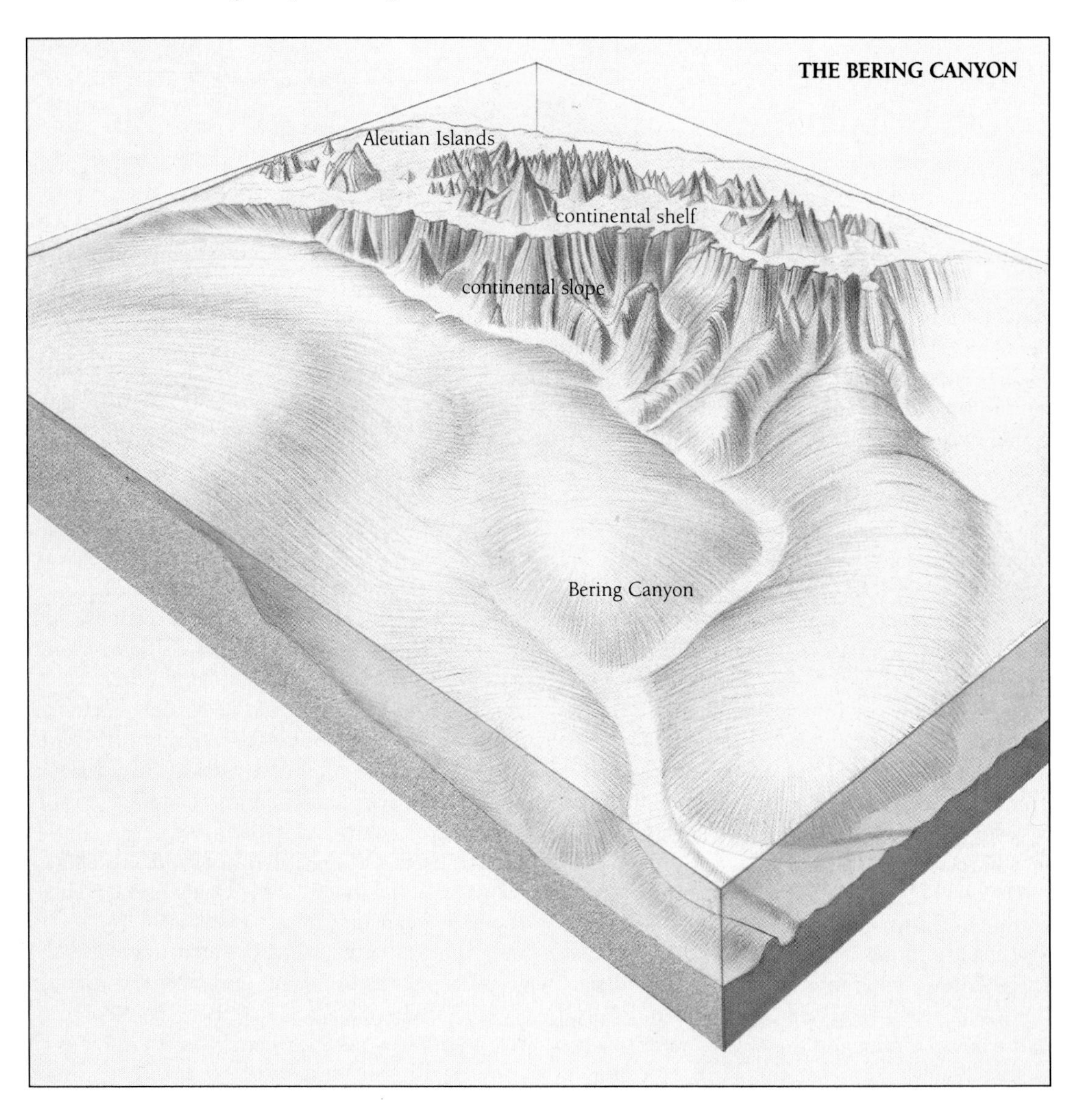


▸ Some continental margins are dissected by steep-walled canyons, scoured out by turbidity currents as they carry sediments to the ocean basins. The longest submarine canyon in the world is the Bering Canyon, north of the Aleutian Islands. The canyon emerges from the continental slope, and makes its sinuous way through the continental margin for more than 1,000 kilometers (680 miles).

Jeffery L. Rotman

▲ Jagged cliffs and irregular boulders surround a deep rockpool in the Galapagos Islands. Rockpools allow many plants and animals to live higher on the shore than they normally would, in addition to providing a refuge between the tides for some mobile species.

than today. During such periods, the canyons off Corsica and southern France could have been cut to near their present depths.

Nearly all canyons can be traced to the base of the continental slope where at their mouths are fans, similar to the alluvial fans at the base of mountain ranges on land. Canyons are worldwide, on all slopes. The sediments on the floors are usually coarse sands and gravels with many shallow-water animal and even land plant remains in the layers There is an adjacent land river more often than not, and the heads of active canyons have access to a rapid influx of unstable sediment that is carried seaward periodically by flows and slides. All these conditions indicate that the canyons are active conduits through which sediments are carried to the deep-sea floor, sometimes with enough velocity to erode any soft or unstable material on the canyon walls.

One means of carrying the debris down a canyon and creating some erosion on the way is by turbidity currents—masses of water with suspended sediments, the combination of which is heavier than the adjacent clean water. The sediment-laden water will descend a slope, at speeds up to 25 centimeters (10 inches) per second, displacing the clean water as it goes along. Yet most of the sediments on the floors of canyons are of sand and gravel, only occasionally mud. So, although turbidity currents must play a role, they are unlikely to be the whole story. If, however, they are combined with sand falls and other gravity-induced slides, and with strong currents extending throughout the water column resulting from high tides and exceptional surf onshore, the three processes together probably provide most of the action necessary to keep the canyons going.

ROBERT E. STEVENSON

Conrad Limbaugh

◄ Where submarine canyons abut coastlines, such as here in Baja California, they act as collecting points for land-based sediments. Sand piles up on the canyon wall, or perhaps is subjected to heavy wave action, whereupon it flows down the wall (as can be seen in this picture). Periodically, the huge sediment deposits that build up on the floor at the head of the canyon are flushed to the mouth to form vast submarine fans, which resemble terrestrial river deltas.

# ROCKY SHORES

Intertidal rocky shores occur in several different forms: vertical cliffs; and sloping shores, including platforms cut by the action of waves and areas that are littered with boulders. These varied shores often support different sets of plants and animals, nearly all of which are fully marine in origin. Very few terrestrial species are adapted for life between the tides on rocky shores.

## INFLUENCES ON ROCKY SHORE ORGANISMS

Two major physical influences act on the organisms in rocky intertidal habitats. First are the periodic and predictable patterns of the tidal rise and fall of water. In most areas of the world, the tide rises and falls roughly every 13 hours. As a result, marine plants and animals at the top of the shore are out of the water for a relatively long

# KELP FORESTS

Kelps, a form of brown algae, are conspicuous marine plants that usually grow on rocky reefs in temperate water (5–22° C/ 41–72° F), and are common along stretches of open coast in many parts of the world. The best-known species is the giant kelp *Macrocystis pyrifera*, which grows in cold temperate waters of both the Northern and Southern hemispheres. Other familiar kelp plants include *Ecklonia radiata*, from the Southern Hemisphere, and *Laminaria* species from the Northern Hemisphere.

Although there are several different species of kelp, two general growth forms can be recognized: those that have a simple long trunk (thallus) reaching heights of between 50 centimeters (20 inches) and 2.5 metres (8 feet), with a frond branching from the top of the thallus; and those, like the giant kelp, which grow to extraordinary lengths and have fronds appearing all the way up the thallus. Giant kelp may grow to over 50 meters (165 feet); to scuba divers it looks,like a cathedral of enormous plants. Some kelps, including the giant kelp, have floats or gas-filled chambers at the base of their fronds to maintain buoyancy, while others remain upright without floats. All plants are attached to the bottom by a root-like structure called a holdfast.

Kelps have a fascinating life cycle. They begin life as microscopic spores produced in special tissue on the mature adult plants, or sporophytes. When released, these spores develop into tiny male and female gametophytes. The male gametophytes fertilize the eggs produced by the females, which develop through embryonic and juvenile stages into the large marine plants. Adult kelps produce thousands of spores, but the vast majority never complete the life cycle. However, those that survive can grow very quickly: giant kelp, for example, often exceeds 30 centimeters (12 inches) per day, which makes it the fastest growing plant in the world. Individuals of some species have been known to live for more than 11 years.

Because kelps often form dense canopies, either on the surface (like giant kelp) or a meter or so above the sea floor, groups of plants are often referred to as "forests" or "stands". In terms of their interaction with their environment, they do behave rather like forests on land. Not only do they provide shelter and food for numerous species of mobile animals, including fish, sea-urchins, crabs, and sea-otters, they also reduce light levels to the sea floor, thereby helping to determine which other plants and sessile animals will grow beneath them. Beautiful animals such as sponges and sea-squirts can often be found growing beneath kelp forests, and studies have shown that the biological relationships among the organisms are very complex.

Thus kelps are not only visually dominant components of reef environments, they have an important role in structuring their environment. Humans too, play a significant part in the story of kelp. Many kelps are cultivated or harvested for food, fertilizer, and as a natural source of products such as algin and potash. Moreover, because kelps grow close to the shore they are adversely affected by the common human habit of disposing waste products in coastal waters.

MARGARET ATKINSON

Mark Conlin/Planet Earth Pictures

The kelp forest is an important coastal habitat for a range of marine creatures.

period before the tide rises again. They must therefore withstand periods of increased or decreased air temperature and desiccation during their time out of water. Organisms lower down the shore have to withstand only short periods out of water before their marine habitat is restored.

Coupled with this pattern is the regular fortnightly progression from neap to spring tides and back again. The water rises to higher levels and falls to lower levels during spring tides than during neap tides. In a period of neap tides, plants and animals high on the shore may not be covered by water for several days at a time; at the bottom of the shore, they may be under water continuously. In either case, prolonged exposure to either of these conditions may impose severe physiological hardships.

The second major physical factor is the force of waves. On wave-exposed shores, the effects of severe wave-shock are such that only flexible plants that are not damaged when hit by waves, or tough animals such as limpets and barnacles that are well stuck down, are able to survive. Wave-exposed shores are often covered by spray, thus reducing the possible deleterious effects of being out of water during low tide.

The effects of these processes vary according to the type of rocky shore. Vertical cliffs offer no shelter from waves, and plants and animals at all heights are subject to extreme wave-force on exposed shores. In contrast, on boulder-fields there are numerous spaces between the boulders and underneath the rocks that provide shelter from the impact of the waves.

## LIFE BETWEEN THE TIDES

Plants and animals living at the top of the shore must be resistant to prolonged periods of drying out during low tide. The plants at these high levels are able to survive even when they are dried to a crisp at low water. When the tide eventually comes in, these seaweeds rehydrate and begin to function normally.

The most common animals of the high shore are small grazing snails, which have a number of characteristics that prevent them from becoming dried out. Often they remain immobile when the tide is further down the shore, moving to feed only when the tides are high. Many of them also move around and feed when it rains, to take advantage of the dampness. When immobile, they commonly withdraw right into their shells, which are glued to the rocks by a ring of dried mucus or slime. Often the snails seek shelter in cracks or crevices in damp rocks.

At the bottom of the shore, plants and animals are larger, there are more species, and they must cope with different problems. Many species, such as large barnacles and sea-squirts, are sessile—they are stuck to the rocks and cannot move. Others, such as fleshy, branching seaweeds, sponges, and sea-anemones, are soft-bodied. Rarely do these plants and animals need to survive prolonged periods of drying out. Even during spring low tides, there is commonly splash and spray drifting over the lower parts of the shore, except during the calmest weather, and even then the air is usually humid because of the proximity of the sea.

These plants and animals can grow quickly and large. The plants grow profusely because they are under water without physiological stresses. Filter-feeding animals (barnacles, sponges, mussels, and sea-squirts) predominate. They can feed for extensive periods when the tide is in, for they are bathed in water that is rich in planktonic food which they filter out and swallow. As a result, there is often intense competition for space as the animals grow.

Toward the bottom of the shore, competitive overgrowth of some seaweeds by others, and swamping of barnacles or mussels by seaweed, are common. As a result, some species are reduced in numbers or coverage of the shore. There are, however, situations where one type of seaweed provides a favorable habitat for others by reducing the harshness of the environment. Tall plants that form a canopy cast shade on the substratum underneath their fronds, providing a refuge for many other species that cannot tolerate the unshaded conditions during low tide.

Low on the shore, plants and animals are subject to predation or grazing by animals coming upshore with the tide. Biological processes of competition and predation are much more important than physical factors as controls on the types, variety, numbers, and sizes of plants and animals at low levels.

Between the top and bottom of a rocky shore are many types of plants and animals that are patchy in distribution. Unlike the few, but

**TIDAL INFLUENCES ON TIME OUT OF WATER**

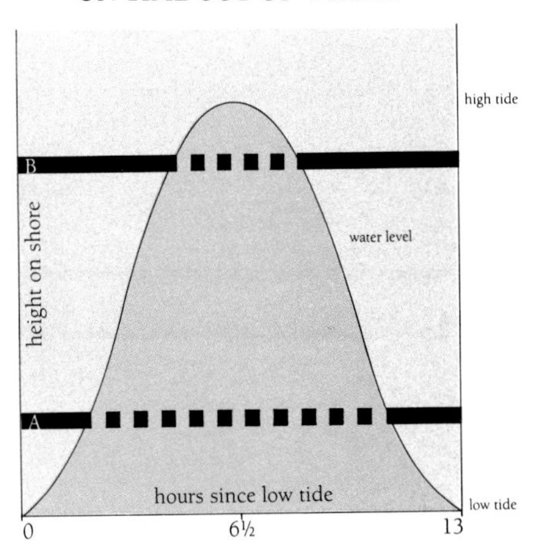


▲ The pattern of tidal rise and fall influences how long plants and animals must spend out of the water at low tide. At point A, low on the shore, plants and animals are out of the water for short periods. Higher up, at point B, they are out of the water for longer and under water for shorter periods.

▼ The multi-hued eight-rayed starfish *Patiriella calcar* is usually found low on the shore or in rockpools.

Kathie Atkinson

Kathie Atkinson

▲ Low tide at De Witt Island, off the southwestern coast of Tasmania, Australia. Few organisms live high on the shore; midshore areas are dominated by patches of barnacles and gastropods; and near the low-tide line macroalgae (in this case, bull kelp) are abundant.

widespread, species at higher levels and the many, but space-dominating, species lower down, many midshore species live in clumps, either alone or in mixtures. There are also patches of bare space and areas that are occupied by encrusting plants.

## A PATCHY HABITAT

Rocky shores on many of the world's coastlines are patchy habitats—a feature caused by a wide variety of processes. Patterns of grazing or predation are linked with features of topography such as holes or crevices in the rocks. Often, grazing or predatory molluscs shelter in cracks or crevices to avoid predators such as crabs or fish, or to reduce the intensity of physical stress (cracks and crevices are shaded and damp). Sheltering animals emerge at high tide to feed on surrounding organisms, but cannot move far from shelter because they must get back into a crevice before the tide falls or before the next period of inclement weather. Thus, when suitable shelters are sufficiently far apart, foraging causes patchy reductions in numbers, or even the complete disappearance, of prey around these shelters. Such "haloes" are widespread on rocky shores.

Opposite patches are formed when prey species manage to grow in, or move to, areas that are inaccessible to their consumers. Thus, some species of algae thrive in the middle of large patches of barnacles because their grazers do not have enough room among the barnacles to reach them. These refuges create a patch of prey surrounded by bare space.

Predation by large predators that come up the shore with the tide can also be a direct cause of patchiness. As the tide rises, some starfish, fish, and crabs move into intertidal areas to feed on seaweeds or animals inhabiting the shore. In Chile, for example, there is a clingfish that arrives with the waves at mid- to high levels on the shore. The fish attaches its sucker to the rocks and rasps off animals and plants from a sizeable area of the shore before releasing itself and returning to the water to reappear elsewhere.

Another major cause of patchiness is physical disturbance by waves, particularly during storms. For example, if waves overturn a boulder on a boulder-field, many of the plants and animals on the boulder will die. Some will be scraped off as the boulder moves against the rocks. Moreover, the organisms on the underside are now out in the open and those on the top are now in the dark: both are in a different habitat and may die as a result of the new environment, or from predation by predators and grazers that can now reach them.

Thus, new cleared surfaces become available for colonization by other plants and animals. If the boulder is not disturbed again for a long time, there will usually be competition for space among the organisms on each surface, leading to the eventual re-establishment of a uniform, non-patchy appearance throughout the boulder-field. Disturbances by waves affect different boulders at different times and boulders of various sizes at different rates. The result is a mosaic of patches—boulders at various stages of development, from newly overturned to completely undisturbed.

The primary cause of patchiness in the numbers and species of plants and animals on rocky shores is, however, the process of colonization. Most intertidal species have a life cycle that includes a dispersive larval stage. These organisms reproduce by shedding their spores or larvae into the sea, where they develop—often through a complex series of larval stages—before

Jeff Foott/Survival Anglia

◀ When threatened, a purple Californian shore crab *Hemigrapsus nudus* brandishes a large pair of powerful chelae (pincers) before retreating to a crevice in the rocks. Rocky shore crabs are both predators and scavengers, using their chelae to crush, rip, and tear their food.

being washed back into an intertidal habitat. During this period of development, the planktonic offspring may be washed many kilometers along the coastline by winds, currents, and tides. Large numbers are eaten by predators, ranging from other planktonic larvae to fish. Usually only a small percentage survives, and these are presumably widely scattered. When their development is complete, they can colonize a suitable empty patch of shore. The animals hastily attach themselves to the rocks and metamorphose to take on the adult body form.

The plants and animals that can colonize a patch will therefore depend upon which larvae and spores are in the water when the patch is cleared. This can result in different mixtures of species in different numbers, from one place or time to another, thus maintaining the patchy appearance of the shore.

Rocky shores are characterized by dynamic, variable processes of recruitment, competition, predation, disturbance, and physical stress. These vary in rate and intensity from one height on a shore to another, along a gradient of exposure to wave-force, and, patchily, from place to place and time to time. The result is a diversity of interactions among the species and, usually, a changing structure that provides endless fascination for the observer.

A. J. UNDERWOOD AND M. G. CHAPMAN

▼ Crabs, like most intertidal animals, reproduce by releasing their larvae into the sea, where they develop through several larval forms before returning to an intertidal habitat. Larvae may be dispersed over considerable distances; only a small percentage survives to colonize a rocky shore.

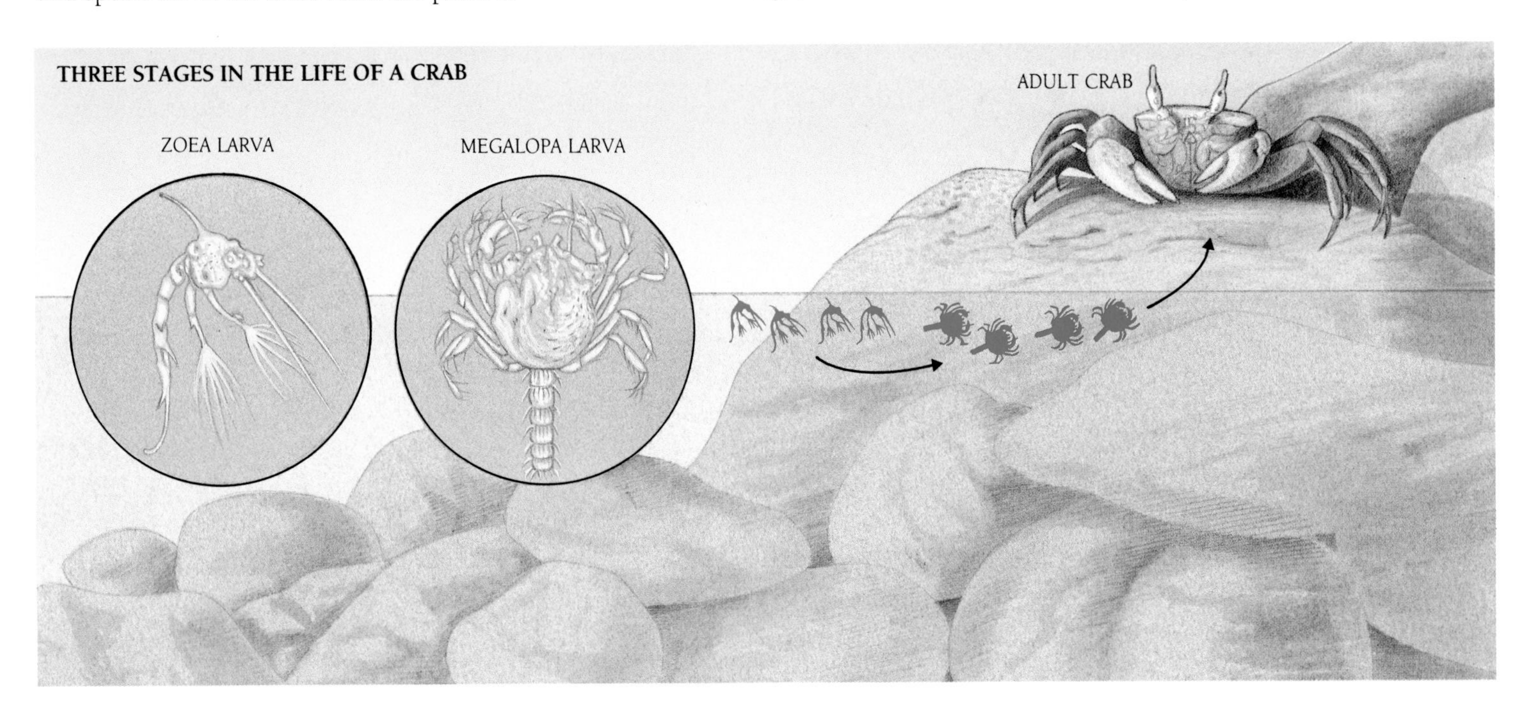

Peter Johnson/NHPA

▲ A beach undergoes constant flux, its form dependent on the availability of sediment and the force of the waves. Clearly, on this desert coast of Namibia there is no shortage of sediment. On shallow sloping beaches such as this, much of the movement of sediment is simply an exchange between offshore bars (evidenced by waves breaking far off the beach) and the dunes (berm) at the top of the beach.

# SANDY SHORES

Sandy shores make up some 75 percent of the world's ice-free coastlines. The occurrence and structure of these extremely dynamic environments depend on movements of air, water, and sand. Patterns of water movement determine whether a stretch of coast will be predominantly eroding or depositing. In the latter case the force of the waves determines the size of the particles deposited; minimal wave action is likely to result in a muddy shore, greater wave energy in a sandy beach.

Most sand consists mainly of quartz (or silica) which originates from inland erosion and reaches the sea via river systems. Some, however, has its origin in the skeletons of marine animals, the shells of molluscs, or cliff erosion; this sand consists predominantly of calcium carbonate. Other materials which may contribute to beach sands include heavy minerals, basalt, and the mineral group feldspar.

## SANDY BEACH SYSTEMS

A range of beach types may be distinguished, the extremes being *reflective*, in which the beach slopes steeply both intertidally and offshore, and *dissipative*, where the beach shelves gently, with a system of bars and shallow troughs offshore. In the former case, waves break on the beach itself, the wave energy being reflected back to sea. On fully dissipative beaches, on the other hand, the waves break on bars or sandbanks well out from the beach, so that their energy is largely dissipated before they reach the intertidal zone. The waves then give rise to a relatively gentle swash running up and down the beach. Dissipative beaches thus have broad surf zones and are of far greater recreational value than reflective beaches.

The width of the beach, and to a large extent its slope, are determined by cycles of deposition and erosion. In those few places where deposition consistently exceeds erosion, the beach marches steadily out to sea. For example, Hastings in southern England, which was situated on the sea when the Normans invaded in 1066 (and for a long time after), now lies several kilometers inland. It is more usual, however, either for cycles of deposition and erosion to balance each other or for erosion slightly to exceed deposition, so that the beach slowly retreats. Erosion occurs chiefly during storms, when sand is removed from the intertidal zone and dumped in the surf zone. It slowly returns to the beach as conditions return to normal. The beach and its surf zone may thus be regarded as a wave-dissipating apron, absorbing wave energy through the movement of

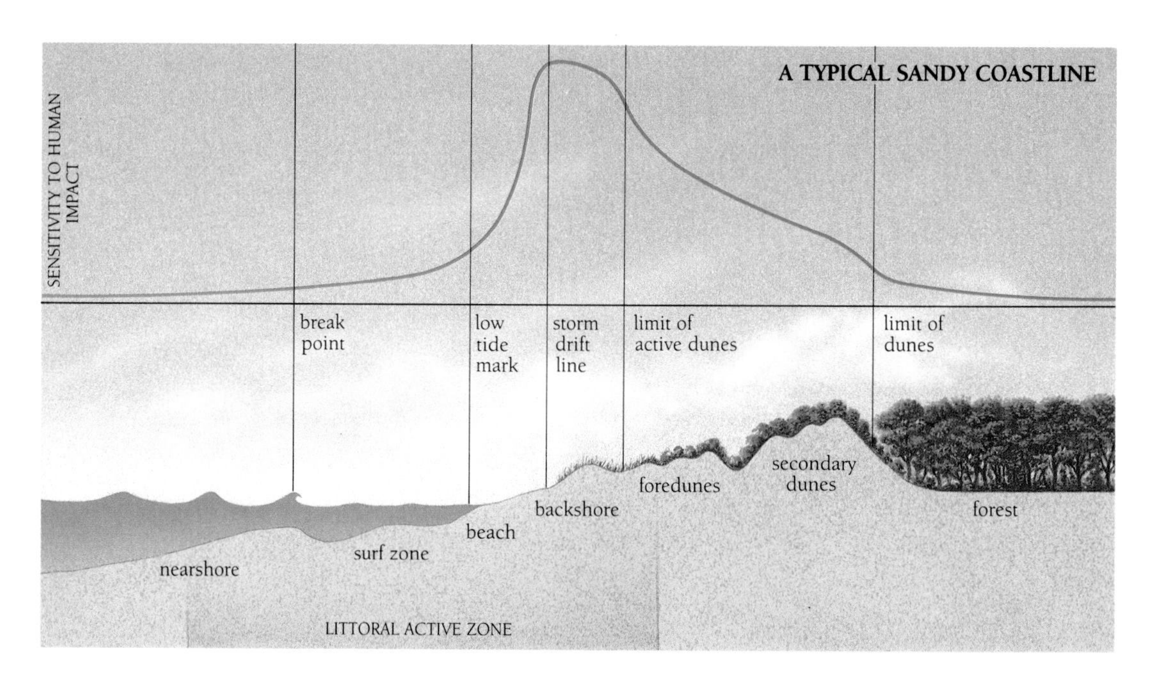


► A cross-section of a typical sandy coastline. The sensitivity curve indicates the areas most sensitive to human impact.

sand and thus protecting the land behind.

Wave action, the size of the sand particles, and beach slope are all interrelated. In general, the greater the wave action, the greater the slope and the greater the average particle size. However, this is complicated by the fact that cycles of deposition and erosion are more extreme on exposed than on sheltered beaches, so that exposed beaches show greater variation in slope over a period of time. Cusps are a common feature of sandy beaches and result from the complex interaction of bars, troughs, wave action, and rip currents.

Sandy beaches act as giant filters. Much of the water in the swashes running up the beach percolates down through the sand until it reaches the water table; this creates a hydrostatic pressure which forces water out at the bottom of the beach. There is thus a flow of water through the beach, most marked when the sand is coarse, and particles in the water are trapped between the grains. These particles are largely organic and form a food supply for bacteria, which metabolize them and return enormous quantities of dissolved nutrients to the sea. Sandy beaches have therefore been described as huge "digestive systems".

Sand moves not only up and down the shore but also along the shore, particularly in the surf zone, through which large amounts are transported during storms. Although the longshore movement of sand may change direction according to circumstances, its net annual movement is usually in one direction only, a phenomenon known as "net littoral drift". This creates problems for coastal planners and marine engineers, as any structure which impedes longshore sand movement results in the deposition of sand on its updrift side and erosion downdrift. A classic example is Madras harbor in India, the construction of which resulted in changes to the sandy shoreline for several kilometers as well as severe shoaling of the harbor entrance, so that the entrance had to be moved and an outer quay constructed.

## DUNES AND DUNEFIELDS

Sand brought ashore by the waves may be pushed higher and higher up the beach until it comes under the direct influence of the wind. When this is onshore, sand is blown shoreward beyond the driftline, becoming available to form dunes. Dunes originate in two ways: vegetation may trap windblown sand or, in the absence of vegetation, the sand may form ripples which slowly expand to form more extensive systems. The amount of sand moved by the wind depends on wind speed, the moisture content of the sand, and the grain size. As fine sands undergo greater transport than coarse sands, dunes tend to consist of relatively fine particles. Although they form slowly, dunes can grow up to 80 meters (260 feet) high. Often they are stabilized by salt-tolerant vegetation, but if this does not happen, sand continues to be transported up the windward face of the dune and over its crest, to be deposited on the slip face. In this way unstabilized dunes advance at rates between 1 and 10 meters (3–33 feet) a year, eventually creating extensive dunefields. The Alexandria dunefield, in Algoa Bay, South Africa, for example, covers 120 square kilometers (46 square miles) and is still advancing.

Dune sands are alkaline because of their high calcium carbonate content. Initially poor in organic materials, they evolve over time, with their organic content increasing and alkalinity decreasing as calcium carbonate is leached out of the sand. They thus become increasingly hospitable to plant, and therefore animal, life. Dune ecosystems are extremely fragile and must be protected from human activities far more rigorously than the intertidal beach.

## SANDY BEACH FAUNA

Although sandy beaches teem with animal life, most species are only just visible to the naked eye. They live between the sand grains, never willingly leaving the substratum, so are rarely seen by the casual observer. Most animal phyla are represented in this interstitial fauna and some groups are found nowhere else on Earth. Although small, most interstitial animals are very mobile,

**HOW SANDY BEACHES ACT AS FILTERS**

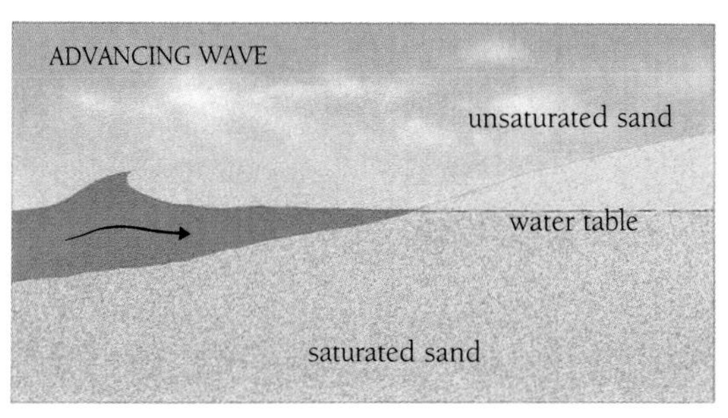


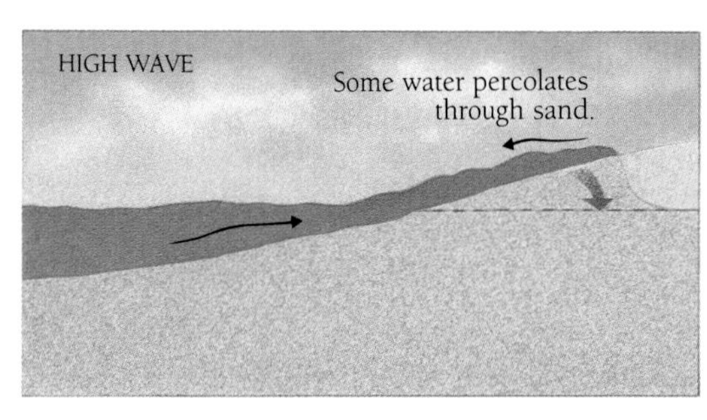


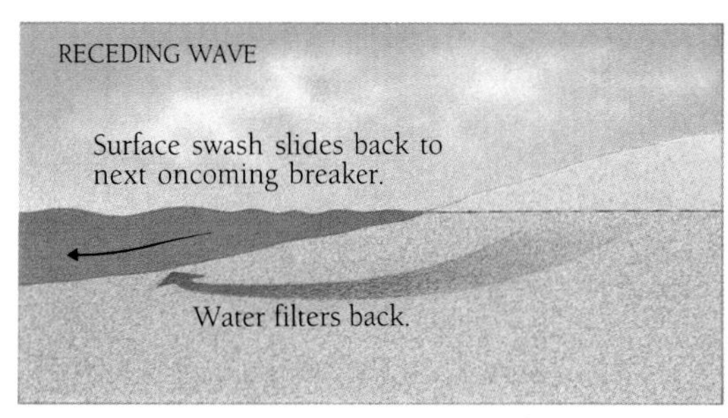


◀ As waves constantly break on the beach, water percolates through the sand, ensuring a constant flow of nutrients through the beach.

Kathie Atkinson

▲ *Ocypode ceratophthalma*, the tropical ghost crab, might almost be mistaken for a land crab, living as it does above the high-tide mark, were it not for the fact that it must return to the sea in order to breed. Ghost crabs are fast-moving, well-camouflaged crustaceans that can burrow more than 1 meter (3 feet) into dry sand.

migrating up and down through the sand with the tides or in response to other factors, such as temperature or light.

There are fewer species of relatively large animals, although individual species achieve high densities under suitable conditions. Polychaete worms, such as lugworms and bloodworms, are much in evidence, especially on relatively sheltered beaches, while clams and whelks can attain vast populations on more exposed shores. Crustacea are also important: burrowing prawns are common on sheltered beaches, while on more exposed beaches smaller forms, such as sand lice (Isopoda) are more likely to predominate. On tropical beaches, mole crabs *Emerita* sp. often occur in considerable numbers. Crustacea are also common at the top of the shore, around the driftline, where they have adapted to a semi-terrestrial existence. Sand hoppers (talitrid amphipods) are important here, particularly in temperate climes, while in the tropics the driftline is often dominated by ghost crabs *Ocypode* sp.

The surf zone supports a rich fauna and acts as nursery areas for many fishes. Among the invertebrate surf-zone fauna, swimming prawns and mysid shrimps predominate, together with smaller crustaceans such as copepods. The surf-zone fauna must be considered an integral part of the sandy beach ecosystem, as these animals invade the intertidal zone as the tide rises and many are predators on intertidal invertebrates. Just as the beach is invaded by surf-zone animals with the rising tide, so it becomes available to terrestrial forms during ebb. Many birds feed on the intertidal fauna during the day, and some even at night, when they may be joined by other predators and scavengers, including mammals, reptiles, and spiders.

A characteristic of all intertidal sandy beach invertebrates is their ability to burrow into the sand. This protects them not only from predators but also from desiccation and extremes of temperature. Some adjust their depth to prevailing conditions, burrowing deeper during cold or hot weather, or when storms threaten to wash them out of the sand.

Burrowing is often accomplished with surprising rapidity: the mole crab can bury itself completely in less than 1.5 seconds. At the other end of the scale, the lugworm *Arenicola* may take several minutes. There is some correlation between speed of burrowing and exposure to wave action, for under exposed conditions, if it is not to be swept away by the next surge of water, a burrowing animal must be able to obtain a firm anchorage in the sand between swashes. *Arenicola* can thus burrow only in quiet water; however this is consistent with the fact that it lives in a semi-permanent burrow which it can keep open only where wave action is slight. On beaches more exposed to wave action, the sand is too unstable to support burrows, and the animals must be more mobile and agile if they are to maintain their position on the beach.

Most of the larger species migrate up and down the beach with the tides, often surfing in the waves. By doing this, the animals maximize their food supply, which is often most abundant at the water's edge, and also make it difficult for predators to reach them. An example of this behavior is provided by the beach clam *Donax*, which surfs up the shore with its large foot and siphons extended to take full advantage of the waves; arriving in the swash zone, it buries itself rapidly and uses its siphons to filter feed, emerging as the tide rises to surf once more in the swash zone. It avoids the upper part of the shore, where it might be stranded, and surfs down the

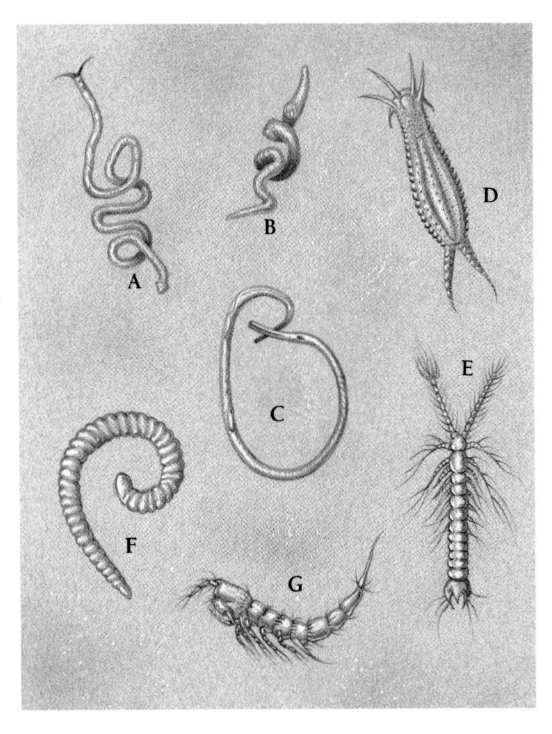


► Sandy beaches are home to a range of microscopic creatures. The illustration indicates a representative sample:

**A** tubellarian *Diascorhynchus* (1.5 mm)
**B** archiannelid *Poligordius* (2 mm)
**C** nematode *Nannolaimus* (1 mm)
**D** gastrotrich *Xenotrichula* (0.5 mm)
**E** mystacocoid *Derocheilocaris* (0.7 mm)
**F** oligochaete *Marionina* (2 mm)
**G** copepod *Hastigerella* (0.6mm)

slope again as the tide falls. Whelks such as *Terebra* and the plough snail *Bullia* also surf by using the fully expanded foot as an underwater sail. The mole crab, on the other hand, allows the waves to transport it by rolling up into a tight ball.

Sandy beaches provide an extremely harsh habitat, not only because of wave action and sand movements, but also because the instability of the substratum precludes colonization by attached plants. Intertidal animals are thus deprived both of shelter and of a resident primary food source, and must rely on food imported from the land or sea. Although debris and insects may be blown in from the dunes, and the bodies of birds and other terrestrial animals may end up on the beach, nearly all the food of intertidal fauna is ultimately of marine origin.

There are three intertidal food source systems: one based on phytoplankton (single-celled algae) washing up from the surf zone; one based on an input of kelp or wrack; and one dependent on carrion such as stranded jellyfish. These systems are not mutually exclusive but the major food source determines to a large extent the type of animal that can colonize the beach. Where the food consists mainly of phytoplankton, filter feeders such as *Donax* or *Emerita* dominate the beach; where carrion is the chief input, scavengers like *Bullia* and crabs come into their own; if kelp or wrack is the main food source, the center of gravity of the community moves toward the top of the shore and the animals consist predominantly of semi-terrestrial herbivores such as talitrid amphipods and large isopods.

The survival of sandy beach animals depends on one thing—adaptability; the ability to adapt to changing conditions of wave action, to cycles of deposition and erosion, to changing temperatures, and above all to an erratic and varied food supply. They have little room for specialization and, like us, are born opportunists.

ALEC C. BROWN

# The Estuarine Environment

Estuaries are the interface between rivers and the ocean, an environment where fresh water from a river runs through a low-lying coastal plain and meets oceanic waters in a bay or channel that is semi-enclosed but has a connection to the open sea. The two types of water mix to produce a habitat with reduced salinity.

## ESTUARIES

Estuaries can be formed through several geological processes. Some, such as the Severn and Thames estuaries in Britain and the mouth of the Amazon River, are drowned river valleys. These are extensive shallow estuaries formed when the sea rose (for example, since the last ice age) so that sea water entered and swamped the river system. The large amounts of sediment deposited by the rivers form the bottom of the estuary. A second form of estuary occurs where wave action creates a build-up of sand across the mouth of a river, eventually forming a partial barrier, behind which the fresh water is trapped and mixed with sea water. An example is the extensive Waddensee in the Netherlands. Fiords, common in Norway, are a third form of estuary, created where a river ends in a deep area of water isolated from the sea by a "sill" or "step" at the oceanic end. The sill is a rocky barrier creating only a shallow opening to the outside sea. Finally, movements of the Earth's crust, during earthquakes or volcanoes, can cause depression of an area of coastline, which sinks to form a bay with a narrow opening to the sea. San Francisco Bay is a well-known example.

Because their mouths are relatively narrow, estuaries are usually sheltered from the force of waves. They are, however, subject to periods of rising and falling tide. A rising tide brings in oceanic water, making the estuary more salty. When the tide falls, water leaves the estuary and the influx of fresh water from the river reduces salinity. Plants and animals in an estuary therefore live in an environment of cyclic change in the water. There can also be unpredictable changes. After periods of great rainfall, the rivers flow more rapidly and more fresh water enters the estuary. Moreover, after rain, rivers can bring considerable amounts of silt into an estuary, making the water cloudy or turbid.

## MANGROVES

Mangroves are a feature of many tropical estuaries and can form extensive forests. There is a wide variety of mangrove trees, ranging from small

D. Parer & E. Parer-Cook/Auscape

▲ At sunrise, the estuary at King Sound in northwestern Australia, where the Fitzroy River enters the Indian Ocean, has a calm and lonely beauty.

▲ The stilt roots of *Rhizophora stylosa* are a characteristic feature of mangrove forests on the leeward side of many coral cays.

▼ The complexity of the estuarine food web is indicated by this generalized diagram. Most estuarine fauna are marine species adapted to withstand the changing salinity that accompanies tidal ebb and flow.

shrubs to large trees. Despite their obvious differences, they have several features in common. They can all tolerate living in soft, waterlogged, often anoxic mud (mud without air), and can cope with the daily inundation of salty water.

Mangroves, like all trees, take up water via their roots, but large amounts of salt are toxic. They cope with the toxicity in different ways. Some prevent the salt from entering their roots with the water by physiological cellular processes. Some absorb the salt and then secrete it out of small pores in their leaves. Others collect the salt in particular parts of the tree, sometimes old leaves, which they then shed.

Mangrove trees have aerial roots to assist with the exchange of gases with the air, which is difficult because the mud is often anoxic and waterlogged. Some roots (stilt-roots) arise from the stem of the plant, well above the ground, and help to support the plant in addition to helping it to breathe. Others (pneumatophores or peg-roots) come up through the ground from the shallow, horizontally spread root system.

Because of their widespread matted root system, mangrove trees trap and consolidate silt, creating a stable habitat for many other plants and animals and ensuring an appropriate environment for the trees. Suitable soil is so important that several species of mangrove trees develop their seedlings to an advanced stage before they are released from the parent tree. As a result, seedlings are able to establish themselves quickly before they are washed away from the soil. Other mangrove trees produce floating seeds which can be washed to a different part of the forest before they sink to the bottom and develop.

Mangrove forests provide shady and moist habitats for a diversity of plants and animals. Some move into the mangroves from the land;

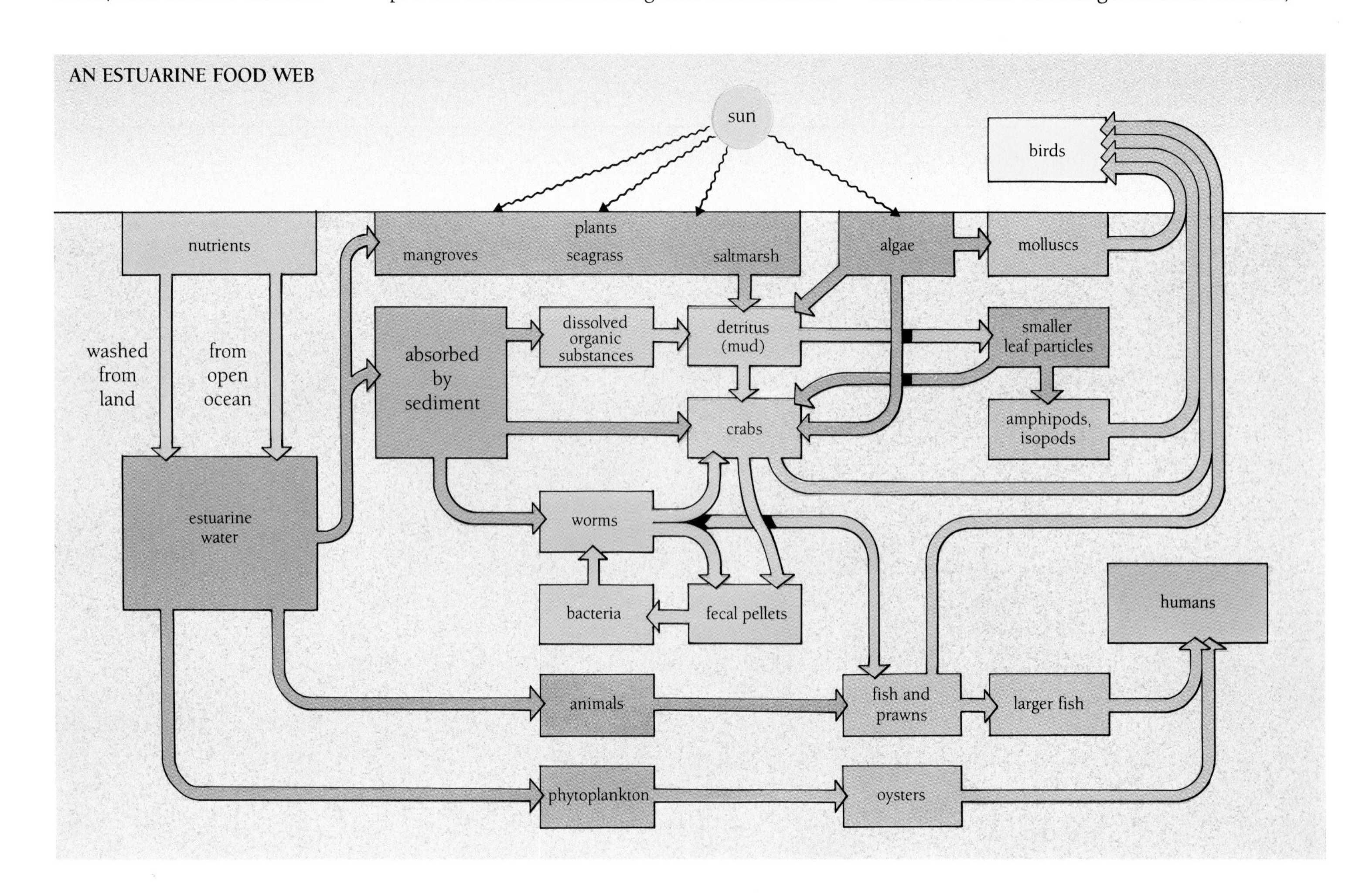

others come in from the sea, using the forests as an extension of their more usual habitats. Other species are specialized inhabitants of mangrove forests and do not occur anywhere else. Some mistletoes, for example, occur almost exclusively on a few species of mangroves.

The upper branches, leaves, and trunks of mangrove trees are a terrestrial environment. Many spiders, insects, and birds live on the trees, just as they do in a terrestrial forest. The lower parts of the mangrove trunks and the aerial roots themselves often serve as a home for marine animals such as barnacles and oysters, which pack onto these hard surfaces because they cannot survive on or in the mud. Other marine animals, however, live on the surface of the mud, or burrow into it. Particularly striking in many parts of the world are the vividly colored, busy crabs that move over the mud to feed on algae, but dart back into their burrows when disturbed or to avoid being eaten by predators such as fish, which come into the forests with the rising tide.

## SALTMARSHES

Saltmarshes are also commonly found in estuaries. If there are mangrove forests, the marshes are behind them. If there are no mangrove forests, saltmarshes can form large habitats and serve to consolidate sediments. Typically, they are characterized by grasses and shrubby or prostrate plants rather than large trees. Like mangroves, however, the plants in saltmarshes must cope with salty water regularly arriving with the tide. Interestingly, many saltmarsh plants have succulent leaves, a feature more commonly associated with arid habitats. Although the plants are surrounded by water for much of the time, the water is saline and it is thus difficult for the plants to take it up into their tissues. It is as if they were living in a very dry habitat.

As in mangrove forests, numerous species of insects are found in saltmarshes, although most are also widespread in other habitats. There are some animals, such as snails and crabs, that are similar to, or the same as, those in mangroves. Saltmarshes often form important habitats for birds, particularly migratory species such as ducks and geese, that feed on the marsh plants.

## SEAGRASSES

Seagrasses are marine plants derived from terrestrial forms, but living mostly under water in estuaries, where they form extensive meadows. Unlike seaweeds, which usually attach to hard rocky surfaces by means of a holdfast (a specialized sucker at the base of the plant), seagrasses have proper roots which anchor them in the mud. They produce flowers and seeds that float to new sites and germinate to develop into new plants. New vegetative growth also arises from rhizomes—specialized branching stems that run under the sand putting up new shoots. These shoots bear elongated leaves similar to those on grasses, although seagrasses are not related to true grasses.

Seagrasses are associated with an array of other plants and animals because their leaves provide stable, hard substrata for the recruitment of seaweeds and small animals, which cannot live in soft sediments. Many other animals burrow around their root systems, which offer protection from predators and make burrowing easier.

In some areas, intertidal regions of estuaries are not stabilized by plants and form continuous mudflats. These often support large numbers of marine or estuarine animals, particularly small crustaceans, worms, and some snails and bivalves that feed on the abundant microalgae growing on and in the mud. These animals, in turn, provide food for many species and large numbers of wading birds.

A. J. UNDERWOOD AND M. G. CHAPMAN

Richard Vaughan/Ardea London

▲ Saltmarshes in the coastal regions of Europe's Low Countries are significant habitats—in the area they occupy and in the habitat they provide for other species. These Dutch saltmarshes provide an important source of food for several species of geese.

# 6 EXPLORING THE OCEANS

The story of oceanic exploration is one of reckless adventure and meticulous planning, honored success and tragic failure, accident and persistence, endurance and exploitation, plunder and scientific enquiry. Oceanic explorers were—and still are—motivated by forces as complex as the individuals themselves. Curiosity and commercial advantage, the thirst for knowledge and the quest for trade, patriotic pride and personal aggrandizement: through the ages, all these motives have inspired humans to cross and chart the world's oceans.

David Hiser/The Image Bank

▲ The earliest and simplest forms of ocean-going vessel were probably canoes or rafts. In many parts of the world, canoes remain an important means of transport for fishing and traveling short distances. Here a dugout canoe is being constructed in the Truk Islands in Micronesia.

## THE FIRST EXPLORERS

Most of the explorers of the great oceans are anonymous. The dates and purposes of their voyages, the vessels they sailed in, and the hazards they faced are all unknown to us. But we do know that in remote prehistory humans were capable of making long sea-crossings.

### CROSSING THE OCEANS

The first Australians were aboriginal only from the perspective of the Europeans who encountered them two centuries ago: they themselves had arrived by sea at least 40,000 years earlier. Archeologists can document the spread of early peoples from Asia along the island chain of the East Indies to New Guinea. There, axes and other stone artefacts have been found sealed under a layer of volcanic ash deposited about 40,000 years ago. Within 12,000 years (and perhaps considerably earlier) groups of people had crossed the 160 kilometers (100 miles) of open water that lie between New Britain and the Solomon Islands. Such voyages would have required quite sophisticated craft, capable of bearing the vagaries of wind, wave, and current in these unpredictable seas. Perhaps more significantly, their instigators would have had to break through the psychological barrier of voyaging out of sight of land, and the intellectual one of navigating without fixed landmarks.

Once these momentous barriers had been broken, the traversing and exploration of the wider ocean was just a matter of time. Certainly by 1500 BC those who had colonized the western margin of the Pacific and developed farming there had brought their root crops, livestock, and cultures deep into Oceania. By the beginning of the Christian era voyagers had reached the Marquesas. A thousand years later all the major Pacific islands lying in the great Polynesian triangle bounded by New Zealand, Easter Island, and Hawaii had been colonized by masterly seafarers—surely the most remarkable feat of sustained oceanic exploration in history.

Human beings arrived in America at least 12,000 years ago via the Bering Strait. Twice over the past 38,000 years lower sea levels have created a "land bridge" across the 85-kilometer (53-mile) wide passage, and some may have crossed on those occasions. But it is likely that many came by sea. When Columbus reached America in 1492 water transport was in widespread use throughout the continent and its coastal margins from Alaska to Tierra del Fuego.

### ANCIENT VOYAGERS

Maritime exploration in the old world, radiating from the great civilizations of the Mediterranean and Mesopotamia, was mainly driven by a desire to trade. In 1750 BC Pharaoh Mentuhotep III sent an expedition down the Red Sea to Punt (modern Somalia). Two centuries later Queen Hatshepsut sent five ships there "to bring back all goodly fragrant woods . . . heaps of myrrh resin, with fresh myrrh trees, with ebony and pure ivory, with

◂ Polynesian seafarers in impressively seaworthy vessels were among the first oceanic explorers.

▾ The ancient Egyptians traveled extensively by sea to trade with countries bordering the Aegean, Mediterranean, and Red seas. The ships shown here were powered by rowers with oars, but by 3200 BC the Egyptians had invented sails.

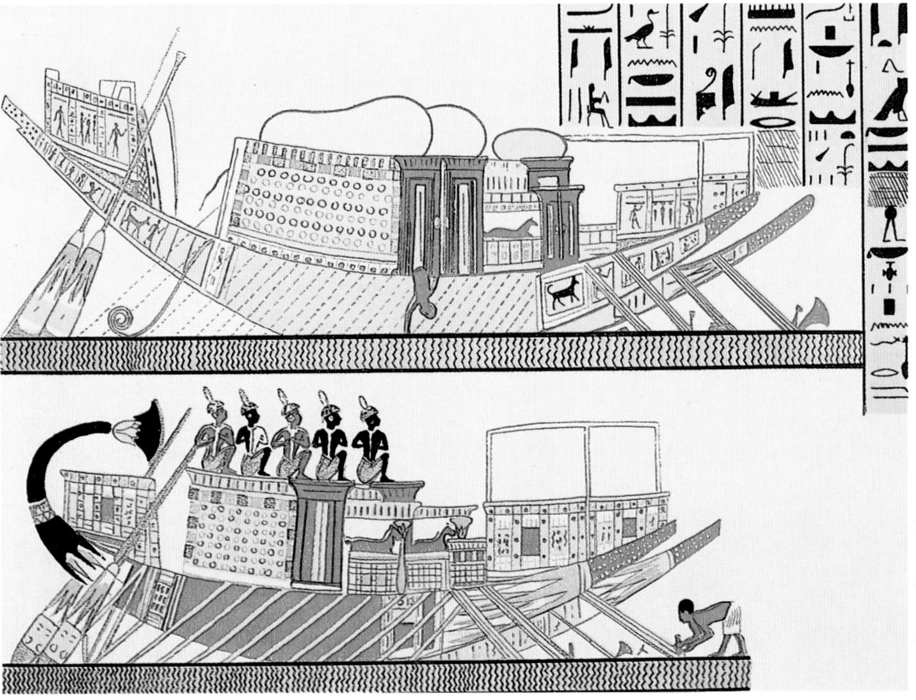

Mary Evans Picture Library

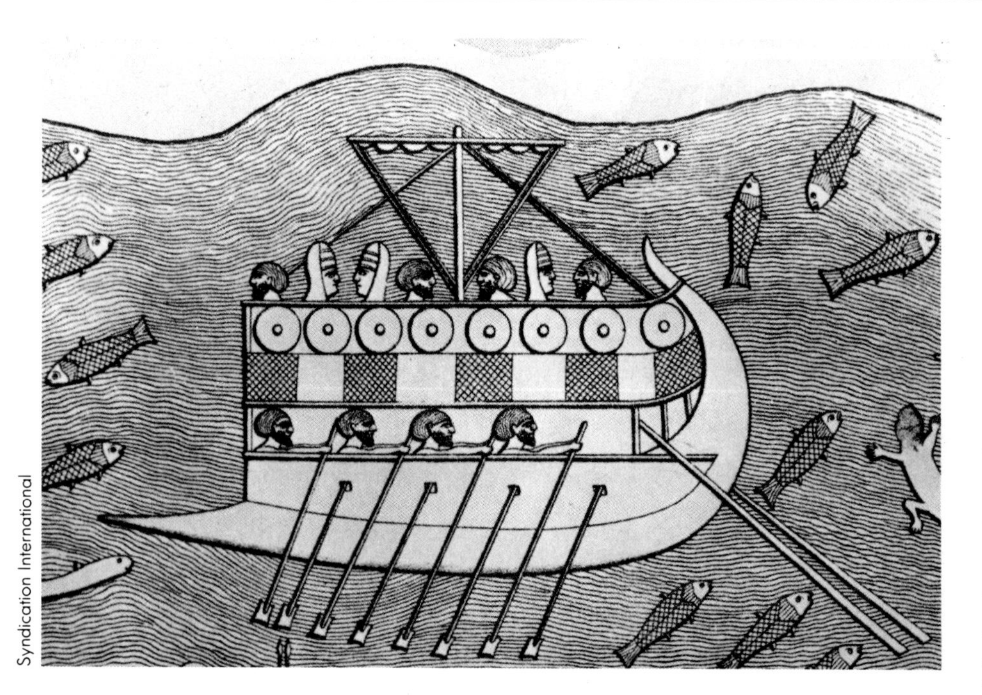
Syndication International

▲ The Phoenicians were great seafarers of the ancient world. Their navigational skills enabled them to be the first to explore the length of the Mediterranean and beyond—including a circumnavigation of the coast of Africa. This illustration represents a Phoenician warship.

green gold of Emu, with cinnamon wood, Kheyst wood, with two kinds of incense, eye cosmetic, with apes, monkeys, dogs, and with the skins of the southern panther, and with natives and their children". One of her trading ships is depicted on a relief from a tomb at Kenamon, Thebes, and the accurately observed frieze of Red Sea fauna swimming below it inspires confidence in the representation of the ship itself.

According to the Greek historian Herodotus, Pharaoh Necho (610–594 BC) sent an expedition manned by Phoenicians to circumnavigate Africa from the Red Sea to the Mediterranean. A century later a Carthaginian expedition, attempting the voyage in the opposite direction, may have reached the Guinea coast. Another attempt was made about 470 BC by Hanno of Carthage, and although he too failed to complete the journey he does seem to have reached a point in equatorial west Africa and founded colonies there.

Northward from the Pillars of Hercules (Straits of Gibraltar) Phoenician influence centered on Cadiz, perhaps from as early as 1000 BC. By 700 BC this great port was handling tin and amber from the almost unknown lands of northern Europe. Herodotus, writing around 460 BC, speaks of these commodities as coming from "the ends of the earth"; in the same passage he mentions the "Tin Islands" which some have equated with Britain's tin-rich Cornish peninsula, though the identification is far from certain.

About 325 BC Pytheas of Marseilles sailed to Britain by way of the west coast of Spain, visiting and reporting upon the Cornish tin mines before going on to complete a circumnavigation of Britain, visiting Ireland and perhaps touching on the coast of Norway before returning home. Four hundred years later Julius Agricola, during his attempt to bring Britain's furthest extremities under Roman control, sent his fleet on a voyage of exploration around Scotland. At one stage it was accompanied by a Greek scholar, Demetrius of Tarsus, who later met the writer Plutarch at Delphi and told him of his adventures in remote Caledonian seas.

### EAST AND WEST

Elsewhere in the world trade by sea between the Persian Gulf and the Indus had been established by the third millennium BC. Some two thousand years later extensive seafaring was also taking place along the coasts of Southeast Asia. India was the natural link between these two great spheres of maritime activity, and by the later classical period long-distance trade between west and east was flourishing. Roman coins and pottery have been found in southern India, and the spread of Graeco-Roman objects extends from east Africa to Indo-China. When the Europeans began their so-called "Age of Discovery" in the fifteenth century, most of the world had already been discovered by seafaring travelers. It remained only for the various cultures to discover one another.

COLIN MARTIN

► This plate depicts Dionysus, the Greek god of wine and fertility, sailing across the sea as marauding pirates are turned into porpoises. Dionysus is reputed to have traveled widely, perhaps as far as India, teaching people how to cultivate grapes and make wine.

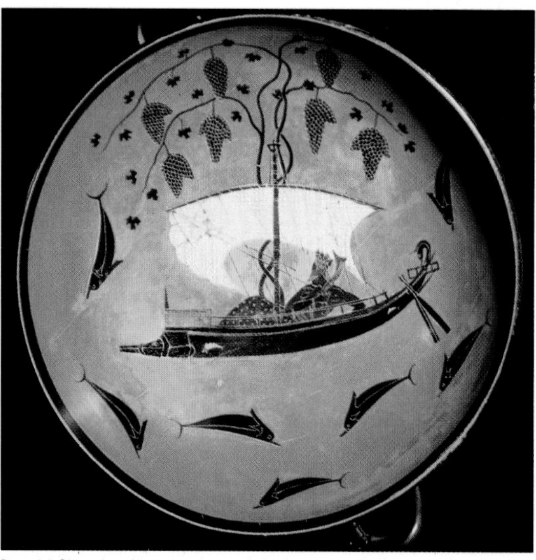
Ronald Sheridan/Ancient Art and Architecture Collection

## CROSSING THE ATLANTIC

The first European explorers of the Atlantic were unknown prehistoric travelers who, in bark- or skin-covered vessels, plied its coastal waters from southern Spain to Arctic Norway. In the early sixth century St Brendan, an Irish monk, undertook a remarkable expedition in a large hide-covered *curragh*, the detailed construction of which was described in a medieval account of his voyage. It seems likely that he and his companions, seeking solitude in a "promised land of the saints", reached Iceland, and possibly even distant Newfoundland.

### THE NORSEMEN

Two hundred years later the Viking expansion westward brought sturdy clinker-built ships into north Atlantic waters. The stepping-stones of Orkney, Shetland, the Faroes, and Iceland carried Norsemen to Greenland by 982 AD, and within 20 years the first Europeans (so far as we know) set foot in North America when Eric the Red and his men landed somewhere in Labrador. Colonization followed, but it was short lived. Archeologists have located the site of one Norse settlement at L'Anse aux Meadows, near the northern tip of Newfoundland. Within a few years of its establishment, however, the site was abandoned. Perhaps the colonists were decimated by Indians; perhaps they simply could not survive so far from their parent culture.

Werner Forman Archive

▲ Eric the Red discovered Greenland around 982 AD and shortly afterward persuaded several hundred Icelanders to establish Viking colonies there. Archeological evidence suggests that all colonies were abandoned a few years later. This site at Brattahlid in Greenland is the remnant of one such settlement.

## COLUMBUS AND HIS FOLLOWERS

It fell to Iberian explorers to conquer the Atlantic, and with it the vast resources of the new world. Christopher Columbus was a Genoese of humble birth who made his momentous voyage of discovery on behalf of the Spanish crown in 1492. He had hoped to discover a westward route to Asia; instead he found the islands of the Caribbean. Three further voyages followed, the last in 1502-03, still with the riches of China as the ultimate goal.

Columbus's voyages were the inevitable outcome of the advances in science, technology, and rational thought that characterized the European Renaissance. Many of these advances, however, were rooted in the achievements of earlier civilizations. From China came the compass and gunpowder; from Arabia the skills of mathematical computation and navigational instruments—particularly the astrolabe, by which latitude could be gauged. Maritime traditions that had matured for thousands of years in the

Archiv/Photo Researchers

◄ "The first landing of Kolumbus", a print by Theodore de Bry (1528–98) showing Columbus on the beach at Fernandez Bay in the Bahamas. Columbus named this region the West Indies, as he mistakenly believed that he had sailed as far as India.

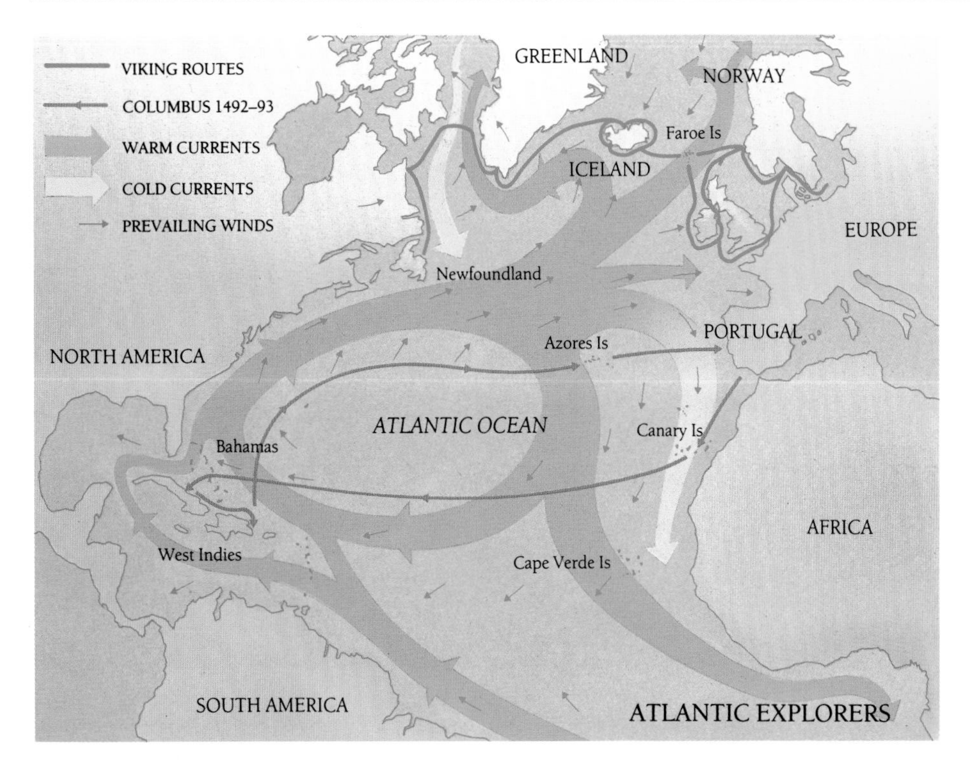


▲ While the Vikings pioneered long-distance voyages across the northern Atlantic, it was the discoveries of Columbus that challenged the European world-view and enabled future seafarers to gain mastery of the ocean.

Mediterranean, refined and adapted to Atlantic conditions by seafarers (from the Carthaginians onward) who had sailed beyond the Pillars of Hercules, provided a sound pedigree for the workaday Iberian fishing boats and traders which, duly adapted, were capable of trans-Atlantic voyaging. After Columbus the old world and the new were inextricably linked.

Between 1499 and 1505 as many as 11 small Spanish fleets followed in Columbus's wake, and colonization of the Caribbean basin began. Their outward route exploited the summer winds that sweep southward down the coast of Africa to the Canaries, and then curve westward toward America. Home-bound ships could pick up the returning winds which carried them north of the Azores and thence to the Iberian coast. But many did not return. The true explorers were not the handful of individuals whose names grace the history books but the legion of anonymous voyagers who, from their tenuous bases on what must have seemed to be the edge of the world, spread out to explore, exploit, and colonize the unknown.

What may be the remains of one of their vessels has been found stranded on a remote reef in the Turks and Caicos Islands. Nautical archeologists have recovered fragments of the hull of a small, sleek vessel, heavily armed with wrought-iron guns. Dating evidence indicates that the ship went down before 1513. It probably belonged to a group of tough, freebooting Europeans intent on carving out their own slice of the new world's riches. Like modern astronauts they must have lived very close to the limits of their resources. They even carried with them the means to manufacture their own cannon-balls, underlining the one overwhelming advantage they had over the indigenous peoples of the Americas—firearms. This particular group evidently failed, but the combined efforts of their contemporaries and those who followed them brought about—for good or ill—Europe's mastery of the Atlantic.

▶ Magellan crossed the Atlantic in 1519, and in 1520 eventually found the passage to the Pacific for which he had been searching. This fanciful engraving from about 1585 depicts Magellan and his fleet in the Straits of Magellan near the southern tip of South America.

The Granger Collection

MASTERY OF THE ATLANTIC

Other Europeans had pioneered different routes to the Americas. In 1497 John Cabot, sailing from Bristol, reached Newfoundland, thus reopening the northern route which had lain dormant since the days of Eric the Red. Amerigo Vespucci, in 1499, discovered South America, and over the following three years explored much of the coastline as far as Rio de Janeiro and possibly beyond. These discoveries were rapidly consolidated. Spain soon monopolized the Caribbean and Central America, and the mid-Atlantic routes that served them, while Portugal held sway over the eastern coastal tracts of South America.

In the north the great fishing and whaling resources off the Newfoundland coast were exploited by European adventurers, particularly the Basques, while Dutch and English explorers, vainly seeking a northwest passage to Asia, discovered and charted the closed Arctic seaways of Baffin and Hudson. In the late sixteenth century the Arctic regions of northern Europe were explored by the Dutchman Willem Barents, who was attempting to find a northern route to the Far East. He reached Nova Zemlya in 1596, where his vessel was stranded in the ice, and died the following year while returning to Holland in an open boat.

COLIN MARTIN

Coo-ee Historical Picture Library

▲ This early etching (circa 1741) depicts the "whale fishery and killing the bears". The scene is probably set in Greenland. Before the sixteenth century the north Atlantic fishing resources were exploited by the Basque whalers, who were later joined by the Dutch and the British.

# THE CHALLENGE OF THE PACIFIC

Many of those who followed Columbus to America still sought a passage to the Far East. It was Vasco Núñez de Balboa who, in 1513, first crossed the Isthmus of Panama to reach the Pacific coast. Exploration of the western coasts of Central and South America soon followed, and in 1527 Alvaro de Saavedra crossed the Pacific to the Moluccas. But he was not the first European to navigate these waters. In 1519 Ferdinand Magellan, a Portuguese navigator with wide experience of Indian and Far Eastern waters, set out from Seville with the intention of circumnavigating the globe on behalf of Charles V, king of Spain.

MAGELLAN'S VOYAGE

The voyage was fraught with difficulty. The fleet of five ships took almost six months to reach the Rio de la Plata, and the travelers were forced to winter at Port St Julian, at a latitude of 41°S. They spent much of the following year trying to find a passage through to the Pacific. One ship disappeared in the attempt, and morale, already low, plummeted. Eventually the tortuous strait which now bears Magellan's name was discovered, and the little fleet broke through into the ocean beyond. Another ship was lost, and many of the men wanted to turn back. But Magellan persisted, and for three months the ships ran with the steady and gentle southeast trade winds which Magellan named "Pacific".

By now the voyagers were close to despair, with food stocks so low that they were forced to eat ox-hides, sawdust, and rats. But on March 6, 1521 they reached the Marianas, and the following month landed at Cebu in the Philippines. They were back in charted waters. Unhappily, Magellan was shortly thereafter killed, and the expedition was reduced to a single ship, the *Victoria*, under the command of Sebastian del Cano. When they

The Granger Collection

◄ Ferdinand Magellan (1480?–1521) led the first expedition to circumnavigate the world. Although he died before the journey was completed, the return of his ships provided the first proof to Europeans that the world was round. This engraving dates from 1695.

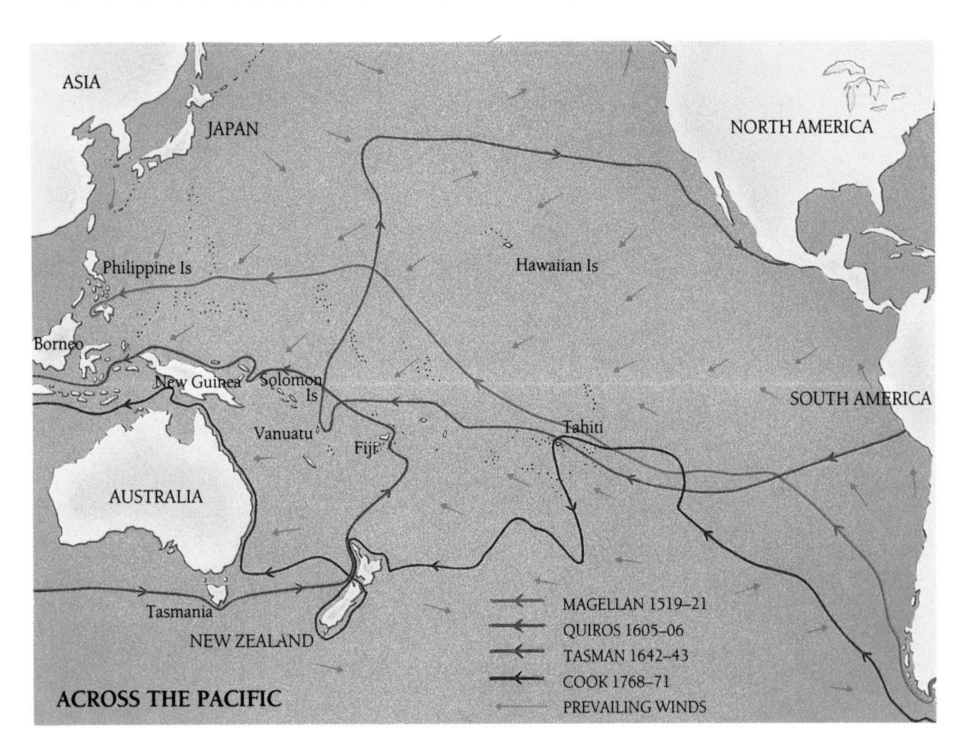


▼ This color engraving, from a drawing by Dutch explorer Abel Tasman, portrays his ships and a Polynesian outrigger canoe at an island landfall during his 1642 voyage.

reached the Moluccas a crewman who had been born there became, technically, the world's first circumnavigator. *Victoria* eventually reached Spain in late 1522 with only 19 remaining of the 280 voyagers who had set out. Del Cano received the honors which were rightly due to Magellan, who had achieved what Columbus had striven for—the linking of continental Europe with Asia via the western ocean.

## SPANISH EXPLORATION

In 1565 the Spaniards pioneered the east-west Pacific route to link Asia with the Americas. This was more difficult, because the prevailing winds blow from the east. However, they discovered that north of the Philippines eastbound ships are aided by the North Pacific Drift toward a landfall in the vicinity of California. From there, the California Current runs southward to Central America. This was to become the route of the famous Manila galleons which, each year from 1574, sailed from the Philippines to Acapulco in Mexico, bringing oriental silks and porcelain in exchange for American silver, to the profit of the Spanish crown. This journey could take up to seven months, but the outward voyage was much quicker, as it exploited both the prevailing winds and the westerly set of the powerful North Equatorial Current.

Other Spanish navigators such as Lope de Aguirre (1561) and Pedro de Quirós (1604-05) sought knowledge and opportunity in the Pacific, and in particular the great southern continent that was reputed to lie there. But they were not encouraged by what they found. The vastness of

The Granger Collection

the ocean, and its apparent lack of continents or large islands to exploit, provided little inducement. Nor did the seafarers of other nations show much interest in the Pacific. Even Francis Drake's circumnavigation of 1577–80, during which he sailed up the west coast of America as far as Vancouver, landed in California, and then crossed the Pacific to the Philippines and East Indies, was a voyage not of exploration but of plunder, although he was the first to round Cape Horn, thus defining the southern end of the American continent.

## THE EUROPEAN DISCOVERY OF AUSTRALIA

Mystery surrounds the European discovery of Australia, the great southern continent. It is not inconceivable that Arab or Chinese seafarers at times encountered its coasts, whether by accident or design, while a Portuguese chart of 1536 has been held to depict the Australian coastline from King Sound on the Timor Sea right around the Pacific seaboard to the southern coastline. But even if Portuguese navigators did chart these seas, they evidently found little to detain them.

The Spaniards came next. Luis de Torres, who had sailed with Quirós from Peru in 1604 but later became separated from him in the New Hebrides, continued his voyage westward and probably reached the Great Barrier Reef at about latitude 21° S. He then headed north to round Australia's northern tip, probably sighting Cape York but failing to recognize it as part of the great southern continent for which he was searching.

From the early seventeenth century Dutch East Indiamen bound for Batavia regularly sighted the

Coo-ee Historical Picture Library

▲ Luis Vaez de Torres sights Cape York, the northernmost tip of Australia.

▼ "Natives of Otaheite attacking Captain Wallis the first discoverer of that island", a watercolor by an unknown artist.

Rex Nan Kivell Collection/National Library of Australia

▸ These storage jars were recovered during a joint Thai–Australian archeological excavation of a wreck site in the Gulf of Thailand. The excavated ship, a small Thai trading vessel, is thought to date from the sixteenth century. The jars probably contained food such as oil and sugar.

Australian coast, and some were wrecked there. But the rugged western coastline was of little interest to the pragmatic Dutch merchants, who regarded it simply as a navigational mark and a hazard to be avoided. In due course, however, higher authority demanded fuller information about the continent, and in 1642 Abel Tasman was sent by the Dutch governor of Batavia to gather it. His route took him across the Indian Ocean to Mauritius; then, doubling back to reach Australia, he sailed too far south and found Tasmania instead. Continuing eastward he discovered New Zealand before returning to Batavia via Tonga and New Guinea. In 1643-44 he investigated the northern coast of Australia from the Gulf of Carpentaria westward to the North West Cape.

COOK'S VOYAGES

The greatest European explorer of the Pacific was Captain James Cook. In the genuine pursuit of geographical knowledge he conducted three major voyages between 1768 and 1779, mainly in the Pacific. During the first he charted New Zealand, observing the strait which separates the North and South islands. Then, in 1770, he surveyed the entire east coast of Australia. His second voyage aimed to establish whether—as some believed—another great continent lay in the south Pacific. In the course of proving that it did not exist he visited, charted, and determined the exact positions of Easter Island, the Marquesas, Tonga, Tahiti, and the New Hebrides. He also discovered New Caledonia, Norfolk Island, and the Isle of Pines.

Jeremy Green/Western Australian Maritime Museum

Cook's third voyage resulted in the discovery of Hawaii, and the exploration of the northern coastline of America to the Bering Strait and beyond. He then sailed southward along the coast of Asia before returning to Hawaii where—like Magellan before him—he was killed in a skirmish with natives.

COLIN MARTIN

# EXPLORERS OF THE INDIAN OCEAN

The coastal fringes of the Arabian Sea, from the Horn of Africa to the southern tip of India, were well known to the ancient world. But the hegemony enjoyed by classical seafarers passed to the Arabs with the fall of the Roman empire, and with the rise of Islam in the seventh century Muslim merchants rode the monsoon winds and brought their *dhows* to India, Malaya, Indonesia,

▾ Portuguese ships from the time of Henry the Navigator. His school of navigation and encouragement of seafaring expanded Western knowledge of the world beyond Europe.

National Maritime Museum, London

China, and east Africa. Such voyages were the inspiration for the legendary journeyings of Sinbad the Sailor, whose stories date back to the ninth century.

CHINESE EXPEDITIONS

Although coastal watercraft in Southeast Asia have been used from the remote past, oceanic voyaging was a much more recent phenomenon. By the ninth century Chinese navigators possessed information about the Gulf of Aden and the Somali coast, and 200 years later they knew of Zanzibar and Madagascar. Marco Polo, the Venetian traveler, described the Chinese ocean-going junks he saw in the 1290s. They had four masts, and some also had two auxiliary ones which could be raised and lowered as required. The crews might be up to 300-strong and the largest ships could carry 6,000 baskets of pepper.

Although Chinese introspection and mistrust of foreigners discouraged exploratory voyages, a remarkable series of maritime expeditions took place in the fifteenth century. They were conducted by Cheng Ho (also known as the Three-Jewel Eunuch), admiral of the Ming emperor Yung Lo, who between 1405 and 1433 led seven expeditions to visit 37 countries on the coasts of Indo-China, the Indian Ocean, the Persian Gulf, Red Sea, and east Africa. Among the more exotic acquisitions he brought back to the emperor was a fully grown African giraffe. But Yung Lo's successors closed the exploratory doors opened by Cheng Ho, and the road to the east via the Indian Ocean was left open to others.

PORTUGUESE EXPLORATION AND COLONIZATION

Perhaps more than anything else it was the existence of the Moslem empire that encouraged Europeans to bypass it by opening maritime routes to the east. The main inspiration came from Portugal, where Prince Henry ("The Navigator") had set up his celebrated school of navigation in 1416. By the middle of the century mariners like Alvise da Cadamosto and Diego Gomez were probing the west African coast, and by 1487 Bartholomew Diaz had reached the Cape of Good Hope.

Ten years later Vasco da Gama set out from Lisbon on his epic voyage to India, touching Cape Verde before heading west to a point some 960 kilometers (600 miles) from the South American coastline to avoid the currents in the Gulf of Guinea which had deterred navigators from Carthaginian times. He then doubled east to make a landfall just north of the Cape, from which he proceeded via Mossel Bay, Natal, and Mombassa to Malindi. There he took aboard a local pilot to convey him on the long-established Arab route to Calicut. His return trip, sadly, was one of plunder rather than exploration. The way was now clear for the Portuguese and their successors to discover and exploit the resources of Africa, India, and the Far East.

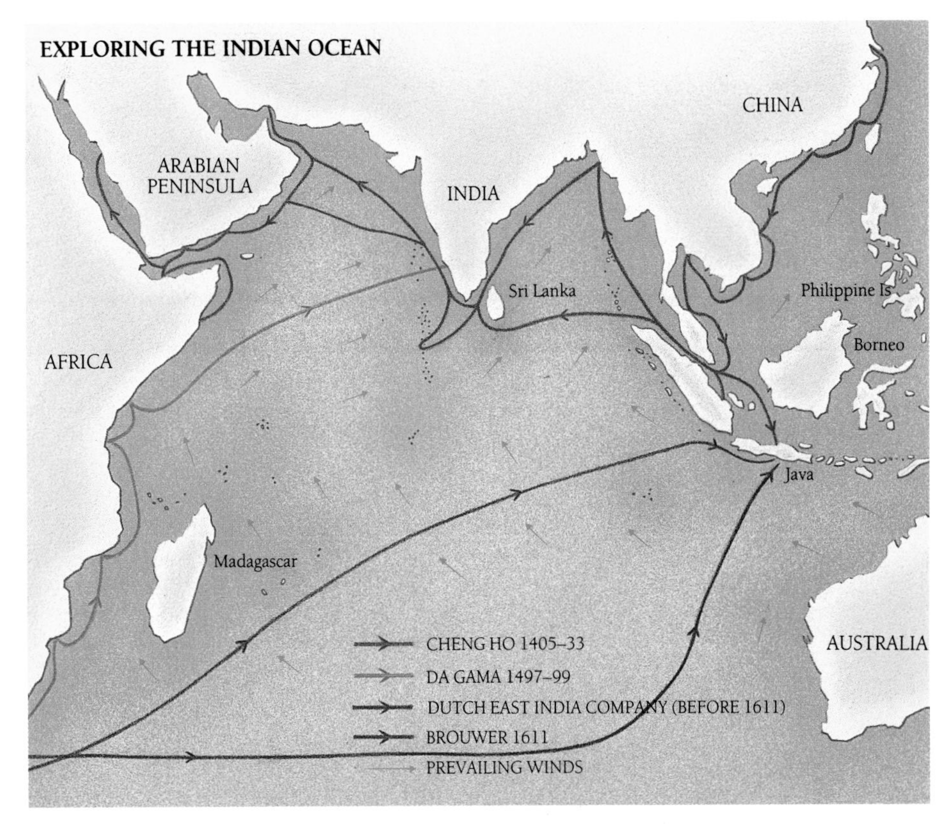


▲ Chinese, Portuguese, and Dutch explorers all contributed to knowledge of the Indian Ocean. The Dutch route after 1611 made use of the strong oceanic winds known as the Roaring Forties.

This colonial process accounts for practically all the European seafaring, and hence exploration, that took place in the Indian Ocean and beyond over the following three centuries. First came the lumbering Portuguese *náos* of the *Carreira da India*, sustaining by their wearisome and dangerous six- to eight-month voyages the farflung trading stations along the east African coast, at Goa, Calicut, and Cochin on the Indian subcontinent, and thence via Colombo and Malacca to Macao on the Chinese mainland. By the end of the sixteenth century this vast and cumbersome trading empire, now united with the American and Pacific conquests of Spain, was at the zenith of its power and prosperity.

DUTCH AND ENGLISH EXPLORERS

But there were rivals. Dutch and English seafarers, inspired by new religious ideology as well as by commercial aspirations to break the Spanish-Portuguese monopolies, and encouraged by maritime successes in their own waters, were eager to develop and control trade with the Far East. The tools with which they sought to achieve this were the great incorporated East India companies, founded in 1600 (English) and 1602 (Dutch). Their sleek ships, built in the tradition that produced the ocean-going warships that had resisted the Spanish Armada in 1588, proved ideal for the run to India and beyond.

Dutch ingenuity in navigating the Indian Ocean led to the European discovery of Australia.

▲ From the sixteenth century, whalers and fishermen were exploiting the resources of the Arctic. A more scientific approach to exploration began in the eighteenth century.

In 1610 Henrik Brouwer pointed out that the usual route taken to Java was unsatisfactory, since it involved crossing in a tropical latitude, where the wind, if not contrary, was often absent and where the heat caused pitch to melt, cargoes to rot, and men to fall sick. Far better, he argued, to follow a temperate latitude from the Cape for as long as possible, heading north only when the longitude of Java—about 105°E—had been reached. This point is only 800 kilometers (500 miles) from the western coast of Australia.

Brouwer himself pioneered this route with conspicuous success, and thereafter it was routinely taken by Dutch East India Company vessels. It was only a matter of time before one of them discovered the Australian continent. The honor fell to Dirk Hartog of the *Eendracht*, who in 1616 reached the island off Western Australia which now bears his name. Others were less fortunate, and were wrecked on these inhospitable coasts. One, the *Batavia*, drove into the islands of the Houtman Abrolhos in 1629, and its remains are now on display in the Western Australian Museum.

COLIN MARTIN

# POLAR EXPLORERS

The last great unexplored tracts of the world's oceans lay in the polar regions. As long as the compelling motives for discovery were colonization or exploitation, these barren and climatically hostile areas held little appeal. But the sixteenth-century seafarers who penetrated the northern polar region in search of a northwest passage were soon followed by fishermen and whalers who seasonally visited the fishing grounds off Newfoundland and the whaling stations of Labrador and Spitzbergen. Throughout

▶ An engraving by J. Brandard (based on a drawing by Sir John Ross) shows two Boothia Peninsula Eskimos giving directions to Ross during his 1829–33 expedition to the Canadian Arctic. Between 1818 and 1834, Ross commanded six Arctic voyages in search of the elusive northwest passage.

Mary Evans Picture Library/Photo Researchers

Royal Geographical Society/The Bridgeman Art Library

▲ Sir Allen Young's vessel *Pandora* "nipped in the pack" in Melville Bay in 1876. The Arctic was a harsh and difficult environment for European explorers.

much of the seventeenth century these waters were the scene of fierce rivalry between the fishing nations, particularly the English and the Dutch, and discoveries were often jealously guarded as commercial secrets.

## THE ARCTIC

By the eighteenth century geographical enquiry—though not always untouched by political motives—was more frequently conducted for its own sake, and the polar regions began to be explored. In 1728 the Dane Vitus Bering, on behalf of the Russian crown, discovered that the continents of America and Asia were separated by the strait that now bears his name. A few years later the Russian navigator Chelyuskin explored much of Arctic Russia's northern coastline. In 1806 William Scoresby, after traversing the Arctic east coast of Greenland, reached a latitude of 81°30'N in the vicinity of Spitzbergen. Twelve years later Sir John Ross sailed up the west coast of Greenland into Baffin Bay, and on a later expedition, from 1829 to 1833, he explored the islands and seaways of the Canadian Arctic.

These hazardous journeys sometimes ended in disaster. In 1846 Sir John Franklin, searching in the same area for the elusive northwest passage, became trapped in the ice. His party's fate was later revealed by rescue attempts and, subsequently, by archeologists. A search party aboard the barque *Breadalbane* became icebound in 1853 and, although its members survived, the ship was lost. At length, in 1859, the remains of Franklin's party were discovered on King William Island, together with a diary recording their terrible ordeal.

More than a century later, in 1984, archeologists discovered some of Franklin's men buried in the permafrost, their flesh and features preserved and their clothing intact. About the same time the wreck of the rescue ship *Breadalbane* was found and photographed, 100 meters (330 feet) below the Arctic ice.

A northwest passage was finally achieved by the Norwegian Roald Amundsen, who in 1903 sailed from the Atlantic via the McClintock Strait to emerge through the Bering Strait into the Pacific.

## TO THE NORTH POLE

The final prize of Arctic exploration was the pole itself. Inspired by Cook's successes in the Pacific, Britain's Royal Society commissioned two ships in 1773 to make an attempt on the north pole. But the expedition, led by C. J. Phipps, failed to find a passage north of Spitzbergen, and narrowly escaped disaster. It was on this voyage that the

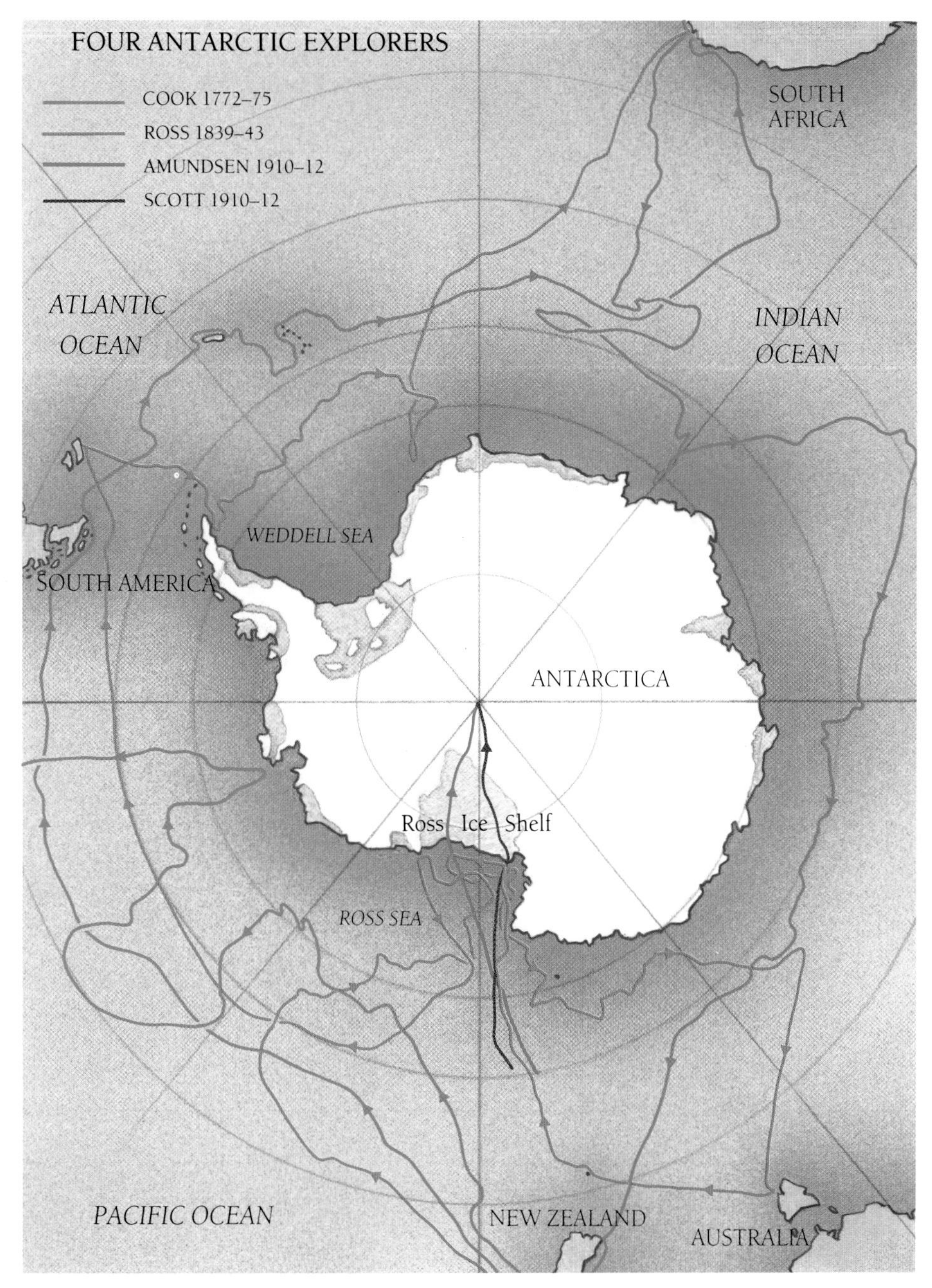

Syndication International

14-year-old Horatio Nelson had his celebrated encounter with a polar bear. In 1827 another attempt was made by William Parry, an outstanding seaman and explorer of Cook's calibre, who by sledging boats over the drift ice which lay between stretches of open water reached a latitude of 82°45'N.

Following an attempt by the Norwegians Nansen and Johansen in 1895, during which they came within four degrees of their goal, the pole was finally reached in 1909 by the American Robert Peary, accompanied by Matthew Henson and four Eskimos. Recently this claim has been challenged, but a mathematical resolution of the shadows cast in Peary's photographs has demonstrated that he came as close to the true pole as his instruments could determine.

## THE ANTARCTIC

During his second voyage, in 1773, Cook became the first man to cross the Antarctic Circle. In trying to determine whether there was a continent in the far south he was turned back by pack ice when he was no more than 160 kilometers (100 miles) from the coast of Queen Maud Land. Later

▲ The versatile James Cook was the first to cross the Antarctic Circle in 1773. Exploration of the fringes of the continent followed, but the south pole itself was not reached until 1911. Sir Robert Falcon Scott (1868–1912) and his team at the south pole, January 18, 1912 (*right*). On the return journey all five men died from hunger and cold.

▶ Sir James Clark Ross's ships *Erebus* and *Terror* caught in the ice in Antarctic waters in 1842. Ross led expeditions to Antarctica between 1840 and 1843. He discovered the ice shelf that bears his name as well as Victoria Land and two active volcanoes.

The Bettmann Archive

that year, from the Pacific side, he reached a latitude of 71° S. Almost half a century later, between 1819 and 1821, the Russian navigator Fabian von Bellingshausen circumnavigated Antarctica, and discovered Alexander I Island. On his return jouney he encountered seal hunters who had just discovered the tip of the Graham Land peninsula.

Exploration along the fringes of the continent followed: Weddell in 1823, Biscoe in 1831, and Dumont d'Urville in 1840. But the major investigation was that conducted by John Clark Ross in a series of voyages between 1840 and 1843 with the ships *Erebus* and *Terror*. Having sailed from Tasmania, they penetrated the great sea (now the Ross Sea), largely free of ice, which pushes into the continent. Ashore they discovered a mountainous coast and two active volcanoes, which they named after the ships. A second season was devoted to further survey in this region, and a third was spent exploring the fringes of the Weddell Sea.

The Ross Sea provided explorers with access to the continent at a latitude of 78° S—less than 1,600 kilometers (1,000 miles) from the pole. A tragic race for the prize began in 1910 between a Norwegian team under Roald Amundsen, who planned to negotiate the final leg with dog-sleds, and a British expedition led by Captain Robert Scott, who proposed that the sleds would be hauled by the expeditioners themselves. Amundsen reached the pole on December 14, 1911, and returned safely. A month later Scott got there too but, with his companions, perished during the return journey.

COLIN MARTIN

Scott Polar Research Institute/The Bridgeman Art Library

▲ Scott led his first expedition to Antarctica in 1901, sailing from New Zealand in the *Terra Nova*, shown here in Antarctic waters.

# MODERN OCEANIC ADVENTURERS

Although there are no new seas or continents to be discovered, the oceans continue to challenge explorers. Their goals vary. Scientists pursue knowledge on the fringes of the sea and in its extreme depths. Others, often benefiting from the fruits of scientific discovery, exploit the ocean's vast but fragile and often finite resources. Some strive to understand and find ways of mitigating the destructive effects of exploitation. And individual men and women, driven by a multiplicity of motives, rise to the challenges presented by the ocean itself.

## AROUND THE WORLD, ALONE

Perhaps the greatest of these challenges is solo circumnavigation. Such a challenge has many components—designing and fitting out a vessel; selecting a route; anticipating the emergencies likely to arise; and of course developing and utilizing high skills of seamanship and navigation. Above all, the challenge tests the determination and endurance of the sailor, who explores not so much the oceans as the deepest recesses of his or her inner self.

As far as is known, the first successful single-handed voyage around the world was that of Captain Joshua Slocum, an American sea-captain of Nova Scotian birth, between 1895 and 1898. His voyage sprang from earlier misfortune. In 1886 his trading barque *Aquidneck* was wrecked off Brazil and Slocum, left stranded and destitute with a wife and two sons, built a 10-meter (35-foot) craft in which they eventually reached New York. While writing an account of this adventure Slocum was given an ancient and neglected oyster sloop, *Spray*, which he rebuilt with his own hands. He tried using it as a fishing boat but, in his own words, "had not the cunning to bait a hook", and resolved instead to sail around the world.

Leaving Boston, Slocum progressed via Gibraltar, the Magellan Straits, Australia, and South Africa to Rhode Island. He paid for the voyage and supported his family by giving lectures at his ports of call, and his book *Sailing Alone around the World* became a nautical classic. In 1909, at the age of 65, Slocum set out on another lone voyage, from which he and *Spray* never returned.

Another North American sea-captain, the Canadian John Voss, followed Slocum's example in an even more unlikely craft. His boat, the *Tilikum*, was a converted Indian dugout canoe. He set out from British Columbia in 1901, crossed the Pacific to Australia and New Zealand, and then proceeded via the Indian Ocean to the Cape of Good Hope, Brazil, and London. Like Slocum, he financed the voyage by giving lectures.

Many solo voyagers find in the oceans an escape from humdrum life ashore. Harry Pidgeon quit his farm in Iowa before the turn of the century and for several years spent a roving life as naturalist and photographer before achieving his real ambition—to build a boat and sail around the world. After his return in 1925 he lived aboard the vessel, *Islander*, until in 1941 he set out on a second circumnavigation. Three years later, at the age of 75, he married, and with his stalwart bride began a third trip round the world. They were wrecked near Vanuatu. Nothing daunted, Pidgeon built a new boat in which this redoubtable couple lived until Pidgeon's death in 1954.

Some voyages are made in the face of great personal adversity. Francis Chichester first achieved fame as a lone aviator in 1929, when he completed the second solo flight from England to Australia in a de Havilland Gipsy Moth biplane. Thirty years later, while convalescing from lung cancer, he sailed his yacht *Gipsy Moth III* to win the first trans-Atlantic solo race in 1960. In 1966 he set out to equal, single handed, the average passage time of the fast clipper ships that had plied between England and Australia during the golden age of sail. In the now legendary *Gipsy Moth IV*, he reached Sydney in 107 days, and went on to complete a full circumnavigation in a record overall time of 274 days, of which 48 had been spent ashore. On his return he was knighted by Queen Elizabeth with the sword used by her predecessor, Elizabeth I, to knight Sir Francis Drake when he returned from his circumnavigation in 1581.

In 1968 seven competitors set out from Falmouth on a non-stop solo round-the-world voyage sponsored by the British newspaper *Sunday Times*. Only one, Robin Knox-Johnston in the 12-ton ketch yacht *Suhaili*, finished the course, covering 48,450 kilometers (30,120 miles) in 313 days. One of the others, the Frenchman Bernard Moitessier, abandoned the race but continued to complete a non-stop record distance of 49,910 kilometers (31,000 miles) in 10 months.

Round-the-world cruising and racing, solo or otherwise, has become relatively commonplace, though the challenge remains undiminished. It appeals to women as much as to men. In June 1988 the Australian Kay Cottee sailed into Sydney

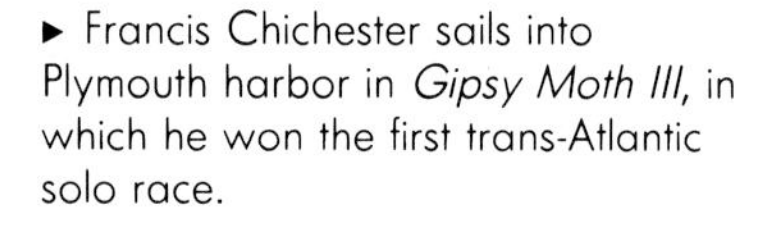

▸ Francis Chichester sails into Plymouth harbor in *Gipsy Moth III*, in which he won the first trans-Atlantic solo race.

Popperfoto

Syndication International

◄ *Kon-Tiki,* the balsa wood raft sailed by the Norwegian ethnologist and author Thor Heyerdahl from Peru to the Tuamotu Archipelago in eastern Polynesia. The voyage was devised to test the theory that Polynesia could have been settled by Indians from South America.

harbor after 189 days at sea, becoming the first woman to sail solo and non-stop around the world without any outside assistance except for radio contact. The accounts of sailors like Clare Francis and Naomi James—and most recently Tracy Edwards, who skippered an all-female crew in the 1989-90 Whitbread round-the-world race—have become seafaring classics.

RECONSTRUCTING THE PAST

Other oceanic explorers attempt to replicate the great voyages of the past using reconstructions of the original vessels. The first such experiment was probably the quatercentennial crossing of the Atlantic in 1892, when supposed replicas of Columbus's fleet set off from Spain for America.

Only the *Santa Maria* made the voyage unaided, though her extremely poor handling qualities suggest that the reconstruction was a bad one; more embarrassingly for their sponsors the other two, *Pinta* and *Niña*, had to be towed across by steamship. A year later a much more successful Atlantic crossing was made by the Norwegian Magnus Andersen in a replica of the ninth-century Viking longship found in an ancient burial mound at Gokstad.

More recently Thor Heyerdahl has made several voyages seeking to emulate the possible feats of prehistoric seafarers. In 1947 his balsa raft *Kon-Tiki* sailed from Peru to the Tuamotu Archipelago in the Pacific, covering 6,920 kilometers (4,300 miles) in 101 days. He made an Atlantic crossing from Morocco to Barbados in 1970 with a reed vessel *Ra*, which was based on early Egyptian pictures. Another reed vessel, *Tigris*, was used to explore possible routes and seafaring techniques of the ancient Sumerians by sailing through the Persian Gulf to Karachi and then across to Aden and the mouth of the Red Sea.

Other replica voyages have been made by Tim Severin, who between 1976 and 1977 sailed a hide-covered *curragh* based on descriptions of St Brendan's craft from Ireland via the Hebrides, Faroes, and Iceland to Newfoundland. In 1980-81 he built and sailed an Arab *dhow* of traditional form from Oman to China. Such experiments indicate something of the capabilities of early craft and seafarers, but they can neither prove nor disprove the actuality of the voyages they attempt to replicate. Some secrets of oceanic exploration will probably remain locked up for ever.

COLIN MARTIN

# 7 Exploring the Depths

Robert Holland

▲ In the past 20 years, technological advances in the design and construction of submersibles have enabled humans to explore the deep ocean floors.

◀ Through storms, misadventure, poor navigation, or maritime battles the ocean floor has become a graveyard for large numbers of ships. Many shipwrecks have become artificial reefs, home to fish and encrusting organisms, and are frequently visited by scuba divers.

Until this century, most of what has been learned about the nature of the ocean underwater has been derived from observations made during shallow dives and the use of nets, bottles, and instruments lowered from the surface. Human beings are, after all, terrestrial, air-breathing mammals with limited breath-holding capability. Some dolphins, seals, and whales make excursions to depths greater than 300 meters (1,000 feet) while holding their breath for an hour or more, but only a few highly trained human divers have attained round-trip dives to 100 meters (330 feet), holding their breath for about four minutes.

## Oceanography: A Modern Science

Exploring the depths has had an appeal that in many ways is comparable to the lure of the skies above. Throughout history, many have yearned to swim and dive like dolphins and whales. There are scattered accounts over the past two thousand years of diving for food, for recovering ships and goods lost at sea, for carrying out military strategies, and perhaps just for the sheer pleasure of exploring.

### The Beginnings

Alexander the Great is said to have been so taken with the notion of penetrating the depths that he had a diving bell constructed with a glass viewing window through which he observed fish and other sea life in 433 BC. The breath-hold diving exploits of the Japanese and Korean *ama* divers, mostly women, have been depicted in artistic renderings for more than 1,000 years, and continue even now with only modest concessions to modern technology.

During the past several centuries, numerous attempts have been made to develop equipment to enhance access to the ocean depths. Various diving bells were designed and used during the 1600s, and in 1715, successful salvage dives to 20 meters (65 feet) were made in a cylindrical wooden barrel equipped with a viewing port.

The use of compressed air for diving began in the late 1700s, coincident with the development of metal helmets and flexible suits. This technique continued during the 1800s, particularly following the appearance of a diving suit and helmet

Mary Evans Picture Library

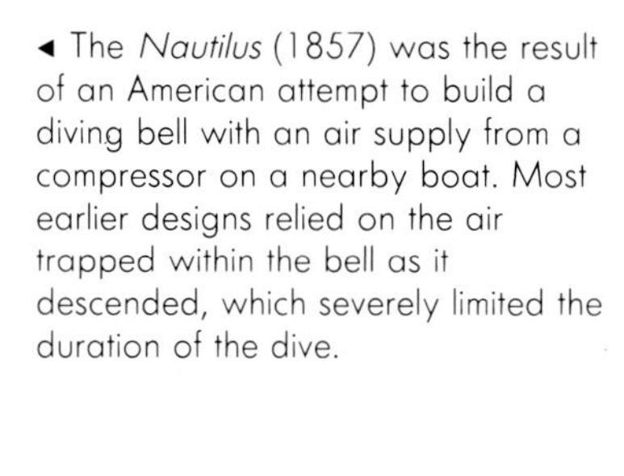

◀ The *Nautilus* (1857) was the result of an American attempt to build a diving bell with an air supply from a compressor on a nearby boat. Most earlier designs relied on the air trapped within the bell as it descended, which severely limited the duration of the dive.

Kev Deacon/Dive 2000

▲ The invention of self-contained-underwater-breathing-apparatus (scuba) was a significant breakthrough in the human quest to explore the oceans. This technology removed the diver's dependence on a hose attached to a surface air supply, and enabled freedom of movement and easy access to depths of up to 50 meters (165 feet).

designed by Augustus Siebe in 1819. In the mid-nineteenth century, Jules Verne wrote convincingly of self-contained diving systems in his classic work, *Twenty Thousand Leagues Under the Sea*. As is often the way, science fiction anticipated science fact, in this case by many decades. Jacques-Yves Cousteau and Emile Gagnan perfected the first practical, fully automatic compressed air "Aqua-lung" in 1943, and self-contained-underwater-breathing-apparatus—"scuba"—soon became an accepted technique for diving. Leonardo da Vinci and Benjamin Franklin both conceived designs for swim flippers many years before an ideal material—rubber—became available for the first practical version in the 1930s.

Although the history of yearning to dive is long, most of the equipment that makes modern diving possible has been developed in recent decades. Millions of sport divers now don masks, fins, snorkels, scuba, and an impressive array of accessories for excursions to 50 meters (165 feet), sometimes more. Scientific, commercial, and military dives are possible to depths over 500 meters (1,640 feet), using exotic mixes of gases, special suits, and chambers for underwater living and decompression. A few experimental dives have even exceeded 600 meters (1,970 feet), but efforts to probe significantly deeper using techniques that expose humans directly to the dark, cold, high-pressure environment of the deep sea are proving elusive. Other approaches involving protective suits and submersibles and a host of remotely operated underwater vehicles are more promising as a means of gaining practical working access throughout the full range of the ocean's depth, even the deepest sea.

THE VOYAGE OF THE *CHALLENGER*

Considering the difficulties of working more than a few meters under water and the relatively recent development of practical diving equipment, it is no wonder that early oceanographic expeditions concentrated on the sea surface, and used equipment that could be lowered from the deck to gather information about the depths below.

Oceanography—the systematic scientific exploration of the sea—began in earnest with the departure of HMS *Challenger* from the shores of England in 1872. A four-year expedition to the major oceans of the world followed, an epic journey that began with such fundamental questions as: "How deep is the ocean?" "Is there life in the deepest sea?" "Where do ocean currents go?" "How salty is the sea?".

*Challenger* scientists returned with many

answers; but many more questions were provoked by the discoveries they made. Nodules of manganese were dredged from the deep-sea floor. Why are they so abundant? How were they formed? Answers are not yet forthcoming. Thousands of species of plants and animals new to science were discovered, and thousands of measurements were made of ocean temperature, salinity, and chemistry. Samples of rocks and mud and sediment were taken from the Atlantic, Pacific, and Indian oceans. They provided many clues but few real answers about the basic physical character of the ocean. In the more than 50 volumes that reported on the extensive findings of the *Challenger* expedition, an underlying message emerged concerning both the great wealth of information and the magnitude of unknowns remaining about the sea. Numerous expeditions from many nations followed, but not until the early twentieth century did a general understanding of the oceans begin to take form.

Mary Evans Picture Library

▲ The scientists and crew of HMS *Challenger* sorting and preserving the contents of a benthic dredge net. Many of the animals dragged from the ocean floor during this voyage had never been seen before and must have seemed truly bizarre to these early oceanographers.

## A DEVELOPING SCIENCE

A comprehensive study of the south Atlantic during the 1925-27 voyages of the oceanographic vessel *Meteor* set a new standard for ocean research. Its scientists were among the first to use an electronic echosounder to measure ocean depths, a technique that provided revolutionary new insight into the configuration of the planet's surface. More than 70,000 echo-soundings clearly showed the rugged nature of the sea floor, and subsequent investigation by many expeditions revealed a startling structure of mid-oceanic mountain ranges, extending like an immense backbone down the Atlantic, Indian, and Pacific oceans. *Challenger* scientists were unaware that one of the ocean's dominant features had escaped their notice. Upon reflection, however, the wonder is that scientists a century ago learned as much as they did, considering the technology available. Even now, oceanographers can be likened to detectives, working with fragments of information, to put together bits of an enormous and ever-expanding puzzle.

Consider the problems. If standard oceanographic techniques were applied to exploration of modern cities or forests or mountain ranges, what would we know about these places? Suppose a net were dragged blindly through the streets of a town and the items fortuitously captured emptied onto the deck of an aerial equivalent of a research vessel. The jumble of shrubbery and cement, perhaps an automobile or confused pedestrian, would give some insight about what occurs below, but real understanding would be as elusive as real understanding is today of the nature of deep-sea communities.

New cameras, submersibles, robots, and other technology are greatly increasing our knowledge of the oceans. More was learned about the sea during the *Challenger* expedition than had been discovered during all preceding history. During the past half-century, it is safe to say that, once again, more has been learned than during all preceding human history. With new voyages and new technology, it is likely that the greatest era of ocean exploration has just begun.

## EXPLORING THE DEEP-SEA TRENCHES

The greatest depths of the sea are in the steep-sided, narrow, deep-sea trenches that form a nearly continuous border not far offshore from the coastal ranges of the Asian and American continents. One of the most striking topographic features of the Pacific Ocean is the ring of trenches along its outer edge. A major trench in the Indian Ocean, the Java Trench, extends northwest from Western Australia. There are two deep-sea trenches in the Atlantic: the Puerto Rico Trench in the Caribbean Sea and the Sandwich Trench offshore from the southern part of South America. But it is in the Pacific that the trenches are most common—and most impressive.

The upper edge of most of these grand canyons of the deep begins at about 6 kilometers (3.7 miles)—about half the maximum depth of the sea. Access to these unique areas has been and still is extremely difficult. Only once has a descent been made to the bottom of the deepest place, the Challenger Deep in the Mariana Trench near the Philippine Islands. Using the bathyscape *Trieste*, US Navy lieutenant Don Walsh and Swiss engineer Jacques Piccard made their historic journey in January, 1960, resolving both the question of whether or not such a descent could be made (it could) and whether or not life occurs in the deepest sea (it does). Looking back at the two explorers 11 kilometers (7 miles) down was a flounder-like fish, at home in a realm where pressure is 110,240 kilopascals (16,000 pounds per square inch).

▼ The interior of this submersible bears a strong resemblance to a spacecraft. This is because humans must operate within a self-contained unit if they are to survive the lack of light and oxygen and the extreme pressure of the deep ocean. The submersible is equipped with the latest in high-technology instruments to enable scientists to observe their watery environment, move around, and communicate with the surface.

Ken Vaughan/Planet Earth Pictures

Using various acoustic sounding techniques, the general configuration of the deep-sea trenches has been charted, but taking samples and even photographs in these great gashes in the sea floor presents enormous challenges. Oceanographer Willard Bascomb is among those who have spent considerable time trying—and sometimes succeeding. His recent book, *Crest of the Wave*, describes his exploration of the Tonga Trench, a formation about 2,400 kilometers (1,500 miles) long, 24–48 kilometers (15–30 miles) wide, and 11 kilometers (7 miles) deep. To sample this magnificent crack, eloquently described as being "a mile deeper than Mount Everest is high" or, viewed another way, "as deep as seven Grand Canyons but with much steeper sides", Bascomb and his colleagues lowered a gravity corer into the depths. The corer is an instrument about 2 meters (6.5 feet) long, weighted with lead, that is used to obtain a cylindrical plug of whatever is on the bottom. When it was retrieved, the sampler was empty, but embedded in the lead weight was a small chip of basalt rock. This represented a slight but important clue from which oceanographic detectives could extrapolate the nature of one of the planet's great deep-sea formations.

Since Walsh and Piccard's visit to the bottom of the Mariana Trench, little attention has been given to direct access. Most deep-sea trenches start at a depth where even the deepest modern submersibles must stop. It seems curious that no present submersible can travel to 11 kilometers (7 miles) for direct access to the deepest sea, thus facilitating exploration of the deep-sea trenches. While constituting only about 3 percent of the total ocean, the depths below 6 kilometers (3.7 miles) represent unique areas with high scientific value. They are also recurrently posed as possible sites for disposal of toxic wastes. Some argue that exploration must precede such use, to avoid possible disruption of systems that may have high value beyond waste disposal. Three percent may sound very small, but this represents approximately 10 million square kilometers (3.8 million square miles), an area about the size of Australia or the United States or Canada. It may be dark and cold and remote with a unique high-pressure environment where strange creatures dwell—but it is a special part of the planet currently inaccessible for scientific exploration or other interests.

Interest in exploration of the deep-sea trenches is growing. In Japan, development is underway for construction of a large tethered remotely operated underwater vehicle that will be equipped with cameras and various sampling tools to operate in depths to 10 kilometers (6.2 miles). In the United States, efforts are focused on small manned submersibles planned as a part of a project known as "Ocean Everest". In an era when travelling 11 kilometers (7 miles) high in aircraft is routine, and spacecraft can go to the moon and beyond, it seems likely that access to the full ocean depths, including direct exploration of the deep-sea trenches, will soon be possible.

SYLVIA A. EARLE

# Modern Submersibles

Modern submersibles may be likened to "inner spacecraft", the undersea equivalent of advanced systems that transport astronauts skyward, beyond the atmosphere familiar to humankind. Similarly, it is necessary to have protective vehicles for those who wish to travel to the hostile atmosphere of the deep sea, or to use remotely operated vehicles, robots, and telepresence—the undersea analogs of unmanned space probes and satellites.

One of the pioneers of deep-sea exploration, Dr William Beebe, was so impressed with the parallels between the sea and space that he wrote in his historic work, *Half Mile Down*, that the only place comparable to the deep sea environment "must surely be naked space itself, out far beyond atmosphere, between the stars . . ." Beebe likened the dazzling array of deep-sea creatures dwelling in the vast darkness, glistening in the lights of his undersea vehicle, to "the shining planets, comets, suns, and stars".

Without submersibles, this magnificent realm would be known only indirectly, through samples brought back in nets and data gathered with instruments and cameras lowered from surface platforms. Access to the sea would be limited to the depths divers can reach. Scuba diving is effective to depths of about 50 meters (165 feet), and other diving methods can be applied for limited access to more than 500 meters (1,640 feet). But it is well to keep in mind that the average depth of the sea is at about the depth where the sunken passenger liner *Titanic* is resting on the sea floor—more than 4 kilometers (2.5 miles) beneath the ocean's surface. The maximum depth is more than 11 kilometers (7 miles) down.

Fortunately, in the half-century since Beebe's early exploration, numerous undersea vehicles have been developed and used to help explore the ocean depths. Two submersibles have visited the site of the *Titanic:* the research vessel *Alvin*, operated by Woods Hole Oceanographic Institution, and the French vehicle, *Nautile*. Although more than a quarter of a century has passed since *Alvin* was launched, the system has been renovated many times, and has maintained the most modern instrumentation. During exploration of the wreck of the *Titanic*, a small remotely operated vehicle, *Jason Jr*, was launched from *Alvin* and, operated with remote controls from within the submersible, was able to probe inside the shipwreck in places too small and too dangerous for access by *Alvin* directly.

Al Giddings/Ocean Images

▲ The *Johnson Sea-Link* submersible can descend to nearly 1,000 meters (3,000 feet) from a surface support ship. Well equipped for collecting samples and taking photographs, the submersible also has a pressurizing chamber which enables divers to use it as a base. The front acrylic dome ensures good visibility.

The combination of a manned submersible working effectively with a small remotely operated vehicle highlights a point of view shared by many who work in the ocean. The issue is not whether manned systems are better than remotely operated or robotic ones; rather, that the best tool for a particular job should be chosen, and sometimes both will work together.

PERSONAL SUBMERSIBLES

Since the early 1970s, there has been a rapid evolution of "personal submersibles" that have greatly altered ocean access. These one-person, one-atmosphere systems began with a renovated

Al Giddings/Ocean Images

▲ In conjunction with the mini-submarine *Star II*, JIM (a self-contained atmospheric diving suit) enables divers to perform many practical tasks in depths of up to 600 meters (2,000 feet). The articulation of the magnesium alloy, pressure-resistant construction allows divers to flex their limbs within the suit.

version of a system called JIM, named for the first person to wear one. Like larger submersibles, JIM has self-contained life support that is maintained at surface pressure—one atmosphere. It is made of articulated metal sections joined by special seals. Normally operated on a tether to a surface ship or platform, one of the 15 modern versions of JIM was used in 1979 for exploratory research dives in Hawaii without the use of a cable to the surface. Nearly 20 related vehicles, called WASP, the "submersible that you wear", were developed for commercial diving, but the system was adapted for use by scientists who used WASP to explore the Santa Barbara channel, thereby revolutionizing previous concepts about life in this seemingly well-known area.

Other tethered one-person, one-atmosphere systems were developed in the 1970s and 1980s, but one that operates without a tether has become especially renowned. Named *Deep Rover*, the vehicle is basically a clear acrylic sphere that looks much like the front end of the *Johnson Sea-Link* submersibles. Two large sensory manipulators are used by the operator as extensions of his or her own arms and hands. It is a splendid example of the interaction between human and machine, and is so "user friendly" that piloting it has been likened to driving a golf cart.

This characteristic, coupled with its small size and lack of a tether, makes it possible to use *Deep Rover* for scientific operations in remote locations inaccessible to large submersibles or even tethered remotely operated vehicles that require complex cable management. For a project in Crater Lake, Oregon, in 1988 and 1989, *Deep Rover* and a small, deep-diving remotely operated vehicle called *Phantom* were flown in by helicopter and operated by a team of scientists to the maximum lake depth of 592 meters (1,942 feet).

Late in the 1980s, increasing interest in ocean research and exploration resulted in the development of several new systems that can dive to 6 kilometers (3.7 miles)—and thus to approximately one-half the ocean's depth, but within reach of about 97 percent of its area. In 1987, the Soviet Union launched two three-person submersibles, *Mir I* and *Mir II*, built in Finland and operated worldwide from a large research vessel. Japan now operates the three-person *Shinkai 6500*, the deepest diving submersible currently available. The French submersible, *Nautile*, and the US Navy's *Sea Cliff*, recently renovated and both capable of diving to 6 kilometers (3.7 miles), complement the more modern members of the global fleet of five deep-diving manned systems.

## THE WAY FORWARD

New materials, new power sources and new incentives for ocean exploration are driving the development of innovations in submersible technology that should revolutionize ocean access. Part of the revolution will be attributable to the rapid acceptance and use of remotely operated vehicles and underwater telepresence noted below, but even for manned systems, the changes are expected to be dramatic.

Until now, most manned submersibles have operated on the same principles that govern the movement of balloons and dirigibles—a variable ballasting system. The submersibles descend with weights, either water that is later displaced with air for buoyancy, or lead or iron weights that are left behind to enable the vehicle to become lighter than the surrounding water, and thus able to ascend. The *Mir* submersibles, *Alvin*, *Deep Rover*, and many more operate this way.

The advent of new materials, new batteries, and new designs has led to another approach, one that can be likened to a small, fixed-wing aircraft. Two small, portable, one-person, battery-powered systems called *Deep Flight* have been constructed of "modern" materials—acrylic, glass fiber, composites, and epoxy—and without ballast tanks. Designed to be literally flown through the water, *Deep Flight* is expected to be used for unique scientific applications such as swimming side-by-side with whales and fast-moving fish, and conducting surveys in remote areas not readily accessible to larger systems. The *Deep Flight* design should also serve as a prototype for deeper, faster vehicles that will use new power sources and various modern non-metallic materials—carbon fiber, glass, epoxy, composites, and others—for systems that will be able to travel freely throughout the full range of ocean depth.

Unmanned submersibles—remotely operated vehicles and autonomous underwater vehicles—have been edging to the forefront of ocean exploration and research. Exotic communities of life and associated hydrothermal vents in the deep sea near the Galapagos Islands were first glimpsed through the television "eyes" of an underwater robot, a towed array of instruments operated from a surface ship. Later, *Alvin* transported scientists directly to the site for first-hand observations and documentation. The *Titanic* was found by a towed system equipped with special cameras and lights and *Alvin* later took observers to see at close range what the remotely operated system had discovered.

The first submersible to dive deeper—and much longer—than divers with air tanks in the Antarctic was not a manned submersible, but one of the nearly 200 portable *Phantom* remotely operated vehicles currently being operated globally in depths as great as 900 meters (3,000 feet). Tetherless systems equipped with computer controls are being touted as the cutting edge of submersible technology, and several variations are being developed in the United States, United Kingdom, Canada, Japan, and elsewhere.

What are the limits of submersibles in the sea? The potential of existing technology has not yet been fully realized, and with the addition of new through-water communications, new power sources, new strong, light materials, and a powerful incentive to understand how the oceans work, the developments should proceed rapidly toward a "space age" in the sea.

SYLVIA A. EARLE

Peter Scoones/Planet Earth Pictures

▲ The *Bruker Mermaid* submersible, seen here on the deck of *Star Pisces*. Unlike large military submarines, small submersibles have a limited power supply and are not designed to travel long distances to deep ocean sites. As a consequence, they rely on a support ship to provide site-to-site transport and for storage between dives.

# LIVING UNDER THE SEA

What is it like to live under the sea? Aquanauts emerging after spending two weeks 15 meters (50 feet) below the surface of the Caribbean Sea in the Tektite underwater habitat described extraordinary experiences—subsea moonlight swims with large, silver amberjack; early morning encounters with parrotfish emerging from their night-time lairs; and leisurely meetings with tiny damselfish, vigorously defending their territories among the coral branches from the alien aquanauts. They also described ordinary things—cooking, eating, sleeping, taking showers, writing letters—in comfortable, motel-like surroundings but with an exceptional view of the ocean—a view from the inside looking out.

## THE PROBLEM OF DECOMPRESSION

During the Tektite project in 1969 and 1970, more than 50 scientists and engineers, in teams of four or five, spent between 10 days and two months living in and working from the underwater habitat, with frequent excursions outside. The Tektite habitat, like several systems that preceded it and many that followed, provided several advantages over traditional diving techniques, most important of which were increased time under water and access to greater depths.

Divers using compressed air are limited to brief visits under water. At 20 meters (65 feet), it is possible to stay for less than an hour; at 30 meters (100 feet), about 20 minutes. Longer stays can be made if time is allowed for decompression—the time needed while ascending to allow compressed gases that have entered the bloodstream to dissipate. For a 20-minute visit at 50 meters (165 feet), more than an hour of decompression is required. At greater depths compressed air can be dangerous, with the nitrogen component causing narcosis and oxygen actually becoming toxic. Deep dives require the use of mixes of gases such as helium or hydrogen, but whatever gas mix is used, decompression time increases with increasing depth. However, once tissues become saturated with gas at depth (a process that may take several hours), the decompression time remains the same, no matter how long the dive lasts. For example, Tektite aquanauts all decompressed for the same length of time—21 hours—although some stayed under water for 10 days, while others remained for 60 days.

## UNDERWATER HABITATS

Several pioneers of underwater exploration developed the concepts underlying saturation diving, and thus underwater living. Edwin A. Link was the first to "saturate" using a breathing mix of helium and oxygen in 1962. Later in the same year, Captain Jacques Cousteau launched an experiment in underwater living in the Mediterranean Sea. Soon thereafter, United States Navy Captain George Bond put his ideas about underwater living into action with a series of experimental dives during projects called Sealab I and II, and, in 1968, Sealab III.

More than 50 underwater research facilities have been developed since Link's first relatively simple system in 1962. Most have now been retired, but continuing sophisticated experimental work by the French scientific organization IFREMER and the diving

Flip Schulke/Planet Earth Pictures

A researcher photographs fish and corals from the Tektite habitat.

company COMEX has culminated in the deepest saturation dives yet made—in excess of 600 meters (2,000 feet).

Most underwater living techniques have been developed and applied by those involved with support of the offshore oil and gas industry and other commercial underwater activities. Dives to 300, 400, and 500 meters (1,000–1,650 feet) are never regarded as ordinary or routine, but they have become reasonably common for difficult deep-water work that requires the presence of human intelligence, coupled with relatively small size, dexterity, and precision not yet matched by robots or other submersibles.

### THE HYDROLAB HABITAT

While not the deepest, a system called Hydrolab provided an underwater home for more scientists for more years than any other. Launched in 1968 in Florida, it subsequently spent many years in the Bahamas and the Virgin Islands before retirement to an honored place at the Smithsonian Institution in Washington, DC, in 1986. One of the most interesting series of missions conducted from Hydrolab involved the use of the submersible *Johnson Sea-Link*. The submersible served as an elegant undersea taxi to transport divers from Hydrolab to the edge of a steep ocean wall, where they swam out for brief excursions in depths of 50 to 80 meters (165–260 feet), then returned to the submersible for decompression and transport back to Hydrolab.

Over the years, more than 300 scientists spent many hours in Hydrolab and nearby waters becoming familiar with lobsters on their own terms, the habitats of grouper and snapper, the behavior of predators and prey, and the interactions between the hundreds of creatures on a coral reef. They also became acquainted with the parallels between sea and space, in terms of living and working in close quarters with others for long periods. Astronauts have an unequaled view of Earth and the heavens above, but the aquanauts of Hydrolab, Tektite, and other underwater habitats were provided with access to the Earth from within, a view of galaxies of bioluminescent organisms at night, and unique opportunities to observe creatures whose history precedes our own by many millions of years.

Official US Navy Photograph

Preparing to launch the Sealab III underwater habitat.

### INTO THE FUTURE

The most modern underwater research habitat is Aquarius, owned and operated by the United States National Oceanic and Atmospheric Administration (NOAA), and sometimes used by scientists sponsored by NASA as an experimental analog for living in space. First used in the US Virgin Islands and later moved to Florida, Aquarius provides comfortable accommodation 15 meters (50 feet) under water.

Science fiction yarns suggest that it may be possible to live under water at any depth, but human physiology appears to set the limit at something close to 700 meters (2,300 feet). An alternative approach that will make the science fiction dreams come true is the use of one-atmosphere dwellings coupled with one-atmosphere suits, vehicles, and robots, just as astronauts plan for the space station and beyond. The materials exist to build such structures at any depth, and so does the skill for design and construction. All that remains to bring these visions to reality is the will, the desire to do so.

SYLVIA A. EARLE

Frans Lanting/Minden Pictures

A rainbow breaks over the Na Pali coast, Kauai, in the Hawaiian Island group.

PART TWO

# ISLANDS: WORLDS APART

# 8 Islands of Fire

The Earth is a dramatic structure, undergoing constant change. Nowhere, perhaps, is this more apparent than in the birth and death of volcanic islands, which have their origin deep within our planet, where molten rock wells up through fissures to the Earth's crust. As the molten rock—magma—cools and solidifies, deposits of lava are left, slowly building up on the crust. Although most of the Earth's volcanic eruptions occur on the ocean floor, the lava deposits may grow so high that they break the ocean surface. Thus an island of fire is born.

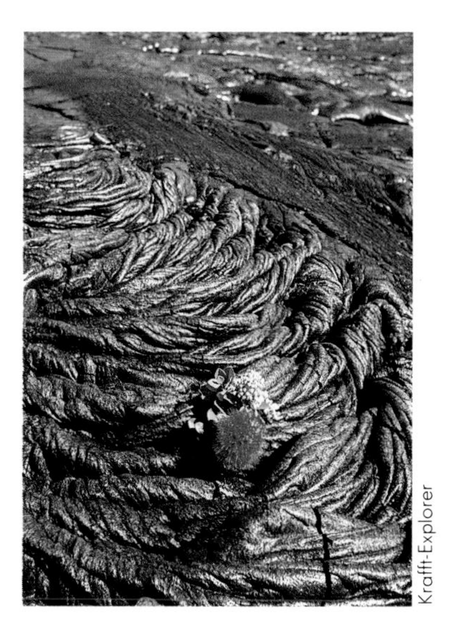

Krafft-Explorer

▲ Intricate swirls and "cords" of basalt typify this ropy *pahoehoe* lava flow on the flanks of Kilauea on the island of Hawaii.

◄ Lava erupts in a fiery fountain and flows as incandescent *pahoehoe* from the main vent of Kilauea.

## Oceanic Volcanoes

The rocks that constitute the surface of our planet are of three principal types, chief of which is igneous: formed from a molten fluid containing all the chemical components necessary to make up the igneous rock upon cooling. This molten parent fluid is called "magma" and occurs deep within the Earth. It reaches the surface through volcanic activity and flows in the form of lava, at a temperature of up to 1,100°C (2,000°F). As the magma cools upon reaching the surface, crystals of minerals grow and interweave to form the igneous rock.

### The Causes of Volcanism

The acceptance of the theory of plate tectonics, which enabled us to explain the process of sea-floor spreading, has in turn given us an understanding of volcanism. The surface of the Earth—both the continents and the ocean floor—consists of rigid plates, and the most active volcanoes lie in belts which coincide with zones of fissuring and collision between these plates. The plate boundaries are bordered by mid-ocean ridges, such as the Mid-Atlantic Ridge, or by transform or thrust faults (areas of stress and fracture where opposing plates meet), or by oceanic trenches created when one plate slides underneath another.

The source of the original magma is located within the asthenosphere, the plastic, partially molten layer lying beneath the lithosphere, the solid, elastic surface of the Earth. This could be as shallow as 60 kilometers (37 miles) at the base of the lithosphere, or as deep as 250 kilometers (150 miles) from the surface within the asthenosphere. The separation of the neighboring plates at the mid-ocean spreading centers leads to a reduction of pressure and an upwelling of magma to the surface. As the magma moves upward, it loses

Greg Vaughn/Black Star

◄ A curtain of fire—the result of lava erupting along a fissure during an eruption of Kilauea in 1986.

Studio R/ Japan Broadcasting Corporation

▶ This computer-generated graphic indicates the lateral movement of cooled magma away from a mid-ocean ridge as continental plates move apart, causing the gradual widening of an ocean basin. This process is commonly known as sea-floor spreading.

▶ These ridge axis cross-sections depict three styles of mid-ocean ridge volcanism and faulting. The arrows indicate magma motion in the chamber and the direction of sea-floor spreading.

**KINDS OF MID-OCEAN RIDGE VOLCANISM**

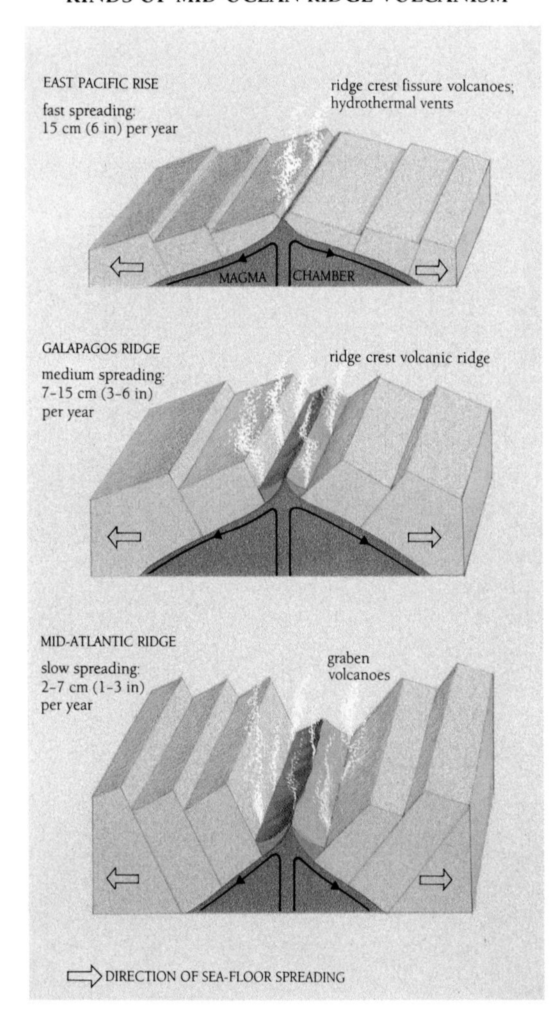


some of its gases and gains some chemical constituents from the neighboring rocks, thus turning into lava which erupts on the ocean floor along fissures and cones dispersed along the deep rift valley which marks the crest of the mid-ocean ridge system.

Slow-spreading sections of the mid-ocean ridge system, such as those found along the Mid-Atlantic Ridge, with spreading rates between 2 and 7 centimeters (1-3 inches) per year, may experience submarine volcanism at any given site once every 10,000 years. Fast-spreading ridge segments found along the East Pacific Rise, with spreading rates between 10 and 18 centimeters (4-7 inches) per year, may experience volcanism at any given place once every few years.

Where magma extrudes along the mid-ocean rift valleys, the asthenosphere has penetrated to the surface of the Earth. At this line of active volcanism, there is no oceanic crust and no lithosphere. As the new crust forms after the lava has cooled, the lithosphere starts to develop. As the lithosphere cools and moves in time away from the ridge crest, it thickens and sinks, attaining a thickness of 60 kilometers (37 miles) over 90 million years.

The depth of the ocean floor is directly related to its age, since it takes time for a segment of the lithosphere to cool along its path away from the mid-ocean ridge, with rates of motion ranging from a rate of 10 kilometers (6 miles) (slow) to 90 kilometers (56 miles) (fast) per million years. The deepest and, therefore, oldest ocean floor (besides that underlying the oceanic trenches) is located along the continental margins of the Atlantic Ocean and in the northwestern corner of the Pacific Ocean. These old sections of ocean floor lie in waters 6,000 meters (19,700 feet) deep (or deeper) and were formed by submarine volcanism along the Pacific and Atlantic mid-ocean ridge segments 170 million years ago.

Overlapping fissures, as well as overlapping rifts, are common features of the mid-ocean ridge segments. Within the central trough (or "graben") of mid-ocean ridges, volcanism initially occurs as a fissure opens along the graben floor. Molten lava extrudes from the fissure and flows along the graben floor. The surfaces of these lava flows are cooled instantly by contact with the cold (2°C/ 36°F) sea water.

## LAVA AND VOLCANIC ROCK

Lava may have a ropy appearance, like that of the Hawaiian *pahoehoe* lavas; or it may have the appearance of folded drapes; or the surface may be broken into sheets similar to plates of broken ice; or it may be rubbly, a lava known as *a'a*. The lava lake eruption from the fissure in the rift axis may be followed by the eruption of slower-flowing lava extruding out of individual fissures on the floor of the rift valley. In this case, the lava is extruded from the fissures in the form of tubes in a manner similar to that of toothpaste being squeezed out of a tube.

Frequently, the tubes on the ocean floor break and the lava flows out to solidify in the form of rounded pillow lavas. One pillow containing molten lava may break out into another pillow and then another. Thus the pillow lava building process occurs as lava flows downslope from one pillow to the next. In this manner, extensive fields of pillows are built up. Indeed, the most common lava form on the ocean floor and the Earth (since the ocean floor covers two-thirds of the world's surface) is pillow lava. The diameter of the pillows and tubes ranges from 1 to 3 meters (3–10 feet), and the ripples, ropes, drapes, and broken sheet structures of the lava lakes may range from a few centimeters to several meters in height. Pillow cones tens of meters high may be built above a particularly active fissure.

Submarine volcanism can generate extensive lava flows on the ocean floor. Mapping by the US Geological Survey of the ocean floor surrounding the east rift of Kilauea has shown submarine lava flows extending over an area of 1,000 square kilometers (386 square miles). Even more surprising is the 25,000 square kilometer (9,650 square mile) lava field mapped over the 4,500-meter (14,750-feet) deep ocean floor 200 kilometers (125 miles) north of the Hawaiian island of Oahu.

When lava solidifies, it forms volcanic rocks, which, because they make up the ocean floors, are the most widespread rock types on Earth. There are three principal kinds of volcanic rock: basalt, andesite, and rhyolite. All are composed of silicon oxide and the metallic oxides of aluminum, iron, magnesium, titanium, calcium, potassium, and sodium. The most widespread volcanic rock is basalt, which consists of 50 percent silicon oxide by weight, and the oxides of other metallic elements. Andesite is formed by the volcanoes of island arcs and oceanic margins, such as the Andes of South America, and is composed of about 58 percent of the oxide of silica. The most siliceous volcanic rock is rhyolite, also formed by island arc volcanoes: this consists of 75 percent of the oxide of silica.

Greg Vaughn/Black Star

▲ Fresh, slow-moving *pahoehoe* lava flows from a fissure eruption on the slopes of Kilauea, Hawaii.

▼ Pillow lava near a hydrothermal vent in the Galapagos region. Lava emitted from such deep-water vents is not explosive because the water pressure at great depth prevents dissolved gases from boiling and expanding. Instead, the gases ooze out, forming these characteristic pillows. Pillow lava can also form in shallow water from terrestrial lava flowing into the sea.

Robert Hessler/ Planet Earth Pictures

Krafft-Explorer

▲ Lava flows from fissures on the flanks of Hawaii's Mauna Loa, while a "sea" of glowing molten basaltic lava surrounds the main crater during an eruption in March 1984.

Basaltic magma produces highly mobile lava at high temperature, as in the flows on Mauna Loa and Kilauea, Hawaii; its eruptions are not very explosive, Rhyolite magma creates viscous and gassy lava of a lower temperature, and produces volcanism that may be highly explosive, such as in the explosion of Krakatoa in 1883 and the 1980 Mt St Helens eruption in the United States. Rhyolites can produce eruptions of incandescent clouds made up of glowing high-temperature particles bathed in steam, moving at more than 100 kilometers (60 miles) per hour. An example of this kind of volcanic force was the explosive eruption of Mount Pelée in Martinique in 1902, which destroyed the town of St Pierre and killed 28,000 inhabitants.

## THE GROWTH OF A VOLCANO

The growth of a new volcano on the ocean floor can be studied using as an example the Loihi submarine volcano, 28 kilometers (17 miles) southeast of Kilauea in the Hawaiian chain. Submarine volcanoes are built on a fissure or rift, along which magma moves to the surface. Loihi, meaning "long one" in Hawaiian, has grown on a fissure system oriented north-south-to-southeast, approximately following the path of the Hawaiian hot spot trace on the ocean floor.

Active volcanic extrusion surrounded by hydrothermal venting is taking place at the summit of Loihi in a water depth of 960 meters (3,150 feet) and at the tip of the southern rift, 21 kilometers (13 miles) southwest of the summit, in a water depth of 5,200 meters (17,000 feet). At the southern end of the rift, the 90 million-year-old oceanic crust upon which Loihi sits is fissured. Fresh pillow lavas and *pahoehoe* lavas have poured out of this fissure, covering the surrounding sediments. Northward of this leading edge of the rift, pillow lava cones 100 meters (330 feet) high have been built along the rift by active volcanism. These lavas are fresh with just a dusting of ocean sediments covering them. The cones are mechanically unstable and mark the sites of frequent downslope slumping of the freshly erupted basalt.

Both the eastern and western slopes of Loihi are subject to slumping processes, and the slopes show slump scars with vertical walls. The volcanic slopes are covered by broken pillow and sheet lava fragments. This slump debris (or "talus") is constantly moving downslope. Episodic submarine landslides generate local shallow earthquakes that are recorded on the seismometers of the Hawaiian Volcano Observatory.

The edifice of Loihi is, therefore, built up largely of talus with interspersed lava flows. Upon

this kind of unstable foundation, giant volcanoes (such as Mauna Loa) are built. It is not surprising, therefore, that in the geological history of the island of Hawaii, and indeed the remainder of the Hawaiian Islands, episodes of large submarine slumping are recorded. During the past two million years, the eastern half of the extinct Koolau volcano, on the island of Oahu, appears to have slumped off onto the nearby ocean floor along a front 50 kilometers (30 miles) wide, with slump debris extending 100 kilometers (60 miles) out to sea.

The summit of Loihi is characterized by pit craters, like those observed on Kilauea, and by active hydrothermal vents with sea water which has been heated to 30°C (86°F) by interaction with the magma contained within the volcano. The mixture of talus and pillow lavas that has built the edifice of Loihi is very porous, and sea water circulates freely through the rock that covers the slopes of the submarine volcano. The interior of the volcano is marked by vertical dykes filled with hot rock or liquid magma. Sea water interacts with the heated dykes, is in turn heated, leaches the surrounding basalt of minerals and then migrates upward in the form of mineral-laden hot water. When this hot fluid comes into contact on the volcanic surface with the cold sea water, the shock cooling of the hydrothermal fluid crystallizes the minerals in solution.

## MID-OCEAN RIDGE VOLCANOES

Current studies indicate that almost all the world's volcanoes—both submarine and aerial—belong to one of three classes: mid-ocean ridge volcanoes, island arc volcanoes, or hot spot volcanoes.

Mid-ocean ridge volcanoes, fed by a persistent, powerful magmatic plume, may form very large volcanoes that grow above the sea surface to create islands. Iceland represents such a plume-driven volcano located astride the northern end of the Mid-Atlantic Ridge, and is an excellent example of how sea-floor spreading works on the ocean floor. Iceland is in a state of east-west tension, one half located on the North American Plate moving westward, the other half located on the Eurasian Plate moving eastward. Tension

Pete Turner/The Image Bank

▲ A dramatic time exposure shows the impressive trajectory of lava and rock fragments in this "explosive" eruption in Iceland.

▼ Most volcanic activity occurs along oceanic ridges, where magma rises through the submarine fissures. The new oceanic crust formed by this process replaces the crust destroyed by subduction on the plate margins. Isolated hot spots create volcanic islands such as the Hawaiian chain. Arcs of volcanic islands form where two plates collide in mid-ocean.

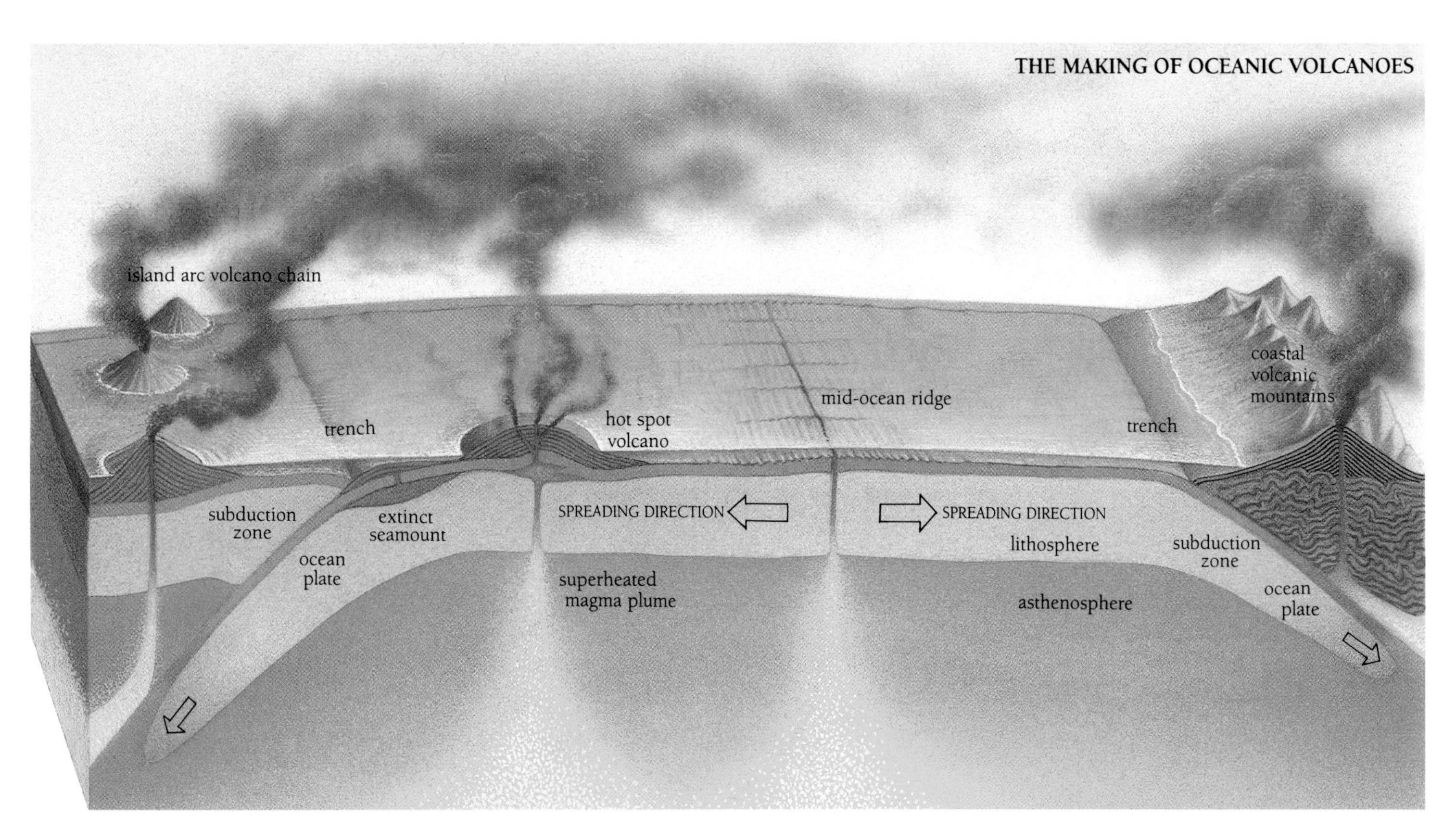

# KRAKATOA: AN ISLAND ERUPTS

Krafft-Explorer

Krakatoa, September 1979.

Indonesia has more volcanoes than any other nation on Earth. Beneath its scattered islands the Australian Plate, slowly moving northeast, is drawn beneath the thicker Eurasian Plate. This subduction zone has produced a chain of volcanoes, from which more than 972 eruptions have been recorded. Because of the region's high population density, 83 of these have caused fatalities.

In this land of volcanoes, the eruption of Krakatoa in August 1883 has acquired a special place. It came at a time of rapid development in communications technology. Telegraph and cable systems linked Europe with the rest of the world, and details of the eruption were cabled in terse, urgent words, to become newspaper accounts around the globe.

Krakatoa is a tiny mountainous island in the Sunda Strait, midway between Java and Sumatra. For centuries the volcano that dominated it had been a familiar landmark to seamen traveling through the strait, along which most of the region's maritime traffic passed. The volcano was believed to be extinct. We now know that it had erupted about 200 years earlier, but rapid revegetation had masked all evidence of the previous explosion.

For three months the people of nearby islands had been aware of rumblings and plumes of smoke coming from Krakatoa. At 1 p.m. on August 26 the rumblings turned to explosions, which continued with mounting intensity. "The ear-drums of over half my crew have been shattered," wrote a British sea-captain 40 kilometers (25 miles) away. Within an hour a huge black cloud 27 kilometers (17 miles) high hung over the strait. By dusk the first of a series of tsunamis (tidal waves) battered the coasts of Java and Sumatra. Lightning flashed through the darkness: the sky was "one second intense blackness, the next a blaze of fire".

All this was merely a prelude: at 10 a.m. on the morning of August 27, the entire island of Krakatoa exploded, and a column of ash, rock, and fire blasted into the sky. A series of deadly tsunamis gathered force and slammed into

Georg Gerster/John Hillelson Agency

The smoking crater is the heart of the island of Krakatoa.

surrounding islands, their massive walls of water, up to 40 meters (130 feet) high, destroying all evidence of human settlement. The explosion was heard on Rodrigues Island, over 4,500 kilometers (2,700 miles) away; ash fell on ships 6,000 kilometers (3,600 miles) distant. For a day the Sunda Strait was in darkness.

When the eruption ended, at least 36,400 people had been killed, most of them victims of the tsunamis, and 165 villages were obliterated. Only one-third of Krakatoa remained above sea level. As one observer telegraphed: "Where once Mount Krakatoa stood the sea now plays." To the north, new islands of steaming pumice had emerged.

The effects of the eruption were felt far beyond the Sunda Strait. The sun became greenish blue as ash, erupted up to 50 kilometers (80 miles) into the stratosphere, circled the Earth. After three months the outpourings of Krakatoa had spread further, causing such vivid sunsets that fire engines were called out in American cities to quench the apparent fires. The fine volcanic dust that created such extraordinary atmospheric conditions also served as a filter for solar radiation, lowering temperatures around the world for a year after the eruption.

The high death toll, combined with the fact that technology made possible the observation and recording of its distant effects, have made the eruption of Krakatoa the most famous volcanic eruption in history. Others have been greater: by most measures the 1815 eruption of Tambora, 1,400 kilometers (840 miles) to the east, was larger, but because of poorer communications it did not elicit the widespread fascination of Krakatoa. From the eruption of Krakatoa, much was learned about the workings of volcanoes; more significantly, perhaps, came an understanding of the global implications of such a phenomenon.

In 1927 a fishing crew reported that Krakatoa was once again smoking. Two years later a new volcano, Anak Krakatoa ("child of Krakatoa"), rose from the Sunda Strait. Eruptions in 1953 raised the cone 60 meters (200 feet) above sea level. Krakatoa is not yet dead.

SHEENA COUPE

Krafft-Explorer

A column of dense smoke billows from the crater.

Krafft-Explorer

▲ Incandescent flows and lava fountains pour from the partially eroded crater of this active Icelandic volcano.

cracks develop, and magma flows up the cracks and erupts as basaltic lava on the surface. In one single eruption in 1783, the Laki Fissure, a fissure 32 kilometers (20 miles) long, opened up and erupted 12 cubic kilometers (3.5 cubic miles) of lava. The magma solidified in the fissure and formed a dyke, a vertical wall of dense basalt.

## ISLAND ARC VOLCANOES

Island arc volcanoes—among them the Aleutians, the Andes, and the islands of the Indonesian archipelago—are the result of one plate plunging beneath another. At the point of contact, the thinner, denser oceanic plate slides under the other to form a deep trench. In this process, called subduction, the oceanic crust is consumed and remelted into the asthenosphere at depths of 700 kilometers (430 miles) beneath the surface. Friction between the downward-moving (subducting) crust and the plate margin leads to heating and partial melting of the subducting plate at a depth of about 100 kilometers (60 miles) beneath the surface. The mixture of remelted basalt and melted oceanic sediments overlying the subducted crust forms a siliceous, gassy magma that migrates to the surface and erupts in the form of lava.

Island arc volcanoes erupt lavas that are more siliceous than basaltic lavas. These lavas are erupted at lower temperatures, are more gassy, and are potentially far more explosive than the basaltic hot spot volcanoes and submarine volcanoes of the ocean basins and ridges.

In 1470 BC, the near-shore island volcano of Thera (Santorini) exploded in a steam and ash eruption, destroying the Minoan civilization then flourishing on the island. In 1815, the volcano Tambora on the Indonesian island of Sumbawa erupted and ejected 150 cubic kilometers (44 cubic miles) of pumice and ash into the atmosphere, leading to an endless winter in North America and Europe during the year of 1816. More recently, the Indonesian volcanic island of Krakatoa was destroyed in a violent eruption in 1883. Tsunamis set in motion by the explosion inundated and killed 30,000 people on the neighboring islands of Java and Sumatra. The explosion ejected 20 cubic kilometers (6 cubic miles) of pumice and ash.

Violent mixing of gassy andesitic and basaltic magmas and sea water in the magma chamber of

► Positioned above a downgoing plate on the subduction zone between the Pacific and Australian plates, Vanuatu is part of an arc of volcanoes and compressed rocks—the "ring of fire" that nearly surrounds the Pacific Basin. Some of Vanuatu's volcanoes are still active, as shown by this "explosive" eruption from Yasu, on Tanna Island.

Peter Hendrie

Krafft-Explorer

▲ Clouds of steam billow when a molten lava flow from Kilauea meets the sea. Because this lava has already traveled some distance across land, it has lost most of its dissolved gases and is no longer explosive. On contact with the sea, remaining heat is lost as steam and pillow lava form.

Krakatoa probably led to the destructive ash eruption. Similar near-shore island arc volcanoes with andesitic or rhyolitic magmas are active in the Mediterranean Sea, in the eastern Indian Ocean, and around all the margins of the Pacific Ocean.

In the submarine environment, these volcanoes also exhibit their typical cone-shaped (rather than basaltic shield-shaped) profiles. They are built of pumice interlaced with lava. The length of the lava flows is shorter in these silica-rich volcanoes and the temperature of the flow lower than 1,000°C (1,830°F), while the content of carbon dioxide, hydrogen sulfide, and sulfur dioxide gases is much higher than that in basalts.

## HOT SPOT VOLCANOES

Thousands of islands and seamounts within the Pacific Basin and the Indian and Atlantic oceans were formed by the process of hot spot volcanism. Some hot spots have been persistent sources of submarine volcanism for as long as 70 million years and, because of the north, and, later northwestward motion of the Pacific Plate, have built long island chains in the Pacific Ocean: the Cook, Austral, Tuamoto, Marshall, and the Hawaii-Emperor island chains.

The process of hot spot volcanism works in the following way. A fixed magma source lies deep in the asthenosphere, well below the moving lithosphere above it. Why the hot spot exists in a particular place in space and time is not known. The heating of the hot spot is probably driven by the decay of the radioactive isotopes of uranium, potassium, and thorium in the Earth's asthenosphere. The raising of temperature of a spot in the asthenosphere produces localized heating. The heated asthenosphere is hotter, and therefore less dense, than its surroundings, which already are at a temperature of 1,800°C (3,270°F). The hotter, less dense material begins to rise in a similar way as a hot air balloon rises through the atmosphere.

The upward rise of the magma is not continuous, but appears to occur episodically, with the heated bubbles of magma rising in a plume to the base of the lithosphere. The lithosphere, as well as the asthenosphere, is in motion: in the Pacific, for example, the lithosphere is moving in a northwesterly direction at a rate of 10 centimeters (4 inches) per year. Following the movement of the lithosphere, the path of the magma curves northwestward and erupts initially as a submarine volcano, and later as an island volcano.

The Hawaiian Islands, therefore, increase in

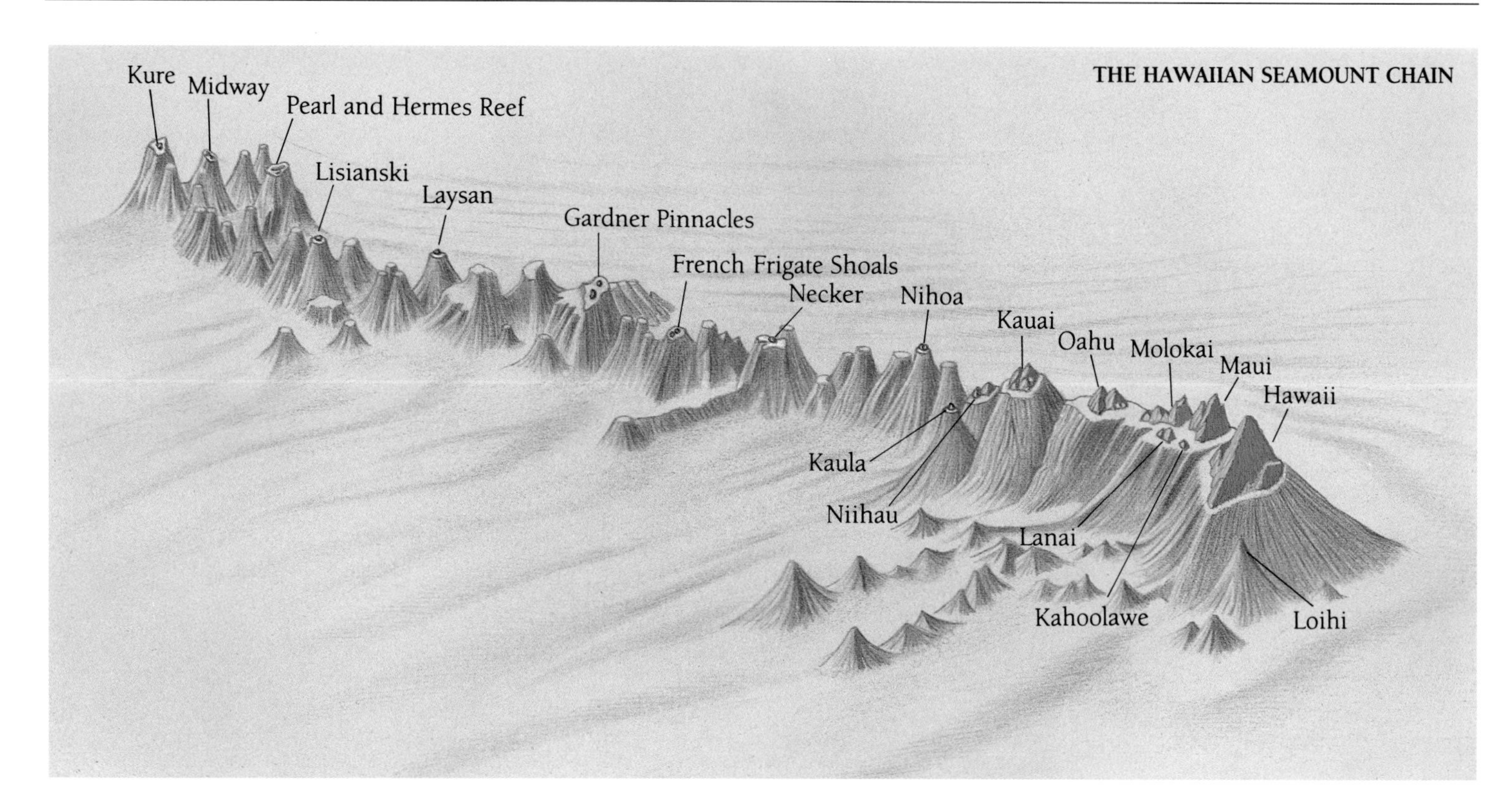


▲ The Hawaiian island chain stretches some 2,400 kilometers (1,500 miles) across the Pacific Ocean, from Kure in the northwest to the island of Hawaii in the southeast.

► Fire fountains of liquid basalt spurt high into the air from one of the vents on Kilauea. Eruptions such as these are spectacular but relatively harmless compared with the more explosive rhyolitic eruptions from volcanoes along subduction zones.

age northwestward away from the current hot spot volcanic activity at the Loihi submarine volcano along the trajectory of motion of the lithosphere. Midway Island was formed by submarine volcanism 28 million years ago at the geographic site of present-day Loihi. Since its formation, the island has moved on the lithosphere in a northwesterly direction for a distance of nearly 2,600 kilometers (1,600 miles). Meiji Seamount is 72 million years old and located 2,500 kilometers (1,550 miles) north of Midway. This data tells us that between 70 and 28 million years before the present, the Pacific Plate was moving north at a rate of 6 centimeters (2.5 inches) per year, or 60 kilometers (37 miles) per million years. Some 28 million years ago, the direction of movement of the Pacific Plate changed to northwest, and the rate to almost 10 centimeters (4 inches) per year. This direction and rate have persisted to the present day.

Since the individual islands along the chain are separated by deep water, it stands to reason that the magma supply from the hot spot sources has waxed and waned over time. At present we

Krafft-Explorer

► *Pahoehoe*, the Hawaiian term for basaltic lava, comes in several forms—smooth, ropy, or billowy. This section of an old lava flow from Kilauea is a good example of glossy, smooth *pahoehoe*.

Krafft-Explorer

cannot explain this phenomenon. As the island volcanoes moved away from their hot spot and lost their magma source and supply, the volcanoes became extinct. They began to sink and erode: Midway was eroded to sea level. Many of the islands developed fringing coral reefs and coral atolls; others lost their living coral caps and sank below sea level, forming flat-topped seamounts or guyots.

THE VOLCANOES OF HAWAII

The Hawaiian Volcano Observatory is located on Kilauea, the most active of the Hawaiian hot spot volcanoes. There are three active and one dormant volcanoes associated with the Hawaiian hot spot, all located within a circle, 100 kilometers (60 miles) in diameter. The three active volcanoes are Mauna Loa, Kilauea, and the Loihi submarine volcano; the dormant volcano is Haulalai. Mauna Loa, Kilauea, and Loihi apparently all have separate conduits along which magma moves up from the asthenosphere through the solid lithosphere to erupt as basaltic lava at the calderas, craters and along the fissures of these volcanoes.

Both Mauna Loa and Kilauea have grown above prominent rift zones; their summits are marked by distinct calderas. In historic times, Mauna Loa (the highest volcano on the surface of the Earth) has been active once every 10 or 20 years, with eruptive phases lasting for up to 18 months. During the past 150 years, lava poured out in volumes of up to 440 million cubic meters (15, 530 million cubic feet) per eruption, an average of 21 million cubic meters (741 million cubic feet) per year. During this period, Kilauea erupted once every few years with an average of 10 million cubic meters (353 million cubic feet) of lava per year. One of the most voluminous eruptions recorded recently has been that of Kilauea in the continuous eruption which started in 1984, producing an average of 150 million cubic meters (5,300 million cubic feet) of lava per year. In 1984, 230 million cubic meters (8,120 million cubic feet) of lava also erupted on Mauna Loa during a three-week period.

Basaltic mid-ocean volcanoes such as Mauna Loa and Kilauea have characteristically mobile lava eruptions, extending as long incandescent rivers of lava moving at speeds of up to 50 kilometers (30 miles) per hour, with very little explosive component. They cool into ropy *pahoehoe* forms or move slowly downslope as *a'a* (block) lavas. Ash cones and ash eruptions are infrequent in these episodes of basaltic volcanism.

The formation of submarine pillow lava can be

▼ An extinct, eroding giant—the crater of Haleakula on the Hawaiian island of Maui. Maui lies slightly northwest of the island of Hawaii, having moved away from the hot spot that fuels the active Hawaiian volcanoes of Mauna Loa, Kilauea, and Loihi.

Stephen J. Krasemann/ Nature Conservancy

Paul Chesley/ Photographers Aspen

observed during the active eruptions of Kilauea, when lava flows into the sea and moves downslope under water. As the lava is chilled below the water surface, new outbreaks pour out incandescent lava, which is immediately chilled and forms pillows. Hydrogen, oxygen, and carbon dioxide escape into the water, with hydrogen exploding in local bursts. There are no major explosive phases in this form of active underwater lava motion.

STUDYING VOLCANOES

Observation of the geological features of mid-ocean ridges and the detailed mapping and sampling of submarine volcanoes has been possible only during the past 10 years, primarily because of the development of three remarkable and complementary technologies.

The first of these came as a result of the US Navy's need for high-resolution ocean floor bathymetric maps. For this purpose, a shipboard swath digital acoustic mapping system was created. In the United States the system is called SEABEAM and is mounted on the hull of the survey vessel. It uses 16 beams to receive ship-emitted, reflected sound waves off the ocean floor. As the ship moves, the system is able to map the ocean depth accurately and automatically along a narrow strip of the ocean floor beneath the ship, moving in a water depth of 5 kilometers (3 miles). The exact water depth for each pixel of the ocean floor within each swath is digitally recorded on the ship and then combined to form a computer-produced bathymetric map.

The second technology is a multi-exposure, ship-tethered bottom camera capable of taking 3,200 photographic frames per lowering. This camera system (originally developed at the Woods Hole Oceanographic Institution and called ANGUS) allows a broad photographic coverage over submarine volcanic features.

▲ Kilauea is one of the most active and most studied volcanoes in the world. Because its eruptions are relatively quiet outpourings of fluid lava, the Hawaiian Volcano Observatory has survived since 1912, perched atop the summit and providing a base from which the volcanic activity can be studied at close range.

Greg Vaughn/Black Star

◀ Liquid basalt lava spews out as a "curtain of fire" from a fissure on Kilauea.

Terry Kerby

▲ Advances in underwater technology enable detailed geological studies of submarine volcanic activity. Here the University of Hawaii's *Pisces V* submersible prepares for take-off from its launch, recovery, and transport platform 20 meters (65 feet) below the surface. The platform and its pilot return to the surface after the submersible has been launched.

The third system is that of deep-diving research submersibles, typified by *Pisces V* of the University of Hawaii. Most modern research submersibles have a depth capability of many thousands of meters. The Soviet *Mir I* and *Mir II* submersibles, for example, are capable of diving to a depth of 6 kilometers (3.7 miles). The research submersible usually carries a three-person crew and is launched from a mothership which, in most cases, is equipped with a multibeam acoustic mapping system and an ocean floor acoustic imaging, or a photographic system that can be lowered to the ocean floor from the research vessel.

The use of integrated shipboard mapping and sampling systems has made detailed study of submarine volcanoes possible. These studies are usually conducted in three simple steps. First, the section of the ocean floor of interest is mapped with the acoustic mapping system. This bathymetric map is used to identify specific targets of interest, such as volcanic fissures, rifts, and craters. In the second step, cameras are lowered from the ship, usually at night, to within 10 meters (33 feet) of the ocean floor to photograph the details of the selected area. The photographic and bathymetric data are then combined aboard ship and analyzed to locate exact targets for daytime submersible dives and investigations.

Recently, another technology has been added to this suite of shipboard systems: the ship-tethered robotic submersible or remotely operated vehicle which is operated by a cable from the ship. Over the past decade we have learned a great deal about the processes of volcanism, but much remains to be discovered: 95 percent of the ocean floor is still to be explored.

ALEXANDER MALAHOFF

# THE INSTABILITY OF OCEANIC VOLCANOES

Charles Darwin was the first to note that the abundance of shallow-water reef assemblages that constitute the bulk of reefs rising from the deep oceans implied a complex interaction between volcano growth, volcano subsidence, and reef development. In his classic model, shown

► Completely engrossed in the primeval power surrounding him, a photographer stands on the magma crust of an old tunnel. Below him flows a new molten stream of lava.

Krafft-Explorer

Volvox/Australian Picture Library

▲ As island volcanoes move away from their hot spot and become extinct, they gradually sink and erode. Crater Island, in the Hawaiian group, is a wisp of its former grandeur.

schematically on page 142 of this volume, the growth of an active volcano above sea level provides a solid foundation for the growth of a reef. Once established, a reef can be remarkably long lived, because of its ability to continue upward growth at a rate about equal to the rate a volcano subsides below sea level. The end result is a circular atoll, the upward extension of an earlier fringing reef. Research drill holes, penetrating deep into the underpinnings of many present-day atolls, generally encounter the tops of subsiding volcanoes, lending credence to Darwin's insightful interpretation relating the growth and death of oceanic volcanoes to the development of atolls.

## GIGANTIC LANDSLIDES

In recent years, however, geologists have learned that large oceanic volcanoes have a far more eventful history than simple growth followed by subsidence. The important missing chapter in this history is "gravitational collapse", the wholesale destruction of large parts of oceanic volcanoes. In a nutshell, volcanic edifices, built mostly of lava flows and loose fragmental material, are notoriously unstable. Not only do they fall apart because of gravitational instability, they tend to be pushed apart by pressures of magma within them. The combined effect has produced some of the most gigantic landslides in the world.

Huge landslides have been recognized at dozens of oceanic volcanoes, but they have been best documented in Hawaii. This may seem surprising, because the volcanoes of Hawaii have very gentle slopes and therefore would not seem susceptible to landslides. However, studies on the Hawaiian Islands and on the nearby sea floor have proved that these volcanoes are literally falling apart. Instabilities have been recognized in mature volcanoes, such as Koolau (Oahu), East Molokai, and Mauna Loa; in actively growing volcanoes such as Kilauea; and in youthful volcanoes such as Loihi, that have not even grown above sea level to form islands.

The active submarine volcano Loihi, whose summit is still 900 meters (3,000 feet) below sea level, has topographical scars believed to be left by large landslides. Most of Loihi is built of pillow lava and related debris, and apparently this material is gravitationally unstable even while it is below sea level. Loihi erupts a few times each decade, and it is estimated that its summit will grow above sea level to form the newest of the Hawaiian Islands in 10,000 to 100,000 years. Obviously, the upward growth of Loihi will be controlled by the offsetting effects of eruptions and landsliding.

Krafft-Explorer

▲ The Kilauea lava lake on the island of Hawaii. Between periods of eruption the top few meters of the lake drain back into the main vent. This causes sections of the lake's lava crust to pull apart, exposing the molten incandescent lava beneath.

THE EXAMPLE OF KILAUEA

Kilauea, on the "Big Island" of Hawaii, is one of the world's most active volcanoes and is in what is called the "shield building" phase of its life. Kilauea has been the site of dozens of eruptions since 1952, and has been erupting almost continuously since January 1883. Despite its rapid growth, and almost paradoxically, Kilauea is in the process of falling apart because of landslides and large slumps. Much of the entire south flank of the volcano is being displaced toward the southern side of the island at very rapid rates. One surveying line, extending from the relatively stable slopes of adjacent Mauna Loa to Kilauea's south flank, has extended more than 4.5 meters (16 feet) since 1970. Recent studies have shown that these large displacements are caused both by gravitational instability of the volcano itself, and by the wedging actions of magma bodies within the volcano that push its entire flank seaward.

For purposes of comparison, the area of Kilauea's south flank above sea level is about 780 square kilometers (300 square miles), and the entire south flank, including its undersea extension, is about 3,890 square kilometers (1,500 square miles). Considering that this mobile flank averages about 5 kilometers (3 miles) in thickness, it becomes apparent that a volume of more than 16,000 cubic kilometers (4,000 cubic miles) of volcanic material is being moved seaward. It is unlikely that this segment of the volcano will slump into the sea in a single catastrophic event; instead, it is more apt to splinter off in smaller increments—perhaps in individual masses of a few cubic miles. In any case, it is clear that dramatic events are unfolding on the south flank of Kilauea.

LANDSLIDES OF THE PAST

Even more dramatic landslides have occurred from other Hawaiian volcanoes in the past several hundred thousand years. Recent side-scan sonar images have revealed jumbled masses of debris that extend outward from nearly all the Hawaiian Islands, as well as from numerous now-submerged volcanoes lying along the Hawaiian chain farther to the northwest. Two of the largest landslides, originating on the northeast side of the island of Oahu and on the north side of Molokai, have coalesced to form a hummocky debris field more than 225 kilometers (140 miles) long containing

► An intensely glowing fire fountain from Mauna Loa (Hawaii's second highest peak) spills over the partly eroded crater to form a river of lava.

Krafft-Explorer

Krafft-Explorer

▲ Luminescent splatterings of molten basalt erupt from a fissure on the slopes of Kilauea, scorching all in their path.

individual blocks of these disrupted volcanoes larger than the entire island of Manhattan. The Alika debris slide, which originated when a large part of Mauna Loa collapsed 100,000 years ago, produced a tsunami (tidal wave) that ran up the side of the adjacent island of Lanai to a height of 280 meters (925 feet). If such an event were to occur today all coastal areas in the Hawaiian Islands would feel its effects; waves up to 30 meters (100 feet) high might roll in upon the city of Honolulu. Clearly, we are dealing with catastrophes of Biblical proportions!

Happily, the probability that such a catastrophic event will occur in our lifetime is extremely small. Nevertheless, the knowledge that giant landslides are important geologic processes should be kept in mind when considering Darwin's classic model for the growth and subsidence of oceanic volcanoes and the development of atolls. As usually presented, this model depicts an erupting oceanic volcano that retains its original geometry as it becomes quiescent and then subsides beneath the sea. The recent discoveries described here suggest instead that oceanic volcanoes commonly lose a significant part of their volume through the process of landsliding. One possible consequence of this is that newly developed fringing reefs, and even young atolls, could be partly destroyed by collapse of the volcanic edifice upon which they are growing. It therefore is possible that imperfections in present-day atolls, such as significant gaps in their normal circumference, might be due to earlier gravitational failures of their volcanic foundations.

RICHARD S. FISKE

# 9 CORAL ISLANDS

Coral islands represent some of nature's finest monuments, often characterized by unrivaled beauty and ecological complexity. Their majestic landscapes and underwater realms house some of the most complex and diverse ecosystems in the world, and they have captured the imaginations of writers, artists, explorers, scientists, and adventurers alike. This chapter explores their geophysical origins, their structure and biology, the processes by which coral islands are colonized by plants and animals, and their vulnerability to natural and human-induced disturbance.

Jane Burton/Bruce Coleman Limited

▲ The flower coral *Eusmilia* sp. is normally expanded during the day. *Eusmilia* is a hermatypic coral, requiring sunlight to live.

◄ Channels through the treacherous coral and pounding surf of fringing reefs enable access to islands such as Moorea, which is bounded by a lagoon. Navigation can still be hazardous because these channels are sometimes subject to strong currents as water fills or drains from the lagoon at the change of tide.

## THE FORMATION OF CORAL ISLANDS

Most coral islands are the product of a rich synergy between geological and biological processes. With the help of symbiotic algae (plant cells) that live within most coral tissues, the coral animal secretes a skeleton of limestone at a rate of about 1 centimeter (0.5 inch) per year. Over thousands or even millions of years coral deposits build massive calcareous reefs sometimes hundreds of meters thick. Uplifted mountains over 300 meters (1,000 feet) high in Australia and China, for example, consist entirely of reef skeletons laid down during the Paleozoic era over 250 million years ago.

### GROWING ISLANDS

Coral islands are usually described as either high or low islands. Most high coral islands are young volcanoes surrounded by reefs in various stages of development, although they may also consist of continental fragments like the Seychelle Islands which are ringed by coral reefs. Low coral islands are nearly flat, and most are made up entirely of the skeletons of coral and other calcareous reef organisms. Atolls are coral islands that consist of lagoons surrounded by islets and barrier reefs.

All coral islands begin with the upward growth of coral from undersea foundations. Because coral growth requires an ocean temperature generally above 20° C (68° F), coral islands are restricted to tropical latitudes between 30° N and 30° S. While coral islands can form on stable foundations, most

▼ The Tuamotu Islands contain many atolls, surrounding a shallow lagoon. A patch reef at low tide can be seen in the foreground.

Nicholas Devore/Photographers Aspen

Nicholas Devore/Photographers Aspen

▲ The rugged beauty of a volcanic coral island is typified by Moorea in French Polynesia. The lagoon surrounding Moorea is filled with a maze of patch reefs and bounded by a fringing reef.

coral atolls have grown upward from the summits of sinking drowned volcanoes. In 1835 Charles Darwin first postulated this as a hypothesis. He had climbed the slopes behind Papeete in Tahiti to view the island of Moorea, which is encircled by a lagoon and barrier reef. To transform such an island into an atoll, Darwin reasoned that if the islands were to sink slowly, the upward growth of the surrounding coral barrier reef would eventually produce an atoll.

Darwin's theory was confirmed over a century later at Enewetak Atoll in the Marshall Islands where two holes drilled into the reef reached volcanic rock at depths of 1,267 meters (4,155 feet) and 1,406 meters (4,612 feet) respectively. Since coral reefs cannot thrive below 100 meters (330 feet) at the maximum, these results proved both the subsidence of Enewetak Atoll over 1,000 meters (3,300 feet) and the constant upward growth of coral during its long history of 49 million years.

## FROM CORAL COLONY TO ATOLL

Coral reefs form around many different kinds of islands in tropical latitudes, not just volcanoes. For example, there are thousands of small high islands in the Indonesian Archipelago, many of which provide shallow shelf environments for the richest coral reefs in the world. Over 80 genera and 500 species of reef-building corals exist in this region—the Indo-West Pacific province—which is the center of global diversity for many tropical plants and animals.

The smallest coral reefs are simply individual coral colonies that coalesce to form large heads or

▶ Atolls and coral reefs occur only in tropical seas. On the sites of sunken volcanic islands, coral atolls and their reefs may extend to depths as great as 1,600 meters (5,250 feet) below the surface of the ocean. Volcanic islands, which may form by intense eruptions (A), are later surrounded by fringing reefs (B). As islands age, they gradually subside, while coral continues to grow upward, forming a barrier that rings the island (C). Finally, the volcanic island disappears altogether, leaving a circular or elliptical atoll (D).

A

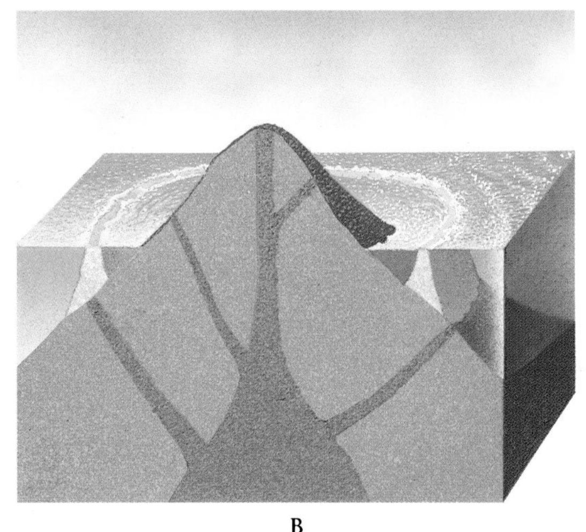

B

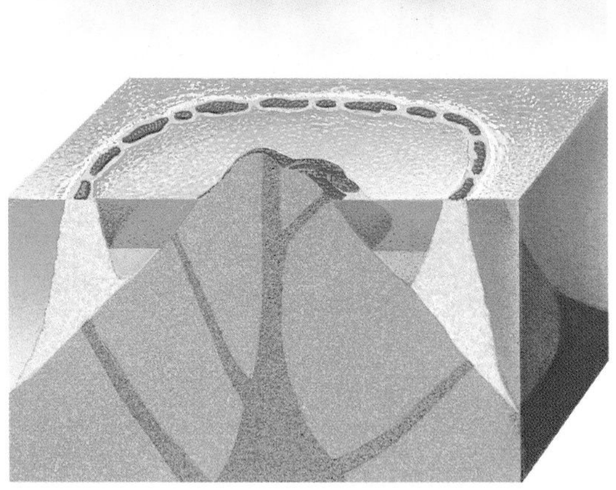

C

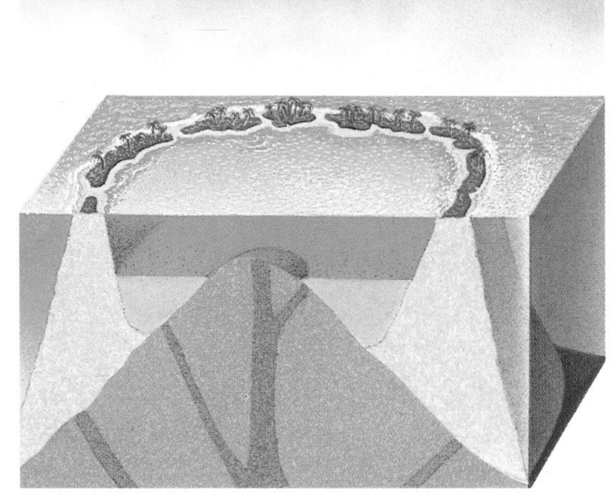

D

▲ A cross-section through a typical atoll, with the vertical scale exaggerated. The reefs provide a variety of habitats that sustain abundant and diverse forms of marine life.

patch reefs several meters in diameter. Micro-atolls are a particular growth form of single coral colonies that result when upward growth is arrested by shallow sea level. When they reach sea level, the corals (usually *Porites*) continue to grow outward. Calcification continues around the outer margins of the colony while in the center coral tissues die off, leaving a saucer-like formation resembling an atoll in shape but only several meters in diameter.

As reefs enlarge, they often coalesce to form patch reefs of various sizes or, along the coastal margins, long veneers of coral that can stretch for many kilometers. These structures develop first into fringing reefs, which are usually shallow enough to cause ocean waves to break but are insufficiently developed to block their shoreward advance. When coral reefs are large enough to dissipate the energy of offshore waves completely, they are described as barrier reefs. Thus a ranking of reef structures from youngest to most mature would be: coral colonies, micro-atolls, patch reefs, fringing reefs, barrier reefs, and atolls.

In areas of optimum reef development, barrier reefs usually develop well offshore and are often extremely important in protecting islands from coastal erosion by waves. Two of the largest barrier reefs in the world are the Great Barrier Reef on the northeast coast of Australia, almost 2,000 kilometers (1,240 miles) long, and the huge

◄ Barrier reefs are formed from the skeletons of a myriad calcareous organisms; however corals provide the primary structure. Kinid Reef in Palau shows an example of an outer barrier, the basic reef and lagoon, with numerous patch reefs.

Colour Australia

Colour Australia

▲ Kayangez Atoll, Palau. Most atolls are found in the Pacific Ocean. Deep lagoons result from the lack of coral growth where sedimentation and sluggish currents combine to retard upward growth.

▶ Crown of thorns starfish *Acanthaster plancii*, which may grow to 70 centimeters (28 inches) in diameter, have a voracious appetite for coral. On reefs where their numbers have reached plague proportions, the starfish have devastated vast areas, leaving only dead coral skeletons which may eventually be reduced to rubble by wave action.

barrier reefs off the east coasts of Honduras and Nicaragua in central America. There are 261 atolls in the world oceans, a large majority of which are in the tropical Pacific.

Andrew Green/Horizon

CORAL GROWTH AND EVOLUTION

To appreciate fully the constraints under which corals live, it is necessary to understand that they depend upon symbiotic algae called zooaxanthellae that live within their tissues. Zooaxanthellae are single-celled dinoflagellates that aid corals by providing nutrients and absorbing carbon dioxide. Both these processes speed the growth of corals by increasing the rate at which limestone is secreted. Corals without zooaxanthellae grow only about 10 percent as fast as corals that contain the algae. Because the zooaxanthellae live only in the well-lit upper levels of the sea, reef-building corals (hermatypic corals) also occupy only these higher levels—the coral euphotic zone. Many other kinds of coral, of course, exist at greater depths, some of the best known being the precious red, pink, black, and gold corals that are used to manufacture coral jewelry. Corals that live in the deep sea all lack zooaxanthellae and are collectively referred to as ahermatypic corals.

While the evolutionary history of coral reefs and islands stretches back over 400 million years when tabulate and rugose corals evolved during the Paleozoic era, most *living* reefs in the world today are only 5,000 to 8,000 years old. This seeming paradox can be explained by the fact that until 8,000 years ago, the sea level was over 30 meters (100 feet) below its present level. About 20,000 years ago, sea level was even lower, almost 130 meters (425 feet) deeper than at present. This drop in sea level was the result of the last ice age, which caused the oceans to shrink in response to global cooling, and the build-up of ice at high latitudes and elevations. Hence, all today's living reefs have formed only during the past few thousand years as rising seas flooded the shelves of elevated landforms. Many of their foundations, of course, were built of older fossil coral limestones which now underlie the latest episode of reef growth. In most parts of the world this veneer is less than 30 meters (100 feet) thick.

The history of rising sea level over the past 20,000 years also explains why many reefs are virtually drowned: that is, they are below the depth (approximately 30 meters/100 feet) where they can successfully regrow to the surface. As we have already seen, most coral reefs grow upward at less than 1 centimeter (0.5 inch) per year, whereas sea-level rise resulting from glacier meltwater and warming has frequently exceeded this growth rate.

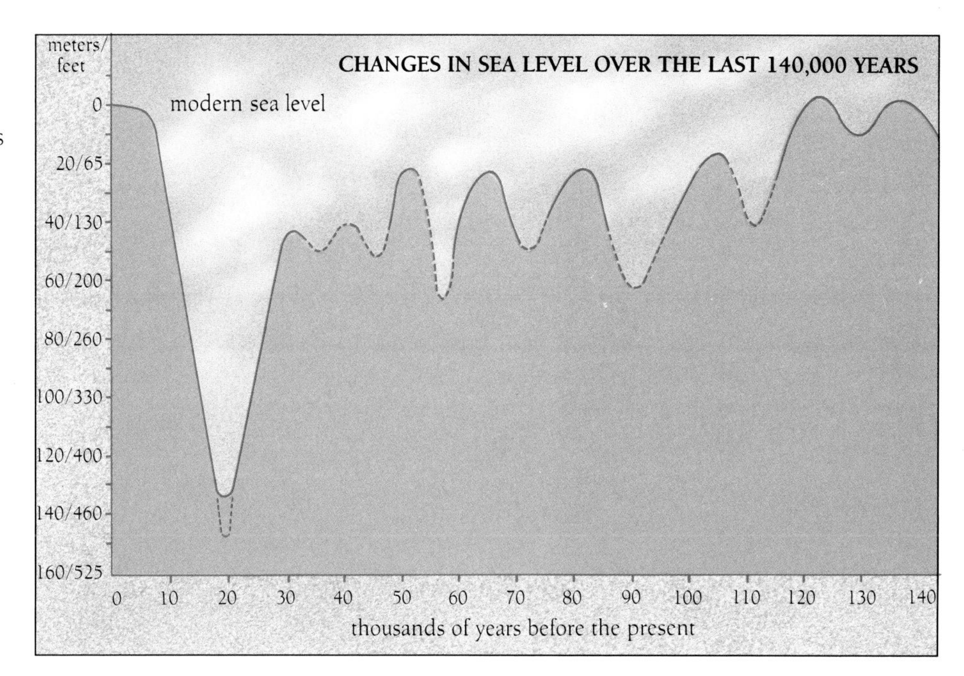


▲ Changing sea levels produce dramatic effects on many coastlines of the world, including the development of coral reefs and coral islands.

## ECOSYSTEMS UNDER THREAT

Coral islands and reefs harbor some of the most complex and diverse ecosystems in the world. There are between 500 and 1,000 species of tropical reef fish, and if all number and kind of algae, invertebrates, and microbes on the reef were to be fully enumerated there could be over 50,000 in any one major zoogeographic province.

Sometimes bizarre in appearance but almost always beautiful, reef ecosystems are often viewed as "delicate" or in "delicate balance" with nature. In truth, this is probably nothing more than a well-accepted myth. Coral reefs have suffered and survived through eons of major Earth change, including episodes of mountain building, continental drift, and massive extinctions like the one 65 million years ago that wiped out the dinosaurs. More recently, they have survived the ice ages and the ravages of tropical storms. In fact, most of today's coral reefs are in one stage or another of recovery from some recent major disturbance, among them storms, sedimentation from river runoff, predation from starfish, or disease. The coral islands that depend on reef ecosystems are equally vulnerable to disturbance.

In spite of all this adversity coral reefs have managed to evolve into a spectacular assembly of

◄ A rare and beautiful red coral *Errina novaezelandiae* from New Zealand, seen here with a tiger snail *Maurea punctulata. Errina* is an ahermatypic coral as it does not form reefs and occurs only in dimly lit environments.

Kev Deacon/Dive 2000

Peter Caine/Horizon

▲ Coral reef ecosystems are particularly vulnerable to natural disturbance from storms, and to exploitation and destruction caused by human activities. Because of this, reef-dependent subsistence industries, typified by this coral seller in Mauritius, pose a serious long-term threat, especially on coral islands with high unemployment and widespread poverty.

highly specialized life-forms. While we now know that corals are quite hardy in evolutionary terms, it is nevertheless important to recognize the limits of coral reef ecosystems—limits associated with humankind: pollution, overfishing, habitat destruction, and exploitation. Coral reefs are slow growing and slow to recover from both natural and human-induced change. A reef may take between several decades and a century to recover fully from a devastating hurricane or typhoon, or the equally destructive practice of using dynamite to harvest fish from the reef, which is widely done in the Philippine Islands.

In many third world countries in the tropics, coral reef ecosystems have been under siege for several decades. Resources are being depleted at an alarming rate. Overpopulation, combined with poverty and the inability of government and society to manage people dependent on reef resources at subsistence level, pose a major threat to coral reefs on a global scale. Here is another example of the "Tragedy of the Commons" where technology may not provide a solution. The threat to coral reefs is part of a larger problem of humans versus nature. The question of who, or which, will prevail holds the answer not only to the future of coral reef ecosystems but perhaps even to the future of our own species.

RICHARD W. GRIGG

# THE COLONIZATION OF CORAL ISLANDS

Coral islands are not always what popular conceptions would have them be. Some are indeed paradises, complete with swaying palms and golden beaches. Others are lonely, spray-swept tracts of barren sand, featureless and shifting. On the Great Barrier Reef of Australia both ends of this spectrum, as well as nearly all conditions between, can be found. Why are islands of the same region so different? Part of the answer lies in an understanding of two influences: time and stability.

Continental islands were once part of the mainland, and their fauna and flora consist partly of species left behind when the sea rose and isolated them. By contrast, coral islands receive all their species by dispersal over water. Some

▶ Sunset at this beachside coconut grove in French Polynesia typifies the image most people have of a tropical island paradise.

Hans-Jurgen Burkard/Bilderberg

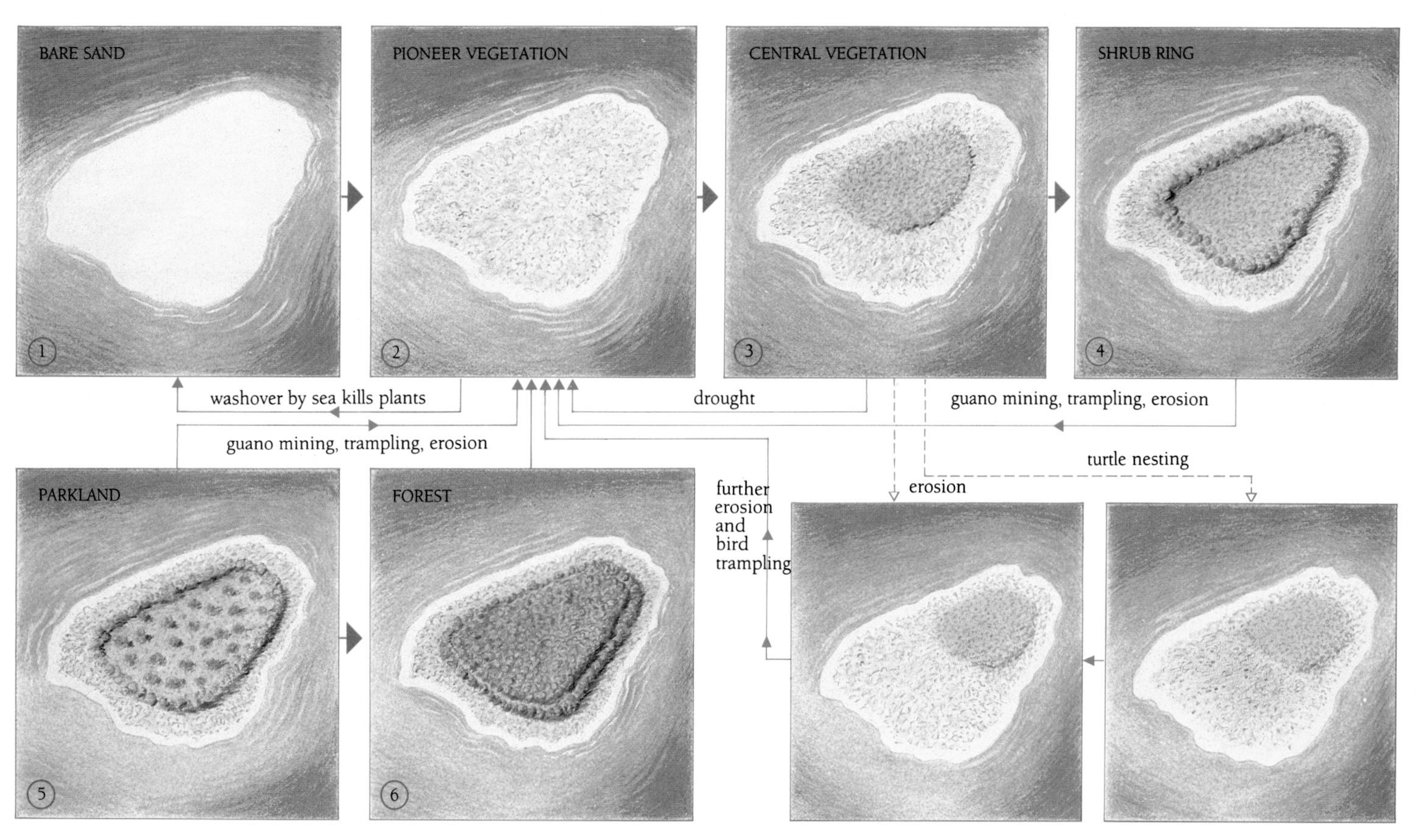


THE DEVELOPMENT AND DEGRADATION OF A CORAL ISLAND

▲ Over time, a bare coral cay can be colonized with a variety of plant and animal species. A combination of factors can create instability, and the island may retreat to an earlier stage of colonization, or even back to its original barren status.

species cross expanses of sea easily; others do so only with difficulty. A newly formed island is colonized quickly by easily dispersed species, and in time accumulates other species that find traveling harder.

The opportunity to accumulate species is not the only influence of time. Islands alter with age, and conditions become suitable for species that could not have survived had they arrived in an earlier period. However, improved conditions for some species are unfavorable for others. If an island is stable, eventually a balance is struck between the establishment of new species and the disappearance of old ones. If it is unstable, numbers of species may either increase or decrease depending on whether the island's environment is improving or deteriorating. The best way to understand these processes is to follow the fortunes of an island from its first appearance above the waters of the reef, to see how it becomes clothed in vegetation and populated with animals, and to witness the dramas played on its shores.

## A BARREN PLATFORM

Sandy coral islands may form gradually as minute particles of coral, broken shells, or the lime skeletons of algae and protozoans accumulate in the quiet waters of lagoons or in the lee of reefs. These sediments build up toward the surface until exposed at low tides. Then the wind heaps them into dunes tall enough that even high tides do not cover them. A submerged sandbar thus becomes an island. Other coral islands, called shingle cays, have a more tempestuous origin. Fierce tropical storms wrench blocks of living coral from the reef-face and hurl them into heaps of coarse rubble on the shallow reef-flat.

Regardless of the texture of an island at its inception, it is devoid of terrestrial life—a dead platform above the sea waiting to become vitalized. There is no shade, and the sands are hot, parched and often laden with a burden of salt from sea spray, or from storm waves crashing over the beach. There are few soil nutrients and little organic matter. Rain quickly evaporates or is diluted by sea water that soaks the ground. Sands, unbound by vegetation, are at the whim of the

Harold Heatwole

◄ Hixon Cay on the Great Barrier Reef is still developing as an island and as yet lacks vegetation. Despite this, the island supports two species of terrestrial animals: a small fly and an earwig, both of which feed on guano and carrion.

▸ The pioneer vines on Bell Cay help to bind sand and protect it from wind erosion. Runners of the beach morning glory *Ipomoea pes-caproe brasiliensis* grow out from vegetation in the center of the island, helping to consolidate a newly forming beach.

Harold Heatwole

wind, being shifted to and fro or blown to sea. It is this hostile environment that welcomes the first terrestrial life cast upon an island's shores.

It is not surprising that most species reaching an island in its infancy do not survive there. Only the hardy few colonize and reproduce. Aptly, they are called "pioneer species". What are they? How do they get to islands?

THE PIONEERS

A knowledgeable person could make a sensible prediction of the order in which species appear on an island. Green plants are the basis of all other life, so one might suppose them to be the first successful invaders. Without them other species could not survive. Next to arrive surely must be animals that eat green plants (herbivores), and then predators and parasites that consume other animals. Somewhere along the line, decomposers and scavengers living on dead bodies or wastes should appear.

As reasonable as these guesses are, they are far from the truth. In fact, animals often become established earlier than plants. Many tiny coral islands on the Great Barrier Reef completely lack vegetation, yet have up to a dozen species of terrestrial animals, including earwigs, slaters, flies, beetles, spiders, and centipedes. Are these islands defying basic ecological laws? No, they merely import food from the marine environment, and ultimately depend on the plants of that habitat for their sustenance. Food of marine origin arrives by several means. An important one is the wash-up of carrion. All beaches have their complements of dead fish, crabs, algae, jellyfish, and starfish. A feast awaits a terrestrial scavenger that might happen to arrive. Indeed, scavengers usually are the first terrestrial animals to colonize an island. Common ones on the Great Barrier Reef are the seaside earwig *Anisolabris maritima* and a small white-winged gnat *Loptocera fittkani*. The earwig burrows under carrion on the beach and eats it from below. The gnat hovers over it, landing to feed and lay its eggs on the decaying flesh.

Seabirds too are an important means of transferring food from sea to island. They eat fish or oceanic invertebrates, and use islands as platforms for roosting and nesting. Scraps of food brought to nestlings, guano, and birds' dead bodies are materials ultimately of marine origin, but become available to insular life through

▾ Seabirds are early residents of newly formed coral islands and because they transfer food from the sea to land, they are important contributors to the food chain. This pair of brown boobies and their naked chick are nesting on Bylund Cay in the Swains Group (Great Barrier Reef).

Harold Heatwole

Harold Heatwole

Harold Heatwole

◂ Flowers and sticky fruits of the pisonia tree *Pisonia grandis* (*far left*). Many coral island plants have developed interesting mechanisms that ensure their dispersion among islands and therefore their long-term survival. *Pisonia* is easily transported to new islands because the fruits stick to bird feathers. Tar-vine *Boerhavia diffusa* in flower (*left*). This is one of the common ground plants growing on small cays on the Great Barrier Reef.

seabirds. For this reason, seabirds are called "transfer organisms".

Fungi and bacteria decompose dead organisms and break them down into finer particles. Terrestrial animals, such as slaters and some insects, feed on this organic matter or on the fungi that grows on it. These, too, make an early appearance on coral islands. As scavengers become abundant, the stage is set for predators to enter the scene. Large wolf spiders *Lycosa* sp. and centipedes *Scolopendra* sp. reach coral islands early and often become part of the food web before the arrival of green plants.

Although green plants usually arrive later than the animals discussed above, many plants are truly pioneers. These species are highly tolerant of salt, drought-resistant, and can grow where there is very little nutrient in the soil. In addition, they cope with a continually shifting substrate. Wind can blow sand away from a plant and uproot it, or cover and smother it. Pioneer plants have two strategies for combating these potential disasters. Some grow rapidly and produce many seeds as very young plants. The future of the species on the island is assured by the abundant seed that has been scattered. A good example of such a plant is swine-cress *Coronopus integrifolius*. During winter, when rain is most abundant, it sprouts from seed left from a previous year, flowers, sets fruit rapidly, and then dies.

Another way of coping with unstable substrates is to send out runners or vines. These spread over the sand and put down roots. Subsequently, if part of the plant is either excavated or covered by sand, it will not die as unaffected parts can continue to supply water and food. Common examples of such plants on coral islands of the Great Barrier Reef are the beach morning glory *Ipomoea pes-caprae brasiliensis*, the beach pea *Canavalis maritima*, and the bird's beak grass *Thuarea involuta*.

Animals disperse to islands by air and by water. Some, such as insects, birds, and bats, can reach islands under their own power. But even within these groups many species are weak fliers and would rarely fly the long distances required to colonize remote islands. These, and even some small flightless animals like mites, may be blown passively by the wind after being wafted aloft on rising air currents. The young of some spiders are particularly adept at this: they spin small parachutes of gossamer that catch the wind and carry them away. Large centipedes and ground-spiders travel by sea, usually rafting in or on a floating log, coconut, or other land debris. Even snakes and lizards have been seen at sea riding on floating objects.

Plants too are transported by sea currents, and beaches are often littered with seeds and fruits. Pioneer species are resistant to sea water and

Jeremy Smith

▴ Drift seeds from trees and lianes, found on the beach at Price Cay, Swains Reefs. Most of these have drifted from as far away as New Caledonia or Vanuatu. None of these species has become established on this cay as many seeds are dead on arrival, killed by sea water en route. Others die as seedlings in the harsh beach environment.

Harold Heatwole

Harold Heatwole

◂ Carrion, such as this dead booby chick on Bell Cay (*far left*), provides food for scavengers on many Great Barrier Reef islands. One of the many insects that find their way to remote islands: a ladybird beetle rests on the leaf of a sea rocket *Cakile edentula* on Bell Cay (*left*).

Walter Deas/Planet Earth Pictures

▲ Heron Island in the Capricorn Bunker Group is a well-established vegetated coral cay. The island has a mixture of forest and parkland and is an important nesting site for black noddies *Anous minutus* and wedge-tailed shearwaters *Puffinus pacificus*.

survive prolonged immersion. On reaching land after a long sea journey they can still germinate. With the establishment of green plants, arriving herbivores have a food base and can become established on the island.

Our tiny island is no longer bare. It has a small community of arthropods, based mainly on a scavenging industry, and a sparse covering of pioneer plants with a few herbivores. What happens next?

## LATER SETTLERS

A small island is essentially all beach. However, as more sediments accumulate, wind-driven sand is piled higher and the beach expands outward. Stormy seas no longer wash over the whole of the interior, and pioneer vegetation there begins to stabilize the soil and reduce wind erosion. The greater accumulation of sand retains rain that floats on top of the salty water in the soil. Plants that are less drought-resistant and less salt-tolerant than the pioneer species can now colonize. Seabirds prepare the soil by fertilizing it with guano.

Almost all the later settlers are dispersed by birds. Sometimes their seeds or fruits are attached to feathers; sometimes they are eaten by a bird, carried in its digestive tract, and deposited in excrement on a distant shore. Many of these plants have adaptations that facilitate their transport by birds. The chaff flower *Achyranthes aspera*, for example, has fruits with sharp hooks that cling to feathers, and the fruits of the tar-vine *Boerhavis diffusa* are covered with a sticky gum that enables them to adhere.

The beaches are still buffeted by waves, swept by wind, and sprayed with salt water, and only pioneer species grow there. But further inland, the changes wrought by seabirds and greater isolation from the marine environment produce a milder, more stable habitat. The island becomes zoned into two habitats: a peripheral one of pioneer vegetation and hardy animals, and a central one of less hardy species. In the central zone, vegetation grows more densely and luxuriantly. Its shade cools the ground and reduces loss of moisture by evaporation. Soil is sheltered from the wind and is stabilized through binding by roots. Dead leaves and stems provide a covering of moist litter. More species of plants now find a favorable seedbed, grow and support a greater number of herbivores.

Our island now has two zones: a beach zone of pioneer species and a milder, central one of later settlers in which herbivores play a greater role in the food web.

## ERECTING THE BARRICADES

So far the island supports only low vegetation such as herbs, grasses, and vines. Further changes depend on the arrival and establishment of shrubs. The most important one on the islands of the Great Barrier Reef is the sea-dispersed octopus bush *Argusia argentea*. Its seeds require soaking in salt water before they can sprout. After a period of exposure to salt, the seeds germinate when soaked with rain, often on a distant beach to which they have been carried by ocean currents for hundreds, or even thousands, of kilometers. The sea lettuce tree *Scaevola sericea* sometimes also contributes to the shrub ring.

At first there are only a few bushes scattered around the edge of the beach. Gradually the gaps close until a ring of shrubs surrounds the island. Following them come arthropods that graze their leaves and imbibe nectar from their flowers. Flies and moths become more abundant, and predators that feed on them become established.

Our island now has three zones: a peripheral one of sparse, pioneer vegetation, and a shrub ring inside which is a central zone of dense herbiage.

## INVASION OF THE TREES

A shrub ring has a profound effect on the interior environment of an island. It serves as a windbreak and screens out salt spray. More species of plants, including additional shrubs and even some trees, become established in this sheltered habitat. These are accompanied by an ever-increasing number of insects and other animals. Occasional wandering land birds are attracted, and species carried as seeds in their digestive tracts become established. Once such plants are numerous enough to provide sufficient food, small populations of land birds become resident. In time, an open parkland of low vegetation interspersed with scattered trees and shrubs replaces the previous central zone of low vegetation. It provides additional shade and a cooler, moister environment. Gradually, as more trees grow, parkland gives way to dense forest.

On the Great Barrier Reef, a common forest tree is the pisonia *Pisonia grandis*. It has a special relationship with seabirds, especially the white-capped noddy *Anous minutus*. The large amount of fertilizer required by pisonia trees is supplied by guano from the noddies nesting on their branches. Dense shade cast by the trees and high levels of guano in the soil prevent the growth of most plants, and the low vegetation is lost. A thick layer of decaying leaf litter accumulates and a richly organic soil develops. This teems with mites, slaters, centipedes, and springtails. With less open space, there are fewer ground-nesting terns, but muttonbirds *Puffinus pacificus* come instead to dig their nesting burrows in the soil.

Our island is now very different from the original desolate sands that first ventured above the sea. It has a number of vegetation zones, each with its characteristic fauna. There is still a peripheral beach of pioneer species, bordered by a shrub ring. Inside the ring, part of the island may be open parkland and part forest. Sometimes forest completely replaces parkland. Ground-nesting seabirds have given way to burrowing muttonbirds, and to tree-nesting noddies and land birds. We have reached our paradise. Can such verdure be sustained?

Harold Heatwole

▲ Octopus bush *Argusia argentea* beginning to form a shrub ring on a cay in the Coral Sea. Shrub rings form an effective barrier against airborne salt, thus enabling more delicate plant species to grow within the circle.

## PARADISE LOST

So far the discussion has centered on one of the two major ecological influences mentioned at the beginning of this discussion: time. Now it is appropriate to consider the second factor: stability. The stages of island development are not irreversible. At any point islands can deteriorate

▼ Sooty terns *Sterna fuscata* on Diamond Cay. Despite the small landmass of many cays, some seabird colonies can number in the tens of thousands, especially when hunting grounds are readily accessible and there is an abundance of suitable nesting sites.

Ron & Valerie Taylor/Ardea London

Harold Heatwole

▲ Verdant open parkland on Heron Island contrasts with the turquoise lagoon beyond. Screw pines *Pandanus* sp. and pisonia trees *Pisonia grandis* form the dominant canopy species in this zone.

and return to previous stages, or even disappear altogether. Among the causes of such change are storms, erosion, seabirds, sea turtles, and the activities of humans.

During severe tropical storms, violent seas pound over beaches, flooding vegetation and animals with sea water, or covering them with sand. Soils on coral islands provide a record of these events. They are often layered: dark bands of organic soil alternate with light bands of mineral sand. Each dark band represents an old surface that was once vegetated, and each light one a smothering by sand. This repeating vegetation pattern tells of many cycles of destruction and re-establishment.

Even in the absence of storms, beach sand is washed from place to place and small alterations of the coastline occur, especially at the narrower ends of an island. A time-lapse movie would show the tip whipping back and forth like a tail. It is this ceaseless activity that keeps the edges of an island at the pioneer stage. Stable islands stay in the same place; erosion and deposition merely affect the edges. However, when there are changes in the currents around an island, the erosion-deposition cycle no longer nibbles around the edges but whole portions are eaten away. Rapid erosion may cut an island back into what had been its central zone. At the same time, on the opposite beach redeposited sand forms new, expanding beaches open to colonization of pioneer species. The zones become skewed, with the central zone offset toward the eroding edge. When one side continues to erode and the other to expand, an island "walks" across its reef. Sand may be washed out to sea and the island dwindle in size, eventually being left only with pioneer vegetation. An island near the rim of its reef may go over the edge and disappear altogether.

Seabirds not only play an important role in the development of an island; they can also contribute to its degradation. Where birds nest in dense colonies, their trampling and high concentrations of guano kill plants. Bare patches develop and expose the sand to wind. Often colonies of birds move from place to place on an island, or from island to island, in subsequent breeding seasons. Such "rotation" allows recovery of vegetation. However, if populations are too large an entire island may become bare, and any plant hardy enough to germinate is destroyed as

► A severe tropical storm threatens Bell Cay. Storm waves can wreak havoc when they wash over low islands such as this, eroding beaches, killing vegetation, and destroying seabird nests.

Harold Heatwole

soon as it appears. Gannet Cay in the Swain Reefs is such an island. In the late 1960s it was large and had a lush cover of plants. Thereafter it began moving over its reef and became progressively smaller. Its dense breeding populations of brown boobies and masked gannets destroyed the vegetation and now the cay is completely bare.

Sea turtles also destroy vegetation. Females dig pits in the sand in which they lay their eggs. Plants growing at the nest site are uprooted and may be killed. Some pioneer species, such as bird's beak grass, beach morning glory, and beach pea, survive better than others. These were mentioned earlier as vines or creepers adapted to shifting substrates. In this case turtles, rather than wind, cause instability, but the result is the same—plants unable to cope with moving sand do not survive. Therefore, turtle-nesting areas in central vegetation revert to a pioneer stage.

Humans can have a great effect on coral islands. Guano miners removed plants and topsoil from islands of the Great Barrier Reef about the turn of the century. This was followed by the introduction of goats that in turn took their toll of the vegetation. Although goats are no longer present and guano mining has ceased, some islands still suffer from these disruptions. With the development of tourism there has been further clearing of vegetation, construction of buildings, and the introduction of exotic plants. Weeds and insects are introduced inadvertently, especially in sand or gravel brought in for construction purposes. Waste disposal can cause acute environmental problems.

Humans may have an indirect, as well as a direct, ecological effect. For example, people may transport weed seeds directly in their gear or attached to their clothes. On the other hand, they may aid the introduction of weeds by attracting gulls. Gulls eat a variety of foods, including weed seeds, and they transport weeds to new localities. They are also scavengers and congregate in the vicinity of humans. Up to three-quarters of the gull population of the entire Bunker-Capricorn group of islands has been drawn at one time to dumps on Heron Island. With a resort, a marine research station, and a park headquarters, this island has a high human population. The number of weed species has progressively increased over the years. It is still uncertain how much of this is due to seeds being carried by people directly, and how much to dispersal by gulls attracted to the human settlement.

Some undesirable changes have occurred through lack of understanding of the ecology of islands. Many, however, have resulted from greed, coupled with a callous disregard for the long-term reef environment. Developers have been quick to exploit before adequate studies were carried out and safeguards established. In some cases governmental requirements have been ignored and restrictions exceeded.

Kathie Atkinson

◀ A female green turtle *Chelonia mydas* coming ashore at sunset on Raine Island to lay her eggs. Vegetation is uprooted and destroyed at turtle-nesting sites.

Not all changes have been bad; there have been some success stories. Lady Elliot Island was perhaps the most ravaged island on the Great Barrier Reef, having been denuded of its vegetation and soil by guano miners and goats. Now it is one of the most pleasant of the islands. It has been carefully restored through an enlightened reforestation program. There are shady groves of coastal she-oaks *Casuarina equisitefolia*, and seabirds are returning to nest in ever-increasing numbers. This miracle was achieved, not by public support and governmental initiative, but by a private citizen wishing to establish a modest, comfortable holiday resort in natural surroundings.

There is a lesson in this. Human interference took a forested, stable island from the apex of its development back to a barren condition. It remained in that state for decades, until humans once again intervened, this time toward restoration rather than degradation. Islands are fragile and dynamic ecological systems. They develop from barren platforms to forested paradises, but they can also go in the reverse direction, and often do. Humans can be a force for improvement or an ecological catastrophe. So far, the track record has been poor, with human activities all too often accelerating destructive forces. By natural processes, islands come and go, develop and deteriorate. Humans have the power to tip this balance. Over-exploitation could lead to deterioration of even the more stable, forested islands, leaving nothing to replace them—truly a paradise lost.

HAROLD HEATWOLE

# 10 Laboratories of Life

Every island—and there are more than half a million scattered throughout the world's oceans—is an individual ecological system. In these isolated worlds—from the tropical islands of the Pacific to the subantarctic islands of the Southern Ocean— flora and fauna have developed and evolved in unique ways. Charles Darwin's conclusions from his visit to the Galapagos Islands in 1835 changed scientific thinking about the evolution of life, and islands are still natural, but vulnerable, laboratories in which we can study the varied, and sometimes complex process that contribute to the adaptation and diversification of life.

Frans Lanting/Minden Pictures

▲ Designed simply to gain attention, the inflated bright red throat sac of the male great frigatebird *Fregata minor* is used in courtship displays. Great frigatebirds have a widespread distribution but are commonly found in the Galapagos Islands: the island of Genovesa has a particularly large colony.

◄ The magnificent cliffs and intersecting valleys of Kauai in the Hawaiian Islands.

## The Galapagos Islands

Arising from a volcanically active hotspot in the Earth's mantle, the oldest islands in the Galapagos Archipelago appeared above sea level some 3-4 million years ago. Since their formation, the islands have remained geographically isolated, never having been much closer to adjacent landmasses than they are today. The Galapagos Islands are 1,000 kilometers (600 miles) west of the South American continent and 5,600 kilometers (3,500 miles) east of the nearest Pacific island chain. Terrestrial immigrants colonized them by rafting (on floating debris or on the feathers and feet of birds), flying, or wind dispersal. Shallow-water marine organisms arrived either directly by swimming or through passive transport of their larvae on ocean currents.

The difficulties of migration imposed by their watery isolation explain in large part the unique flora and fauna of the Galapagos. The only native land mammals, excluding the recent human introduction of cats, dogs, goats, and pigs, are small rice rats. The dominant fauna on land are reptiles—such as tortoises, lava lizards, marine iguanas, and land iguanas—and birds, particularly small finches, mockingbirds, and one species of hawk. Overall, the Galapagos has a remarkably low diversity of organisms, considering the exceedingly rich assemblages of plants and animals found along adjacent South American shores. There are, for example, only 500 native plant species in the Galapagos compared with more than 10,000 species in Ecuador.

### Limited Life-Forms

Why this disparity? There are three major reasons. First, Galapagos is geologically quite young and the terrain is dominated by recent lava flows. Fernandina, the youngest and most volcanically active island, is only about 10 percent vegetated. The arid climate of Galapagos delays the breakdown of rock to soil, leaving much of the shoreline a formidable and hostile environment for both plants and animals. In short, there is a limited variety of habitats.

The second reason for the limited diversity of life-forms can be attributed to differering abilities to survive transportation to the islands. For organisms coming from the adjacent shorelines (South and Central America being the nearest and most direct source of immigration), transport on a raft under the best of circumstances (a current speed of 1kilometer/0.6 mile per hour) would require a journey of 42 days. For most organisms, particularly mammals, the deprivation of food and water over this length of time would almost certainly be fatal. On the other hand, because of their low metabolic requirements, reptiles have the potential to survive without sustenance for long periods of time.

The last reason for the low diversity of organisms is chance. With sufficient time, most events will occur. Given the youthful age of Galapagos, however, there have probably been only a few hundred migrations since the islands formed, and only a small percentage of these would have resulted in the establishment of persistent populations. This conclusion is supported by the results of a recent study of the dispersal of the fruit and seeds of Galapagos plants. It was estimated that a minimum of 413 colonization events were required to account for

◄ The islands of the Galapagos Archipelago. Their geographical isolation has had two major consequences. Firstly, there is a high percentage of endemic forms. Secondly, isolation (and failure to fit the popular notion of a tropical paradise) has prevented much of the damage and exploitation associated with human settlement.

Frans Lanting/Minden Pictures

▲ Geologically, the Galapagos Islands are a recent addition to the archipelagos of the Pacific—they range in age between less than 1 million and up to 5 million years. This youthfulness is reflected in the often barren landscapes: basaltic lava and volcanoes, interspersed with sparse patches of vegetation, as can be seen here at Bahia Sullivan and the Pinnacle.

the floral diversity of the islands. Given the maximum age of the islands as 5 million years, the rate of a successful colonization would be in the order of one species every 12,000 years! This slow rate of colonization, however, is not likely to have been constant. During the earliest stages of island development colonization may well have been even slower, but as conditions for life improved with the establishment of pioneer species, such as lichens and other arid-adapted vegetation, organisms probably colonized the islands more rapidly.

## ENDEMIC SPECIES

Since successful colonization is rare, even when conditions are favorable, most of the species that managed to become established on Galapagos shores became genetically isolated from their parent population. They subsequently evolved independently, responding to a different set of environmental conditions from those in which they originated. Within a very short period of time (believed to be less than 1 million years for most Galapagos species) many species diverged to the point where they are now quite distinguishable, either morphologically or physiologically, from their ancestors.

The product of this isolation is a high percentage of endemic forms—that is, species confined to the Galapagos Islands. About 34 percent of the terrestrial plants, more than 50 percent of the terrestrial animals, and about 25 percent of the marine flora and fauna are endemic. Particularly remarkable is the fact that many of the founding species eventually migrated to other islands within the archipelago and further diverged into a variety of distinct forms—some clearly differentiated into separate species, others only slightly modified in appearance and thus accorded the status of subspecies or race.

The diversity of form within a particular group, such as the tortoises or finches, is believed to have originated from a single common ancestor, a phenomenon known as species radiation. Examples of this radiation occur in nearly every major terrestrial group of organisms—tortoises, iguanas, lizards, land snails, and prickly pear cactus among others. The most famous example is Darwin's finches with 13 recognized species in Galapagos and one on Cocos Island, some 540 kilometers (335 miles) to the north.

## DARWIN AND THE GALAPAGOS

The unique flora and fauna on islands close to South America, and the variation between populations of mockingbirds, tortoises, and plants at several localities within the Galapagos Archipelago, provided Charles Darwin in 1835 with evidence that species were not fixed, but mutable (that is, able to undergo morphological change). These observations supported his hypothesis that species are able to change in response to different environmental conditions, and provided the foundation for the development of the theory of evolution based on the principle of natural selection.

While Darwin was correct about natural selection, he did not recognize the importance of isolation as a requisite for the divergence of species. This is perhaps not surprising: at the time his ideas were formulated, Darwin was unaware of genetics, and did not know that interbreeding populations would dampen the effects of selection on the population as a whole. The importance of isolation, however, was recognized even during Darwin's time. In 1868, only nine years after the publication of *The Origin of Species*, the German zoologist Mortiz Wagner wrote: "The formation of a real variety which Mr. Darwin considers as incipient species, can succeed in nature only where some individuals can cross the previous borders of their range and segregate themselves in nature for a long period from other members of their species."

These early discoveries in Galapagos made a lasting contribution to the development of evolutionary theory, and the islands continue to provide scientists with information about the ecological factors that shape evolution. For example, recent studies of Darwin's finches have revealed that food limitation is the driving force that determines bill shape and feeding behavior. These morphological and behavioral adaptations in turn influence mating behavior, which leads to reproductive isolation and reinforcement of species diversity. Darwin's finches have evolved in response to alterations in their physical environment and to indirect changes caused by their competitors.

**DIVERSITY AND ADAPTATION:** **THE FINCHES OF GALAPAGOS**

ANCESTRAL SEED-EATING FINCH

*Geospiza fuliginosa*
*Geospiza fortis*
*Geospiza magnirostris*
*Geospiza conirostris*
*Geospiza scandens*
*Geospiza difficilis*
*Certhidea olivacea*
*Cactospiza heliobates*
*Cactospiza pallida*
*Platyspiza crassirostris*
*Pinaroloxias inornata*
*Camarhynctus parvulus*
*Camarhynctus psittacula*
*Camarhynctus pauper*

seeds
fruit and buds
insects

▲ In the isolation of the Galapagos Islands, one ancestral seed-eating finch gave rise to 14 species of birds, each adapted to its habitat and food supply. Some of today's finches still eat seeds; others favor insects, grubs, cactus, fruit, or berries. Their size and bill structure reflect this variety of habitats and dietary preferences.

Frans Lanting/Bruce Coleman Limited

◄ Galapagos land tortoises *Geochelone elephantopus* bask in the morning sunlight in a pond on the island of Isabela. These huge creatures gave their name to the islands: *galápago* is an old Spanish word for tortoise.

THE CONVERGING CURRENTS

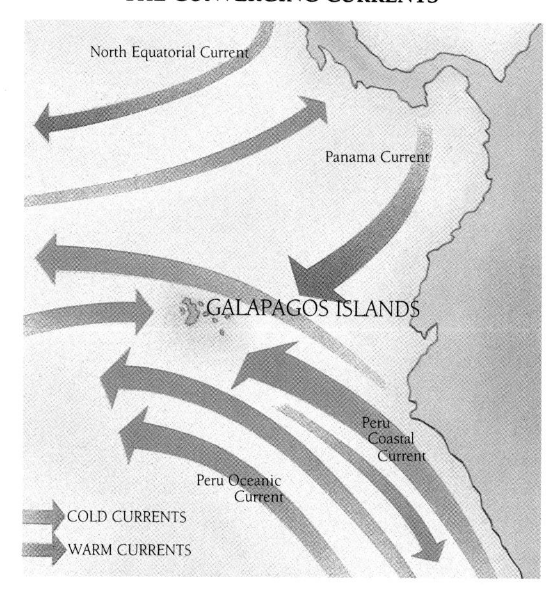


▸ The combination of cool and warm currents around the Galapagos, with their changing nutrient mixes, has created an unusual marine environment to which animals have adapted in a variety of ways.

THE MARINE ENVIRONMENT

The Galapagos Islands have also provided an opportunity to study the evolution of adaptations in the marine environment. Situated at the confluence of a dynamic current system, the islands are seasonally bathed by cool nutrient-rich waters and then by warm nutrient-poor waters. The cool waters have enabled temperate climate organisms such as fur seals, sea lions, cormorants, and penguins to colonize. However, the warm waters that prevail over four to five months of the year pose potential problems in terms of food supply and overheating. One adaptation to minimize overheating is a reduction in body size: the Galapagos fur seal is the smallest marine mammal and the penguin is the second smallest of its kind. In contrast with their temperate water cousins, Galapagos sea lions forage mainly during daylight hours and mate in the water rather than on land. Both of these behavior patterns are believed to represent adaptations that minimize exposure to the tropical heat.

The Galapagos marine iguana has no counterpart anywhere else in the world. In the absence of mammalian predators, this reptile thrives on marine algae along the volcanic shoreline. Thought to share a common ancestor with the land iguana, it has adapted for feeding in the sea. Its tail is laterally compressed to help in swimming; it has a blunt snout to facilitate grazing on the turf-like algae; and its long sharp claws enable it to cling tightly to rocks, thereby avoiding being swept away or battered on the rocks.

There is still much to learn from the Galapagos life-forms, and the islands remain a living laboratory. Newly developed biochemical techniques are currently being applied to reveal the genealogy of various Galapagos species within the archipelago, and identify their mainland ancestors. Knowledge of the time when species diverged, coupled with data on geological and climatic changes in the islands, will yield estimates of the rates of evolution and give us some idea of the environmental conditions that may influence this change.

G. M. WELLINGTON

▾ The marine iguana *Amblyrhynchus cristatus* is unique to the Galapagos. These animals have become largely herbivorous, grazing on marine algae, and have developed long claws and a shortened snout for rasping algae from rocks. They are often found in great densities, basking in the sun to raise their body temperature before venturing into the cold sea to feed.

Ken Lucas/Planet Earth Pictures

Claire Leimbach

▲ Coconut groves stretch back toward the jagged volcanic pinnacles and spectacular bay of Hati Heu on the Marquesan island of Nuku Hiva in the northeastern Pacific.

# THE PACIFIC ISLANDS

When Captain Cook's ship *Endeavour* rounded Cape Horn and ventured into the Pacific in February 1769, a new world of animals and plants was introduced to Europeans. Among the newcomers were tree-living land snails, birds such as the honeycreeper and toothed pigeon, and cultivated plants like the breadfruit. Bearing the labels "South Seas", "Friendly Isles", "Otaheite", and "Owyhee", shells, bird skins, preserved fish, and pressed plants collected on eighteenth- and nineteenth-century voyages were taken to Europe. There, in herbaria, museums, and the curiosity cabinets of the wealthy, they opened new vistas to scientific knowledge and continue today to give impetus to studies of distribution and evolution of plants and animals of oceanic islands.

The islands of the tropical Pacific extend from Palau in the west at 130°E to Easter Island in the east at 110°W, and from 20° north of the equator to 30°S. Included within these boundaries are New Zealand, New Caledonia, Vanuatu, New Guinea, and Palau in the west; the Mariana and Caroline islands, Samoa, Tonga, Kiribati, and the Marshall Islands in the center; and the Hawaiian Islands, Society Islands, Tuamotus, Marquesas, Easter Island, Pitcairn and Sala y Gomez in the east-southeast.

## RULES OF ISLAND LIFE

The explorer-naturalists who traversed the Pacific two centuries ago were keen observers, recognizing not only new species but some of the peculiarities of island life. Johann Reinhold Forster on Cook's second voyage wrote that "The countries of the South Sea . . . contain a considerable variety of animals, though they are confined to a few classes only . . ." and he noted as an example that the only mammals he saw were "the vampyre and the common rat". Adelbert Chamisso, the German poet and naturalist on the Russian ship *Rurik* in 1818, added his observations: "This rich Flora seems to have become more scanty in the islands of the Great Ocean, from the west towards the east" and ". . . the appearance of nature in the eastern islands of the South Seas, reminds us at once of Southern Asia and New Holland, and is wholly dissimilar to America."

The peculiarities remarked on by the explorer-naturalists can be restated as the rules of Pacific island life: the species that inhabit Pacific islands are few in numbers compared with those on equal continental areas; there is a gradual elimination of major groups of plants and animals from west to east across the Pacific; there are diminishing numbers of species from west to east; many islands have species that are endemic, or unique to them; and the animals and plants of Pacific islands are, for the most part, related to those of the west rather than the east.

The accompanying map and chart indicate how rapidly plant and animal groups and numbers of species fall off across the Pacific once the shores of the Philippines, Australia, and the Malay archipelago are left behind. East of the Solomon Islands, which have a marsupial mammal and some rodents, there are only bats, and the 17 bat species in the Solomons are reduced to four in Fiji and one in the Cook Islands. There are neither frogs nor snakes east of Fiji. Even the number of birds falls from west to east, from more than 250 land birds in New Guinea to seven in the Marquesas. Among the higher plants, conifers, bamboos, and rhododendrons do not penetrate beyond New Caledonia and Fiji. Mangroves and large fruited trees of shorelines, such as *Barringtonia*, reach the Marshall Islands but are not found in Hawaii.

## ISLANDS OF THE WESTERN FRINGE

Seven major island groups fringe the western border of the Pacific: New Zealand, New Caledonia, Fiji, Vanuatu, the Solomon Islands, New Guinea, and Palau. Beyond them to the west are the continents of Australia and Asia, the Malay archipelago, and the Philippines, all far richer and more varied in plant and animal species: in the Malay archipelago there are estimates of 3,000 species of plants; in Vanuatu and New Caledonia there are fewer than 1,000 species. The islands of the western fringe are both continental and volcanic; some of them, such as New Zealand, are very old; all are densely vegetated. Their plants and animals are those of Australia, the Malay archipelago, and Southeast Asia, with some elements unique to each island, and often with some major order or family missing.

Tropical rainforests are dominated by massive auricarias and kauris, broad-leaved breadfruits and durians, *Metrosideros*, palms, various species of *Pandanus*, and climbing vines such as *Freycinetia*. Shorelines are rimmed by mangroves. New Zealand, in a more temperate climate, has many tropical elements in its forests, including *Metrosideros* and *Pittosporum*, as well as non-tropical plants such as the southern beech *Nothophagus* and fuschias.

The western islands are a mosaic of presence and absence. New Zealand has three frogs but no snakes, yet both frogs and snakes are present in Fiji. There is a marsupial mammal in the Solomons but no native mammals in New Caledonia. New Zealand is legendary for its now extinct moa, which at 3 meters (10 feet) high was the tallest bird in the world. In New Zealand, too, there is the lizard-like tuatara, and in Fiji an iguana. The kagu is endemic to New Caledonia. It is a remarkable grayish ground bird, about as big as a chicken, with a crested head, well-developed wings (although it never flies), and a call like the bark of a dog. Another singular bird, the megapode or incubator bird, occurs in Vanuatu, the Solomon Islands, and Palau. It too is about the

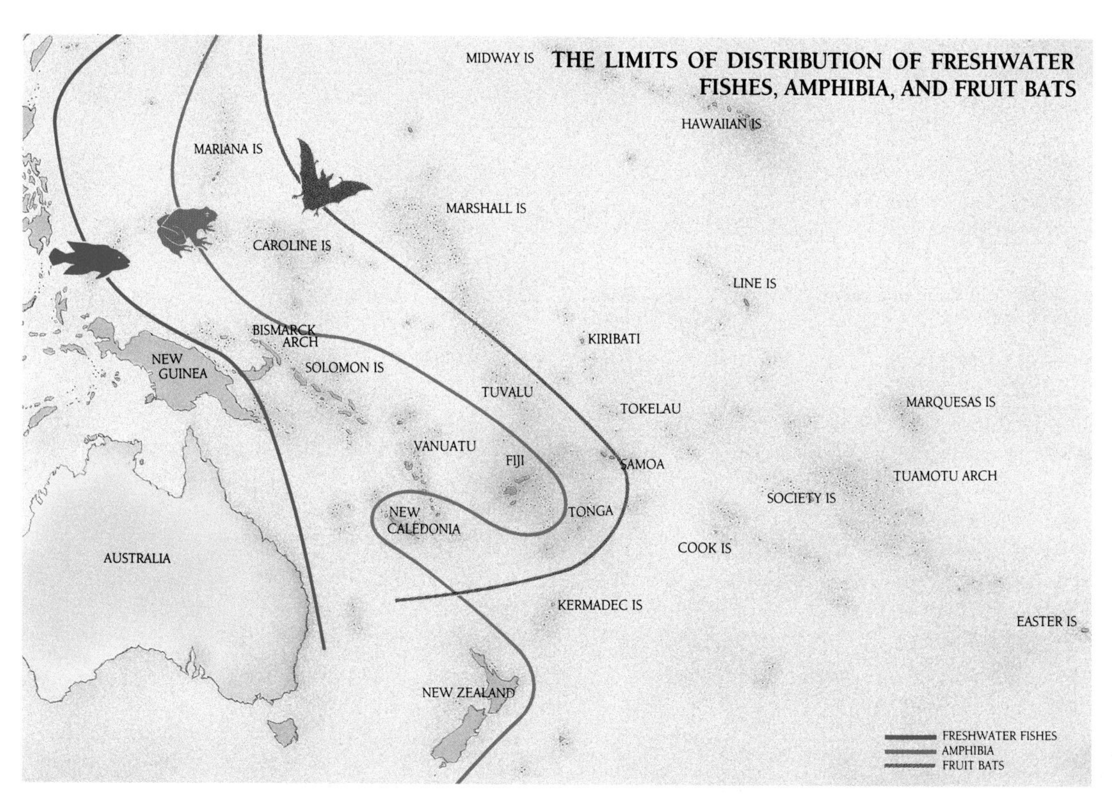


▸ There is a gradual elimination of species from west to east across the Pacific. Freshwater fishes are absent east of the coasts of New Guinea and Australia; amphibians disappear east of Fiji and New Zealand; and fruit bats, which can withstand the rigors of oceanic dispersal more easily than other mammals, disappear beyond Tonga and the islands of Samoa.

size of a chicken except for its feet which are like those of a turkey. Megapodes lay their eggs in piles of leaves or sand, to be incubated by the heat of the composting vegetation. In New Guinea there are tree kangaroos, cuscuses, giant spiny ant-eaters, birds of paradise, and cassowaries.

Insects are big and colorful. Fiji is home to huge stick insects and one of the largest beetles in the world. Many of the beetles of New Guinea are gaudy: golden and rose and green. The flight of the bird-winged butterflies of New Guinea and the Solomons has been described as "poetry in motion". Some of the land snails are also spectacular. In New Zealand *Paryphanta*, a large mahogany brown snail, feeds on earthworms. In New Guinea *Papustyla pulcherrima*, the brilliant green tree snail of Manus Island, lives high in trees in the rainforest.

HIGH ISLANDS IN THE CENTRAL PACIFIC

East of the western fringe are several volcanic islands among the array of islands recognized as Polynesia and Micronesia. Guam and others of the Mariana Islands, and Truk, Ponape, and Kosrae in the Caroline Islands are north of the equator; the Samoas, Tongas, Tahiti, the Marquesas, and Easter Island are below the equator. Because they rise steeply from the sea, often to heights of more than 300 meters (1,000 feet), these islands are called "high islands" to distinguish them from the "low" limestone islands among which they lie.

The woody plants, jungle climbers, mangroves, palms, *Pandanus*, and ferns of the forests are like those of the west, but there are fewer species even in the western islands such as Guam and Samoa, and the number of species rapidly declines eastward: Guam has about 356 species of native plants, Samoa around 320 species, and the Marquesas have fewer than 200 species. A few genera are endemic, among them the *Fitchia*, a relative of the sunflowers, which is unique to Tahiti and nearby islands. On Bora Bora, one species is a shrub growing only on the cloud-shrouded upper slopes; on Rarotonga another species is a large forest tree.

The dominant animals are land birds, small insects, and molluscs; there are no mammals except fruit bats, and no amphibia or reptiles. Each island group has its own subspecies of fruit doves, fly-catchers, fantails, and reed warblers. In the Marquesas there are two species of pigeons and on Tahiti two species of kingfishers. The most remarkable of the fruit pigeons is the tooth-billed pigeon, a large dodo-like bird endemic to only two islands in Western Samoa. In the Mariana Islands there is an endemic crow and on Guam a flightless rail.

A conspicuous feature of the insect faunas of these islands is the absence of large groups, families, and orders common to all continents: there is neither a single native species of scarab beetle nor any endemic leaf beetles. Few if any ants range east of Rotuma, Samoa, and Tonga. Anopheline mosquitoes are absent east of 170°E. The larger part of the insect fauna is made up of weevils. There are comparatively few genera, but some of them have disproportionately large numbers of species. In the Marquesas, for example, a relative of the leafhoppers is represented by eight endemic species.

Pacific island land snails are the only group in which entire families are endemic to the Pacific. The Endodontidae and Partulidae occur on all the high islands of the central Pacific, each island

Frithfoto/Australasian Nature Transparencies

▲ The Emperor of Germany bird of paradise *Paradisea guilielm* is one of the seven sexually dimorphic species of this spectacular group, and is found in the mountain rainforests of Papua New Guinea. Polygamous males such as this one have an extravagant and colorful plumage which they manipulate during ritualized displays, either among males to establish a dominance hierarchy, or to court females.

▲ Dense rainforest on the island of Viti Levu, Fiji. These luxuriant forests are a rich gene pool, teeming with myriad forms of life.

with its own suite of endemic species. There are three genera of partulids: *Partula*, distributed from Palau to the Society Islands; *Eua*, confined to Samoa and Tonga; and *Samoana* found from the Marianas to the Marquesas. *Partula* reaches its greatest diversity in the Society Islands where there are 65 species, most of them arboreal.

### HAWAII: A SPECIAL CASE

The Hawaiian Islands are the most isolated major island group in the world. The island chain consists of low coralline islands in the north and eight major volcanic islands in the south. The landscape ranges from snow-covered mountain tops to shoreline desert, and from bog and rainforest to new lava flows. More than 90 percent of the flowering plants, insects, and land snails are endemic, but the biota is disharmonic: the native flora has no gymnosperms, bromeliads, figs, or mangroves; there are only four orchids, one genus of palms, and two genera of butterflies. There is one endemic family, the land snail Amastridae, and an endemic subfamily of birds. Among the unique flora are lobelias; tarweeds (which include Hawaii's most famous plant, the silversword); and legumes such as *Vicia*, *Erythryina*, and *Canavalia*.

Birds and insects have radiated in remarkable directions. The bills of the honeycreepers range from the delicate curved bills of nectar-sipping birds to the massive mandibles used in tearing off bark and wrenching open burrows of larval insects. There are some 500 endemic species of drosophilid flies, more than in any other part of the world. Tarweeds and beggars ticks in different forms and different habitats range from the top of Haleakala on Maui to new lava flows on Hawaii.

### THE LOW ISLANDS OF THE PACIFIC

There are more than 300 islands between Palau and the Tuamotus, most of which are low islands, either atolls or raised coral islands such as Makatea and Henderson Island. These islands rise only a few meters above sea level and are distinguished not only by their geological base of limestone, but by their sparse and cosmopolitan flora and fauna.

The beach plants of low islands are the same as those on the beaches of all islands throughout the Pacific—morning glory, *Pandanus*, scaevola, beach heliotrope, and the like. On low islands, however, these plants form the flora of the entire island. No two islands have exactly the same plants, and what plants are there may form a jungle where it is rainy and scrub where rainfall is low. The northernmost atoll in the Marshall Islands is in a semi-arid belt and has a flora of nine species; an atoll in the south where it is extremely rainy has perhaps 60 native species.

The birds of low islands are predominantly nesting seabirds which spend many weeks at sea. There are also a surprising number of land birds. Not only does each island have (or had) its own flightless rail, but fruit pigeons and reed warblers are also endemic to specific islands: the Wake rail, the Henderson Island fruit dove, and the Makatea fruit dove are examples.

## HOW DID THEY GET THERE?

What are the explanations for these distribution patterns? Charles Darwin provided some answers in 1859 in *The Origin of Species*, deriving both from the observations of the naturalist-voyagers of the Pacific and from what he himself had seen in the Galapagos Islands in 1853. In Darwin's view, plants and animals colonize islands hundreds of kilometers from continents by migration or dispersal. Only those organisms that survive the long journey from a continental source can survive and reproduce on a far-flung island, and distance from the continents explains why so few kinds of plants and animals inhabit oceanic islands. Because of their common descent, insular plants and animals resemble those on the continents of their origin. Moreover, both the unusual creatures and the endemic species evolve on islands from ancestors derived from a continental source.

## THE SUCCESS OF CHANCE DISPERSAL

Most biogeographers now agree that plants and animals arrived on Pacific islands accidentally, the result of chance dispersal over water: rafting on floating trees, logs, and pumice; as planktonic larvae in ocean currents; as propagules uplifted by air currents; and as seeds and small animals stuck to the feet or feathers of birds.

There is convincing evidence of the effectiveness of chance dispersal. Darwin himself conducted several experiments to see how long seeds would remain viable in sea water. A century later, J. Linsley Gressitt experimented with trapping insects in nets towed by planes flying over the Pacific, and Rudolph Scheltema sieved mid-Pacific water for larvae of marine animals. Gressitt found insects and spiders in the jet stream in the same proportions and of the same types as those on the islands below; Scheltema found larvae of coastal marine snails and worms hundreds of kilometers from land. The efficacy of natural rafts as a means of dispersal has been documented by Paul Jokiel who has found pumice, to which corals and other sessile invertebrates are attached, washed up on island shorelines throughout the Pacific.

FAUNA SPECIES OF SELECTED PACIFIC ISLANDS

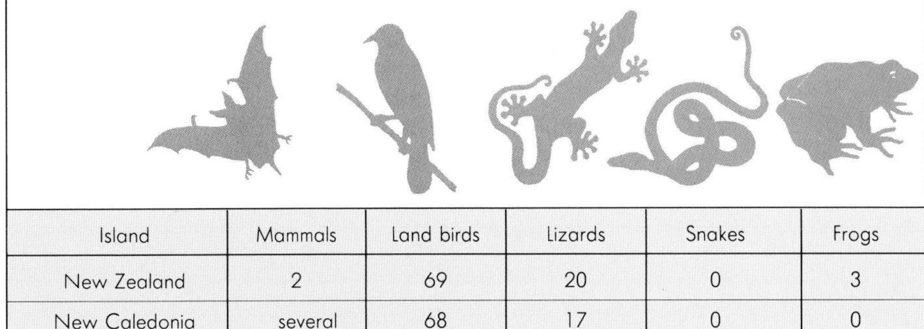

| Island | Mammals | Land birds | Lizards | Snakes | Frogs |
|---|---|---|---|---|---|
| New Zealand | 2 | 69 | 20 | 0 | 3 |
| New Caledonia | several | 68 | 17 | 0 | 0 |
| Solomon Islands | 30 | 136 | 10 | 4 | 10 |
| Fiji | 5 | 60 | 15 | 4 | 2 |
| Samoa | 2 | 33 | 77 | 0 | 0 |
| Tahiti | 0 | 12 | 7 | 0 | 0 |
| Hawaiian Islands | 1 | 44 | 0 | 0 | 0 |
| Marquesas | 0 | 7 | 7 | 0 | 0 |

▲ Pacific island fauna is a patchwork of presence and absence, but the general principle is clear: with increasing distance from a major landmass to the west, the number and diversity of species diminish. The Solomon Islands, for example, which lie about 650 kilometers (400 miles) from New Guinea, support considerable numbers of species of mammals and birds, among other fauna. The Marquesas, on the other hand, some 7,400 kilometers (4,600 miles) from the east coast of Australia, have only small numbers of land birds and lizards.

Darwin argued that chance dispersal was like a filter, that only those organisms capable of dispersal would end up on distant islands. Thus oceanic islands are populated with a limited number of representatives of plants and animals, and the characteristics of Pacific island life derive in part from filtering out most continental species and from the rigors of long-range dispersal. Herein lies the explanation for the absence of all but such agile mammals as bats and seals, of freshwater fishes that cannot survive in the ocean, and of the amphibians and reptiles for which an ocean voyage is not feasible. Here, too, lies the explanation for what is found on Pacific islands: plants with seeds that can be dispersed by wind or birds or in the sea; birds that survive chance dispersal by storm or wind; and insects and snails that are easily transported by wind or raft or bird.

However they might have come, the indications are that only a very few need have arrived. In Hawaii it is estimated that over a period of several million years, only about 272 successful colonists would account for the indigenous flora of more than 1,000 species; 300 founders were sufficient to produce 10,000 species of insects, and 22 to 24 ancestral colonizations gave rise to the 1,000 descendant species of land snails.

## SIMILARITIES AND DIFFERENCES

The frequency and rarity of chance migratory events are reflected in island populations. The cosmopolitan complexion of beach plants on all the islands indicates that beaches and low islands continually receive the seeds of *Pandanus*, morning glory, beach heliotrope, and other plants from the ocean. Seabirds that spend weeks at sea traverse thousands of kilometers of ocean, and the same species—red-footed boobies, great frigatebirds, and white terns—are seen in Fiji as on Pitcairn.

Nevertheless, not all seaborne organisms are exactly the same from island to island. Even on some of the smallest low islands there are occasional endemics: nine species or varieties of plants, three bird species, and several insects and land snails are endemic to Henderson Island. While ocean currents actively distribute plant seeds and the larvae of marine organisms, some suspension of the means of dispersal must occur, and with it the suspension of gene flow and subsequent change in organisms with time. Migration is, therefore, not the only factor that explains the peculiarities of Pacific island life.

One of the most effective disruptions to gene flow is isolation: the distance of an island from a neighboring island or continent. Indeed, the diversity of the floras and faunas of Pacific islands is inversely proportional to the distance between an island and the Malayan region to the west: there is more endemism in the Hawaiian Islands and southeastern Polynesian islands which are further from the Malayan archipelago.

Another factor that appears to ensure diversity is a range of habitats that provide places in which evolutionary changes can take place. There is little endemism on beaches and low islands where there is little topographic relief; rather, it is on the inland areas of high islands, with their ridges and valleys and abrupt changes in climate, that most endemics occur. In Hawaii it has been said that each ridge and valley has its own land snail.

## ADAPTIVE SHIFTS

Some of the unique plants and animals of the Pacific islands have arisen as a result of what are called "adaptive shifts", in which subpopulations of a successful population exploit a totally new resource or habitat, and the resulting plants or animals diverge far from the ancestor. These changes may occur because the niches or habitats in which the new species developed were empty. In the absence of grazing mammals, reptiles, and other predators, for example, the flightless, ostrich-like moa in New Zealand became a giant grazer. One of the most remarkable of these adaptive shifts has occurred in the Hawaiian inchworms: herbivorous caterpillars have become carnivorous, feeding on flies, cockroaches, leafhoppers, and spiders.

Another striking development on Pacific islands is that in which several adaptive shifts have occurred, resulting in a group of organisms occupying a range of habitats. In Hawaii, for instance, in the "silversword alliance", the 28 descendant species of a tarweed include not only the majesterial silversword, but also cushion plants, rosette plants, upright trees, and vines.

## STEPPING STONES, TIME, AND THE HUMAN IMPACT

Biogeographers read the past to explain present distributions. Three relatively recent discoveries point the way to future understanding of the

▼ A relic from the past, the handsome crested iguana *Brachylophus vitiensis* was discovered only in 1979. They are rare, and endemic to the dry offshore islands of Fiji. The male, shown here, is spectacularly colored.

G. E. Schmida/Australasian Nature Transparencies

Claire Leimbach

▲ Bora Bora in French Polynesia, like many other Pacific islands, is renowned for its natural beauty. Here, on the island fringe, mature coconut palms bend gracefully toward the lagoon.

distribution and evolution of the flora and fauna of Pacific islands.

The discovery in the 1950s of chains of guyots or seamounts, from Australia and Asia to Hawaii, has lent support to the concept of continental drift and plate tectonics. These guyots, if they were once at sea level, could provide a means for dispersal by island hopping. Unfortunately the ages of these undersea mountains are inconsistent with the relatively young ages of most of the high islands in the Pacific. But we now know that islands drift away from the hot spots that spawned them, and it has been suggested that these drifting, eroding, and soon-to-be subducted islands could have provided stepping stones by means of which the ancestors of Hawaii's plants and animals colonized islands newly arising from the hot spot.

Naturalists have argued that some of the unique plants and animals of Pacific islands are (or were) relics, surviving from times long past. The moa in New Zealand, the iguana in Fiji, and the gymnosperm araucarias and cycads of the western fringe of the Pacific are representative of ancient groups of organisms. Dating of volcanic islands throughout the Pacific indicates, however, that almost all of them are only a few million years old. Indeed, the island of Hawaii is less than 700,000 years old, which implies that the species endemic to it must also be new. There is every indication that the development of new species on oceanic islands is a rapid process, and that the time span to evolve, for example, flightlessness in rails, may be measured in generations.

The possibilities of rapid evolution have been given impetus since 1970 by the discovery of the bones of a flightless goose on the island of Molokai in the Hawaiian Islands. That discovery has had three significant consequences: the finding of fossil bird bones, many of them representing both recent and extinct species, throughout the Pacific; the recognition that several of these recent island endemics had ranges far greater than we know them today; and the belief that human impact is the most plausible explanation for these extinctions. The surviving birdlife thus poses the challenge of reconstructing the distribution of land birds on Pacific islands in the past, and thereby continuing an ever-changing understanding of the biogeography of Pacific islands.

E. ALISON KAY

Frans Lanting/Minden Pictures

▲ The stunning lighting from a Falkland sunset lends a primeval beauty to a lichen-covered boulder-field in these otherwise bleak and treeless islands.

# THE ATLANTIC ISLANDS

The islands of the Atlantic Ocean are far fewer than those of the Pacific, and their flora and fauna seem much less exotic. Islands such as Newfoundland and the British Isles were connected to nearby continental landmasses during the ice ages, when so much water was tied up in glaciers that sea levels were much lower than at present. The flora and fauna of these continental islands are thus remnants of a larger set of species occurring on the adjacent continents.

Most Atlantic islands, however, are oceanic islands whose native fauna and flora must have crossed over water to colonize them. With the exception of wave-washed St Paul Rocks, home to but three species of seabird and a few insects, all the oceanic islands of the Atlantic are volcanic in origin. Many are on, or were formed on, the Mid-Atlantic Ridge, an underwater mountain range that marks the zone where North and South America pulled away from Europe and Africa to begin forming the Atlantic Ocean some 150 million years ago.

## CONTINENTAL LINKS

As a rule, the flora and fauna of the Atlantic islands were derived from the nearest continent. Thus the affinities of Bermuda are with North America; the islands of Macaronesia (Azores, Madeira, Canaries, and Cape Verdes) with southern Europe and northern Africa; Fernando de Noronha and Trindade with Brazil; and Ascension and St Helena with Africa. The exceedingly remote islands of the Tristan da Cunha group and Gough Island are interesting because, although situated rather closer to Africa, their land bird fauna, at least, seems most similar to that of South America. Presumably this is because the prevailing winds are from the west, so that even today most of the vagrant birds arriving in Tristan are of American origin.

The natural history of islands at the northern and southern extremes of the Atlantic reflects their proximity to the poles. The wildlife of Iceland, for example, is like that of northern and Arctic Europe, whereas South Georgia is populated mainly by Antarctic seabirds, and the Falklands by a combination of Antarctic and South American elements, the most intriguing of which was the Falklands fox, renowned for its extreme tameness, such that Charles Darwin once obtained a specimen with a hammer. As might be expected, the Falklands fox is now extinct because of subsequent persecution by fur trappers and stockmen.

Of the string of volcanoes in the Gulf of Guinea that gave rise to the islands of Bioko (Fernando Poo), Principe, Sao Tome, and Annobon, the first is continental, whereas the others are oceanic and support several species that occur nowhere else, such as a dwarf ibis, a pigeon, a very large weaverbird, and a grosbeak with a truly massive bill. This last species is now presumed extinct, and with the destruction of the museum in Lisbon by fire, the only remaining specimen appears to be the one in the British Museum. But, as would be expected, the overall affinity of the flora and fauna of these islands is overwhelmingly like that of equatorial Africa.

The remaining islands in the south Atlantic are for the most part single, small, and very isolated, whereas in the north Atlantic, with the exception of Bermuda, the islands are all part of larger archipelagos. There is no prehistoric record of human occupation of any of these islands except the Canaries, which were colonized by a people known as the Guanches between 2500 and 2000 BC. Most of the other islands and archipelagos were discovered and inhabited during the great era of Portuguese and Spanish exploration in the 1400s and early 1500s.

## THE HUMAN IMPACT

Human settlement has had a depressingly similar impact on the ecology of all the Atlantic islands. Historical accounts invariably mention the early release of destructive domesticated animals. Goats and rats were among the first, and were usually followed by pigs, dogs, cats, mice, and other pests that either killed native animals directly, or destroyed the vegetation on which they depended. On those islands that were settled by Europeans, habitat destruction continued through burning, clearing for agriculture, and the introduction of noxious plants and insects.

Biologists did not arrive in the islands until long after these perturbations had altered the character of native ecosystems. Thus, the full extent of human impact was not appreciated until systematic searches were made for fossils of extinct vertebrates, beginning in the 1970s. These studies revealed many extinct or extirpated populations of seabirds, land birds, a few reptiles and mammals, and land snails. So far, the search for fossils has included Bermuda, Madeira and Porto Santo, the Canaries, the Cape Verdes, Fernando de Noronha, Trindade, Ascension, and St Helena. Extinct forms have been found on all but Trindade, where the geological environment is not at all conducive to fossil deposition.

Many of these studies are still incomplete and unpublished. All show, however, that the historically known faunas of these islands are often only a pitiful remnant of what was there before the arrival of humans. There are no endemic species of land birds on Bermuda now, yet in the past there was a finch, a woodpecker, an owl, a heron, and several flightless rails, among others. The most distinctive land birds of Madeira are a species of pigeon and a pipit, both shared with the Canaries. But fossils reveal extinct thrushes, finches, a rail, and two species of quail that may have been flightless. Extinct rodents, lizards, a tortoise, and a finch have been described from fossils in the Canary Islands. With these facts in mind the absence of any endemic species of birds in the Azores is quite improbable. These islands are older and more remote than the Galapagos and it is hardly conceivable that no differentiation of birds beyond the level of subspecies took place there.

Rails and gallinules are somewhat chicken-like marsh birds that disperse widely and are often found as vagrants far out of their normal range. For this reason they are very successful at colonizing remote oceanic islands, where they quickly evolve flightlessness. This condition unfortunately renders them extremely susceptible to introduced predators. The only flightless members of this group known historically in the Atlantic are gallinules from Tristan da Cunha (now extinct) and Gough Island, and the tiny flightless rail of Inaccessible Island in the Tristan island group.

Douglas Rogers

▲ A scientist descends into a dormant fumarole (vapor hole near a volcano) in search of the fossilized bones of an extinct rail on Ascension Island. Detailed searches such as this have shown that the extant fauna on many Atlantic islands is but a small subset of the variety of species in pre-European times.

But fossils show that there were probably flightless rails on all the islands of the Atlantic, for in addition to those mentioned above from the north Atlantic, we now know of a flightless rail from Fernando de Noronha, two from St Helena, and one from the harsh and inhospitable island of Ascension, which is hardly more than a giant cinder. This rail was observed alive and accurately described in 1656 by the astute traveler and diarist Peter Mundy. It is thought to have been dependent on carrion and associated insects found in seabird colonies, and indicates the extreme adaptability of rails to various island environments.

## SEABIRDS PAST AND PRESENT

Although terrestrial organisms were hard hit by the arrival of humans, seabirds were dealt perhaps a more devastating blow. Even the most oceanic of birds must come to land to nest. Thus for seabirds, the few tiny islands of the Atlantic were of inestimable importance, far out of proportion to their total land area.

The most renowned of seabirds dependent upon Atlantic islands was the flightless great auk

P. Morris/Ardea London

▲ A nineteenth-century engraving from John Gould's *Birds of Britain* shows the extinct great auk. Once common in the north Atlantic, this peculiar flightless bird (related to razorbills and guillemots) was slaughtered for its oil and as a food source for sailors. By about 1844 it had been wiped out.

*Pinguinus impennis*, an ecological counterpart to the flightless penguins of the Southern Hemisphere. It is known to have nested in historic times on islands from the Gulf of St Lawrence to the British Isles. In prehistoric times it was known from Florida to the Mediterranean, so the species had probably been adversely affected by prehistoric people before Europeans exterminated the species for food and oil: the last known pair was killed in 1844.

Since it surfaced some 30 million years ago, Bermuda was probably an important breeding locality for seabirds. The petrel known as the cahow *Pterodroma cahow* once occurred there by the millions, but the species was reduced to a pitiful remnant of a few dozen individuals by colonists who used it as food and by pigs, cats, dogs, and rats. Fossils show that St Helena was once home to vast numbers of seabirds, at least six species of which have been exterminated since human settlement. The few remaining species exist only in very low numbers on small offshore rocks.

With seabirds, the story is not so much one of complete eradication of species, although this did happen, but of reduction of population sizes by several orders of magnitude. Millions of petrels, terns, frigatebirds, boobies, and other efficient surface predators were removed from the oceanic environment. What the effect may have been on the fish and squid that were their prey, or on nutrient recycling in the waters around the islands where they nested, has never been calculated. Humans have long wrought profound and deleterious changes in the terrestrial ecosystems of Atlantic islands. We must now entertain the possibility that these changes had a significant effect on the oceans themselves.

STORRS L. OLSON

Hakluyt Society

► The flightless rail of Ascension Island was observed alive by Peter Mundy while visiting this inhospitable place (in the south Atlantic Ocean) in 1656. Mundy's observations and his drawing of the rail, together with fossilized bones, are all that remain of this bird.

# ISLANDS OF THE INDIAN OCEAN

Although the islands of the Indian Ocean are not as numerous as those of the Pacific, they are geologically and biologically more varied. Relatively few of them are inhabited. Unlike the Pacific Ocean, archipelagos made up of many small coral atolls are rare—the Maldives and the Laccadives, and the Chagos Archipelago, are the two main groups. However there are many single coral atolls, as well as much larger islands such as Madagascar and Sri Lanka.

## CORAL ATOLLS

Off the northern coast of Western Australia are several uninhabited and isolated coral atolls such as the Rowley Shoals, composed of three atolls, and Scott, Ashmore, and Seringapatam reefs. Although they have little landmass above the water and limited terrestrial scenery, their subtidal environments are often spectacular, with extensive coral reefs, steep drop-offs, and large territorial reef fish. The northern reefs support very dense sea-snake populations, making diving an interesting experience!

Cocos (Keeling) Islands, a group of a few small coral atolls, are roughly equidistant from Australia, Sri Lanka, and Chagos, but their marine fauna is most similar to that of Indonesia. The islands have been inhabited since 1826 and until the 1980s the basis of their economy was the production and export of copra. Human activity has led to vastly reduced populations of turtles, and the bird life too has been heavily affected. The uninhabited island of North Keeling is a world-renowned seabird rookery, with colonies of red-footed, brown and masked boobies, frigatebirds, red- and white-tailed tropicbirds, white, sooty and common noddy terns, and Cocos buff-banded rails.

In contrast, the main island, only 24 kilometers (15 miles) away, is almost devoid of birds. This may have resulted from the felling of hardwood trees for timber, which destroyed the birds' habitat. Birds and their eggs, as well as turtles and

Jan Aldenhoven/Auscape

▲ The endemic red crabs *Gecarcoidea natalis* of Christmas Island undergo an extraordinary annual migration. In early summer, oblivious to all in their path, hundreds of thousands of crabs swarm in a scarlet tide from the rainforest to the seashore to mate. After mating, adults return to the forests while the fertilized eggs drift in the plankton, developing through their larval stages until they return to the intertidal habitat as juveniles. A brief existence beneath boulders follows; then they too head for the inland forests.

their eggs, and shellfish and crabs, were also consumed to supplement the meager diets of the human inhabitants. These factors, combined with the introduction of rats, cats, and exotic plant species, have completely changed the island's flora and fauna, thus illustrating the fragile state of nature on a small island.

## CHRISTMAS ISLAND

Christmas Island is part of the same underwater chain of volcanic activity as the Cocos (Keeling) Islands. In contrast to the low relief of the Cocos Islands, however, Christmas Island rises to a height of 360 meters (1,185 feet). With an area of 13,700 hectares (33,840 acres), it is a much larger island than Cocos, but has a similar surrounding coral reef. Christmas Island is also a major bird habitat with several endemic species, including Abbotts booby, the Christmas Island frigatebird, and the golden bosun. It is particularly famed for its large land crabs—there are more than 20 species—which are thought to thrive because of the absence of large predators on the island. Three species are most common: the robber crab *Birgus latro*, the blue crab *Cardisoma hirtipes*, and the red crab *Gecarcoidea natalis*. The red crab is ubiquitous on the island, and it has been estimated that there are 120 million individuals with a total weight of 8,000 tonnes. At these densities ( more than 1 per square meter/1.2 square yard) there is little debris in the forests or gardens as the crabs remove it all.

## TWO ISLAND REFUGES

The isolation of some Indian Ocean islands has resulted in their becoming a refuge for rare or unique forms of life. The Seychelles and Aldabra, mid-ocean islands north of Madagascar, represent two such refuges from the pressure of changes brought about by humans, and the predators such as rats and dogs they introduced. In the eighteenth century the Seychelles were planted with coconuts as a cash crop, but the islands are now a major tourist destination. Because of their small size and limited variation in habitat, the islands have no native land mammals. However, 15 endemic species of birds, including the world's smallest falcon, the Seychelles kestrel *Falco araea*, and many endemic amphibians have evolved there. The most unusual of the amphibians is a legless burrowing caecilian *Hypogeophis*. Its presence on the Seychelles, while absent from adjacent Madagascar, is difficult to explain.

Aldabra is a low coral atoll of four main islands around a central lagoon. It has about a dozen species of endemic land birds, one of which, the

Adrian Warren/Ardea London

▲ The true Madagascan lemurs, like this ring-tail *Lemur catta* and its baby, have fox-like faces, long bushy tails, and long limbs. The ring-tail is a vegetable eater, adapting its diet to seasonally available food.

white-throated rail *Dryolimnas cuvieri*, is the only flightless bird left in the Indian Ocean. Aldabra is also the only remaining location in the Indian Ocean of the giant land tortoise, which was once widely distributed, especially on Madagascar and the Seychelles. These tortoises are slightly larger than their Pacific counterparts on the Galapagos Archipelago. As with the Galapagos tortoises, they were once used as food by sailors on long ocean voyages, and are now extinct except for those left on Aldabra. Fortunately their future seems more secure now, as the islands have become a focus for wildlife conservation.

### MADAGASCAR

Madagascar lies only 400 kilometers (250 miles) off the southeast coast of Africa and is the fourth largest island in the world. It has a varied climate and topography, rising to a height of 2,880 meters (9,500 feet). Despite its proximity to the African coast, Madagascar's flora is more similar to that of the western Pacific than that of Africa. Its fauna is remarkable, with many endemic species, often relatively small in size. The most famous of these are the lemurs, large-eyed primitive primates, which make up about 40 percent of all mammals on Madagascar. These monkey-like animals usually live in the trees on the wet eastern side of the island, although the commonest, the ring-tailed lemurs, tend to live in rocky crevices in the arid sparsely wooded regions.

There are other unusual creatures: iguanas, whose nearest living relatives occur in South America; tenrecs, primitive insectivores related to hedgehogs, which have interesting social behavior patterns; and half of the world's species of chameleons. There are only 11 species of carnivore, including the cat-like fossa and the smaller fox-like fanaloka. The fauna has a strange balance, which is presumably the result of the processes of immigration and extinction that determine the presence or absence of animals on islands. For example, there are over 150 species of frogs but no toads, poisonous snakes, or freshwater fish! In comparison with the rich avian fauna of southern Africa, the bird population is very impoverished with only 82 species, half of which are endemic.

The impact of humans on the habitats of Madagascar has led to dwindling numbers of these rare and special organisms. One type of tree-dwelling lemur, the aye-aye, is virtually extinct: only a few remain in the wild, and a group has been established on an island reserve to try to protect the species.

DIANA WALKER

# THE SUBANTARCTIC ISLANDS

Between the Antarctic continent and the southernmost shores of Africa, Australia, and South America is a series of eight small and remote islands and island groups. These islands, which lie in the subantarctic zone, are the only land within the vast Southern Ocean. Scattered around the globe between 45°S and 59°S, they are as remote from each other as they are from the continents.

## A RING OF ISLANDS

The islands straddle the Antarctic Convergence, an oceanographic boundary formed from the meeting of the cold Antarctic waters with the warmer waters to the north. Those islands to the south of the convergence (South Sandwich, Bouvet, South Georgia, and Heard islands) are extensively glaciated and predominantly snow covered, while those at the convergence (Kerguelen Islands) or a little to the north (Crozet, Macquarie, and Prince Edward and Marion islands) tend to be free of snow for most of the year. Their position in relation to the Antarctic Convergence is important in determining their climate and hence their terrestrial life-forms. Those islands to the north are green, and their lower slopes are covered with tussock grasses and herbaceous plants (but no trees). To the south, the islands have progressively less land free of snow and ice, and the vegetation is sparse and limited to mosses, lichens, and only a few flowering plants on the most southerly.

The early history of the islands was one of exploitation, first for fur seals for their pelts, and then for elephant seals and king and macaroni penguins for their oil. Later visits by scientific expeditions raised wider interest in the islands as sanctuaries for seals and seabirds, and for the study of the flora and fauna which have developed in isolation in the subantarctic. The remarkable abundance of seals and seabirds in the unexploited or recovered state has fascinated all who have visited the islands. The surrounding ocean is able to provide food for very large numbers of these species but the land to which they must return to breed is limited.

The Southern Ocean, which encircles Antarctica and flows eastward past the subantarctic islands, was formed during the final stages of the break-up of the supercontinent Gondwana when the deep oceanic trenches formed between Australia and Antarctica, and then between South America and what is now the Antarctic Peninsula. Until this time the Atlantic and Indian oceans were separated from the Pacific Ocean, but with the opening of the Drake Passage in the middle to late Oligocene epoch (about 30 million years ago) the oceans began to mix and circulate around the Antarctic continent and the Antarctic Convergence formed. The geographical position of the Antarctic Convergence is moderately constant in present times but was probably further north in geological times, when most of today's subantarctic islands were to its south and were therefore subjected to a more severe climate.

The geological history of each island is quite different, even though their placement in a ring around the globe might suggest a common origin. Since their formation the islands have been subjected to relative changes in altitude as they have risen out of the sea. They have also been influenced by changes in sea level brought about by melting of the polar icecap, and warming or cooling of the surrounding ocean. Since the last glacial maximum some 18,000 to 8,000 years ago, the sea level has risen between 120 and 140 meters (390-460 feet) and the glaciers have retreated, thus exposing more and more land. These processes are responsible for the amount of land available for colonization by plants.

**ISLANDS OF THE SOUTHERN OCEAN**

| Island | Latitude °S | Area sq km/ sq miles | Snow free area sq km/ sq miles | Distance from nearest continent km/miles | Geological origin |
|---|---|---|---|---|---|
| ***Atlantic Ocean*** **South Georgia** | 54 | 3,755/1,450 | 1,500/585 | 2,210/1,370 | Upper Jurassic to lower Cretaceous: 130-50 million years old |
| **South Sandwich** | 56–59 | 618/240 | 85/33 | 1,670/1,035 | Volcanic: 1–4 million years old |
| **Bouvet** | 54 | 50/19 | 4/1.5 | 2,659/1,648 | 250,000 years and younger: Late Pleistocene strato volcano in Mid-Atlantic Ridge |
| ***Indian Ocean*** **Prince Edward and Marion Islands** | 46 300 | 52/20 300/117 | 52/20 300/117 | 1,900/1,178 | 250,000 years and younger: shield volcanoes |
| **Crozet** | 46 | 400/156 | 400/156 | 2,740/1,700 | 800,000 years and younger: strato volcanoes, deeply dissected |
| **Kerguelen** | 49 | 6,500/2,535 | 2,900/1,130 | 4,110/2,548 | Complex volcanic history: 40 million to 40,000 years old |
| **Heard and McDonald** | 53 | 380/148 | 70/27 | 4,570/2,833 | Heard Island is an active volcanic cone: 3 million years old |
| ***Pacific Ocean*** **Macquarie Island** | 54 | 200/78 | 200/78 | 990/614 | Upturned oceanic crust, perhaps related to Miocene (10–12 million years) subduction or older. |

## A LIMITED FLORA

The biological history of the islands is interesting and complex, although the range of plants and animals inhabiting any one of them, or in fact all collectively, is small compared with the southern continents to their north. Throughout all the islands, there are only 24 species of grass, 32 herbaceous plants, and 16 ferns. There are also a minimum of about 250 mosses, 300 lichens, 150 liverworts, and 70 species of mushrooms and toadstools. The few endemic species, and species in general, show close affinities with their nearest neighbor to the west. Many of the flowering plants are also found on the southern continents; there are, for example, botanical links between South Georgia and Tierra del Fuego, Prince Edward and Marion islands and Africa, and Macquarie Island and Australia.

The mosses and lichens tend to be cosmopolitan; many are found also in the colder regions of the Northern Hemisphere. The scarcity of the subantarctic flora is due to the remoteness of the islands, the direction of the prevailing winds, the area of available land for colonization, and the climate. Only a few islands to the north have any fossil record and it seems that very little, if any, of the vegetation survived glaciation. The means by which the plants reached the islands is still debated, but it is generally agreed that the present-day flora became established after the end of the last ice age.

Species of flora probably arrived on the islands when their propagules were carried by wind. This is consistent with the similarities between island flora and those to the west (the prevailing winds circle the globe from west to east at these latitudes). However it seems likely that other methods of transocean dispersal may be involved, including transport by birds with seeds ingested or attached to feathers, and transport by ocean currents.

Survival on land is, of course, as important as survival while being transported over the ocean. The vegetation of the subantarctic islands shows wide tolerance to environmental conditions and in general grows optimally at temperatures higher than those prevailing. Growth is also limited by lack of sunshine and a marked lack of seasonality in the weather. The species are not necessarily well adapted to the environment, and other species could probably grow better were they introduced to the islands.

The vegetation has developed in the absence of vertebrate herbivores and thus without grazing pressure. The inability of much of the flora to withstand grazing has been shown where herbivores such as rabbits, sheep, cattle, and horses have been introduced. These have extensively depleted and modified the vegetation, often denuding slopes and leading to soil erosion. This in turn has destroyed the habitat of some breeding species and exposed others to the depredations of introduced cats.

## ISLAND FAUNA

The truly terrestrial animals of the subantarctic islands, those which spend all their time on land, are small invertebrate organisms such as nematodes, worms, molluscs (slugs and snails), spiders, and insects, as well as the microscopic protozoans, rotifers, and tardigrades. There are no amphibians, reptiles, or fish, and the only vertebrates are seven species of birds, including sheathbills, ducks, and a pipit. A parakeet and a rail once present on Macquarie Island are now sadly extinct.

The best-known inhabitants of the islands,

▸ A wandering albatross *Diomeda exulans* incubates eggs at midnight. Pairs of wandering albatrosses invest a considerable amount of time in parental care: eggs are incubated for two to three months and chicks may take 38–43 weeks to fledge. Consequently, breeding occurs every two years, beginning during the continuous daylight of the Antarctic summer.

Frans Lanting/Minden Pictures

Frans Lanting/Minden Pictures

seabirds and seals, spend most of their time at sea in search of food. They are, however, tied to the land to breed and during the summer months are present in very large numbers. The most obvious and widespread are the macaroni, king, gentoo, and rockhopper penguins, and the wandering, light-mantled sooty, black-browed, and grey-headed albatrosses. Petrels also breed on the islands in great numbers.These birds, which are related to the albatrosses, range in size from the albatross-size giant petrels to the sparrow-size storm petrels and diving petrels.

Because their adaptations to a life at sea also fit them to survive in the polar regions, and because of the virtual absence of vertebrate predators, seabirds have been able to establish breeding colonies on the subantarctic islands. As breeding sites are limited, the colonies are often very large.

Each subantarctic island is unique in its age and geological history and in its fauna and flora, although some elements are shared. Studies of these islands provide valuable insights into the processes of colonization and the development of communities. Unfortunately, very few islands have escaped the impact of humans, and these processes have to varying degrees been modified by the introduction of organisms, particularly grazing and carnivorous animals, but also species of plants that, having been taken to the islands, are able to thrive better than those species which survived the ocean crossings. Active programs to eliminate these alien species are now underway on most islands.

Only a few islands are in a near-pristine state. These are important as baselines for the study of all the subantarctic islands. There are no introduced species on Heard island (although seals and penguins were once exploited). The nearby small McDonald Islands may be pristine, as are the South Sandwich and Bouvet islands. The subantarctic islands are beautiful, wild, rugged, and remote: at once places of beauty and laboratories for science.

KNOWLES KERRY

▲ Juvenile male Antarctic fur seals *Arctocephalus australis* languish on tussock mounds on South Georgia Island. Fur seals, in contrast to short-haired pinnipeds such as sea lions, have beautiful thick woolly coats beneath their matted upper hair. For this reason many species have been hunted almost to extinction.

# 11 ISLANDS AND PEOPLE

Since the beginning of human civilization, islands have represented both a sanctuary and a challenge. In rafts and canoes, and braving enormous dangers, people once made epic voyages to islands, attracted by curiosity, perhaps, or the belief that in isolation there was safety. Using the resources of both land and sea, some island settlements thrived and became centers of culture and ideas; others floundered amid the sometimes overwhelming odds of nature and isolation. Islands have long been the focus of imperialism, the jewels in the crown of colonial nations intent on extending their empires. Throughout history, they have been fought over and staunchly defended. In today's shrinking world, islands can no longer be isolated refuges, and island societies face a new set of problems no less demanding than those that challenged their forebears.

Klaus-D. Francke/Bilderberg

▲ Scarlet, downward-tapering, cypress wood columns reconstructed in the palace at Knossos, Crete. This site of Minoan civilization was discovered by British archeologist Sir Arthur Evans in 1900.

◄ Elaborately painted and adorned, according to the customs of his tribe, a Tabi Bugar warrior from the highlands of Papua New Guinea attends the annual sing-sing. Hundreds of tribes gather for this festival, to celebrate and share their cultural identities.

## CRETE: AN ANCIENT ISLAND STATE

What is now the island of Crete was once the highest part of a long mountain chain that curved toward the east from the Peloponnese to Turkey. In the Pleistocene epoch the Aegean submerged, the mountains were raised, and Crete became separated from mainland Greece and Turkey. With Karpathos and Rhodes and smaller islands it formed part of the southern chain of islands linking Greece and southwestern Turkey.

### THE LAND AND ITS RESOURCES

The island, about 250 kilometers (150 miles) long, has three main mountain groups: far to the west are the White Mountains, high (over 2,400 meters/7,700 feet) and wild; in the center is the group including Mt Ida (2,500 meters/8,200 feet); Mt Dikte lies to the east. North of the White Mountains is the Khania Plain, with rolling country to the east; between Mt Ida and Mt Dikte is the low northern coastal strip around Knossos, with the Messara Plain to the south; east of Mt Dikte is the low-lying narrow isthmus of Hierapetra. Along the north coast there are valleys between the spurs thrown out by the central mountains; along the south coast the mountains run steeply down into the sea.

As well as the fertile coastal plains and valleys, there are many upland plains. The lower, more accessible ones are occupied all year round; the higher ones are often snowbound in winter and are occupied only in summer.

Klaus-D. Francke/Bilderberg

◄ Sheep and goats have provided food for Cretans since ancient times, and sturdy flocks are still taken to graze on the upland plains.

In antiquity the land grew good crops of olives, grapes, and grain, together with fruit and nuts, and the wide range of beans and root vegetables that formed a large part of the ancient Greek diet. Wild bulbs, grasses, and vegetables were also probably harvested. The sea yielded its fish and shellfish, the air its birds, and the hills small animals. Sheep and goats provided cheeses and other milk products. The diet could be enhanced with wild herbs and honey, and varied occasionally with the addition of meat from hunting in the mountains.

From the forests came timber for building houses and ships. The mountains and hills provided limestone, schist, and gypsum for building purposes, and stones such as serpentine and breccia which could be turned into household vessels. Clay for building and pottery was dug from the plains. The principal lack was mineral resources.

The island's natural resources met most human needs. Its fertility and its position marked it out for attention by early humans, and sustained a culture sufficiently prosperous to be able to supply its missing raw materials through trade.

## A FLOURISHING SOCIETY

The earliest settlers appear to have reached the north coast by sea from Turkey before 6000 BC. They settled at Knossos as farmers. By 2500 BC Crete had a number of permanent settlements, especially in the central region. Some of these already exhibited the "agglutinative" character of later Minoan architecture, with rooms added apparently haphazardly to the original structure: one, at Fournou Korifi consisted of some 80 rooms. Such a strong community feeling is also exhibited in the early establishment of sizeable villages and the use of communal tombs. The natural resources of Crete provided the basis for the prosperity of the people, and their skills allowed them to exploit these: their buildings, pottery, gold jewelry, ivory and stone seals, and stone vases illustrate this, and also point to the importing of new ideas and raw materials not available in Crete.

The high point of Crete's early history was the period between about 1900 and 1450 BC. The Minoans—the people of Bronze Age Crete—had already expanded northwest to Kythera; and by 1600 BC their pottery was being exported to Cyprus, Syria, and Egypt; in return they acquired techniques of fine metalworking, and exported metal vessels to Syria. They also spread into Rhodes and other Aegean islands and mainland sites further north, including Miletus, Kos, and Samos, possibly to relieve overpopulation in Crete and to guarantee supplies of raw materials.

The most striking remains in Crete are those of the great Minoan palaces, especially Knossos, Phaistos, and Mallia. They were first built about 1900 BC, destroyed by earthquake around 1625 BC, and immediately rebuilt on a grander scale. In their final form, each was built around a large rectangular courtyard. The palaces provided for court life and the administration of wide territories; for workrooms for craftsmen who served the needs of the palace and produced a surplus for export; for large-scale storage of food, which was needed for the considerable palace population but was probably also part of Crete's export trade; and for the religious requirements of

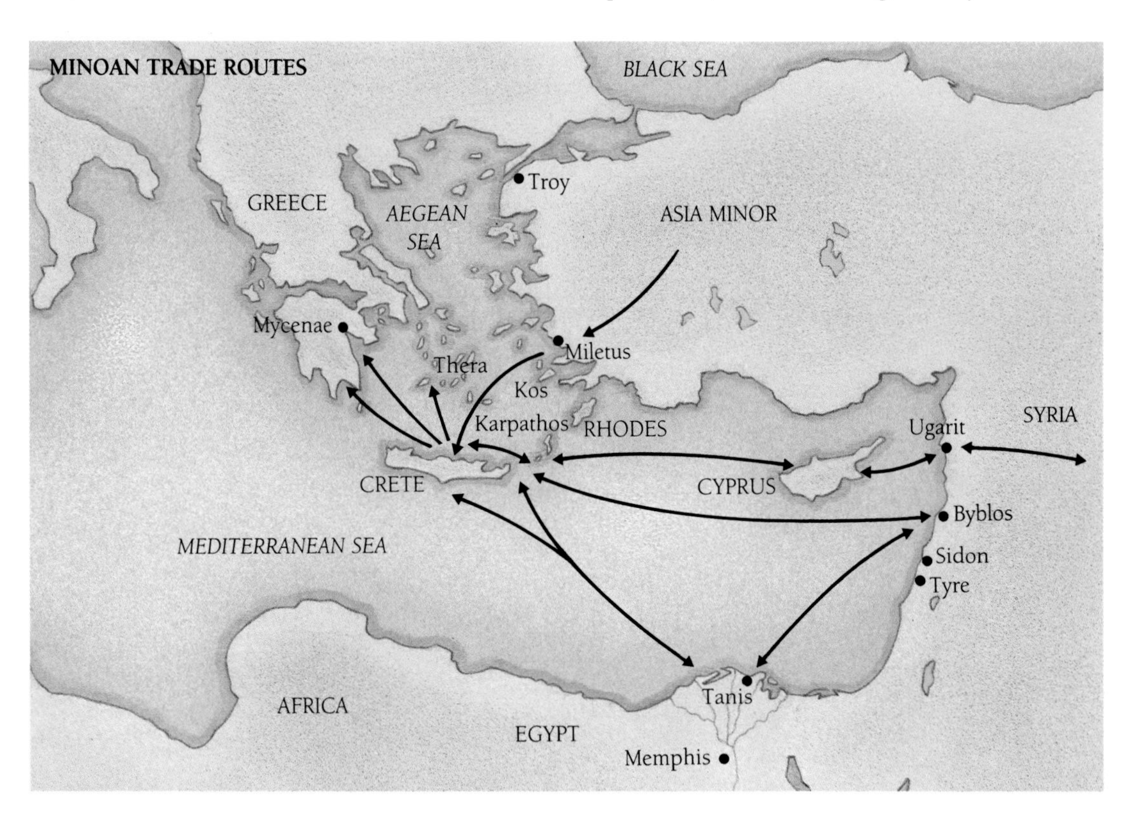


▸ The Minoans were among the world's earliest maritime civilizations. In vessels powered by sail and oar, they mastered the Mediterranean, and ensured that the Aegean Sea was a safe highway for trade. The circular route was dictated by the prevailing northerly winds.

J. Brun/Explorer

▲ The brightly colored fishing fleet of Oulanda in Crete. Small fishing boats operating in nearby waters are an important source of seafood for both local and export markets.

the community. The scale of the buildings, with their lightwells and imposing staircases, the engineering skills, the care and quality of the workmanship all bear testimony to the society that produced them. The construction of such large and complex structures as the palaces indicates a high level of prosperity, and either close political cohesion or centralized authoritarian control.

Minoan art, especially the palace frescoes, provides insights into the daily life of the palaces. Human activities are represented in the processional frescoes of ceremonial gift-bearers and in the packed crowd scenes at public ceremonies. Animal frescoes tell us not only about the animals depicted—monkeys, cats, bulls, partridges, doves, and dolphins—but also about the culture, suggesting a delight in nature which is also represented in the floral motifs and marine decoration used on Minoan pottery.

## THE END OF MINOAN CRETE

The end of Minoan civilization came in two stages. In about 1450 BC the palaces (except for Knossos) and country houses were destroyed by fire. Some were later resettled on a smaller scale, but Minoan social order had apparently collapsed and the population declined. Knossos continued as the principal center, but with new people in charge: they were soldiers (to judge from their burials and their lists of military equipment), which was new in Crete; and they spoke Greek.

The destructions may have been caused by earthquake, or (less likely) by the effects of the volcanic eruption of the nearby island of Thera; but the decline of Minoan power in the Aegean and the occupation by Greek soldiers suggests invasion by the mainland Mycenaeans whose culture the Minoans had influenced strongly. They left records of their administration in the clay tablets preserved when the palace at Knossos was destroyed by fire around 1375 BC—the final stage in the Minoan collapse. The reasons for this destruction are unknown—yet another earthquake, perhaps—but the effects were clear. The Mycenaeans seem to have been displaced; Minoan culture continued on a reduced scale; and the former prosperity of Crete was not recovered until the period of the Roman occupation a thousand years later.

In the subsequent centuries Crete's position as a link between west and east and its prosperity made it a prized possession, contested variously by Saracens, Byzantines, Genoese, Venetians, Turks, and Greeks. Crete was finally united with Greece in 1913.

GRAHAM JOYNER

# The Pacific Islanders

The popular view of the Pacific islanders' lifestyle—perpetual ease amid great abundance in a perfect climate—is impossible to withdraw from general currency, because although we recognize it as fiction, it is nevertheless a fiction we would all like to experience. The truth, of course, is that island life is diverse, and that islanders face the same complexity of pressures and problems as communities anywhere.

## THE COMING OF THE ISLANDERS

Once, nobody lived in the islands: people found their way to them, over time, by various routes. Just where they came from, and when, and the patterns of their distribution, are uncertain, although archeological, linguistic, and other scientific studies are gradually finding answers to these questions. There is wide, if not unanimous, agreement that there was no massive migration, and that the movement into the Pacific islands flowed from the west, through New Guinea, to the east and the north.

New Guinea and Australia were once part of a greater Australian continent. At least 40,000 years ago people had found their way there from Asia, by way of Borneo and Indonesia. It took thousands more years before they pushed further eastward from New Guinea and its associated Melanesian islands into the true insular Pacific. After they had settled Melanesia, others moved into the Micronesian and, finally, the Polynesian islands. There is evidence to suggest that the last wave of expansion was swift, spreading out from New Guinea less than 4,000 years ago.

Great stretches of sea separate the Polynesian and Micronesian islands from Melanesia, and the distribution of people throughout such a huge expanse of ocean was expedited by accidental drift voyages and planned voyages by bold Polynesian navigators in impressively seaworthy craft. Drift voyages are not uncommon today: islanders fishing beyond the reef are sometimes carried away by currents or storms, turning up on islands hundreds, even thousands, of kilometers away many weeks later.

Although the route east from New Guinea was the major one, undoubtedly some people found their way to the Pacific islands from other directions, particularly from South America (a probability that has been reinforced by several modern raft voyages).

▼ Studies by archeologists, linguists, anthropologists, and ethnobotanists have established the broad pattern of settlement in the Pacific. Movement through the island groups was not always sequential, and some stages and dates are tentative.

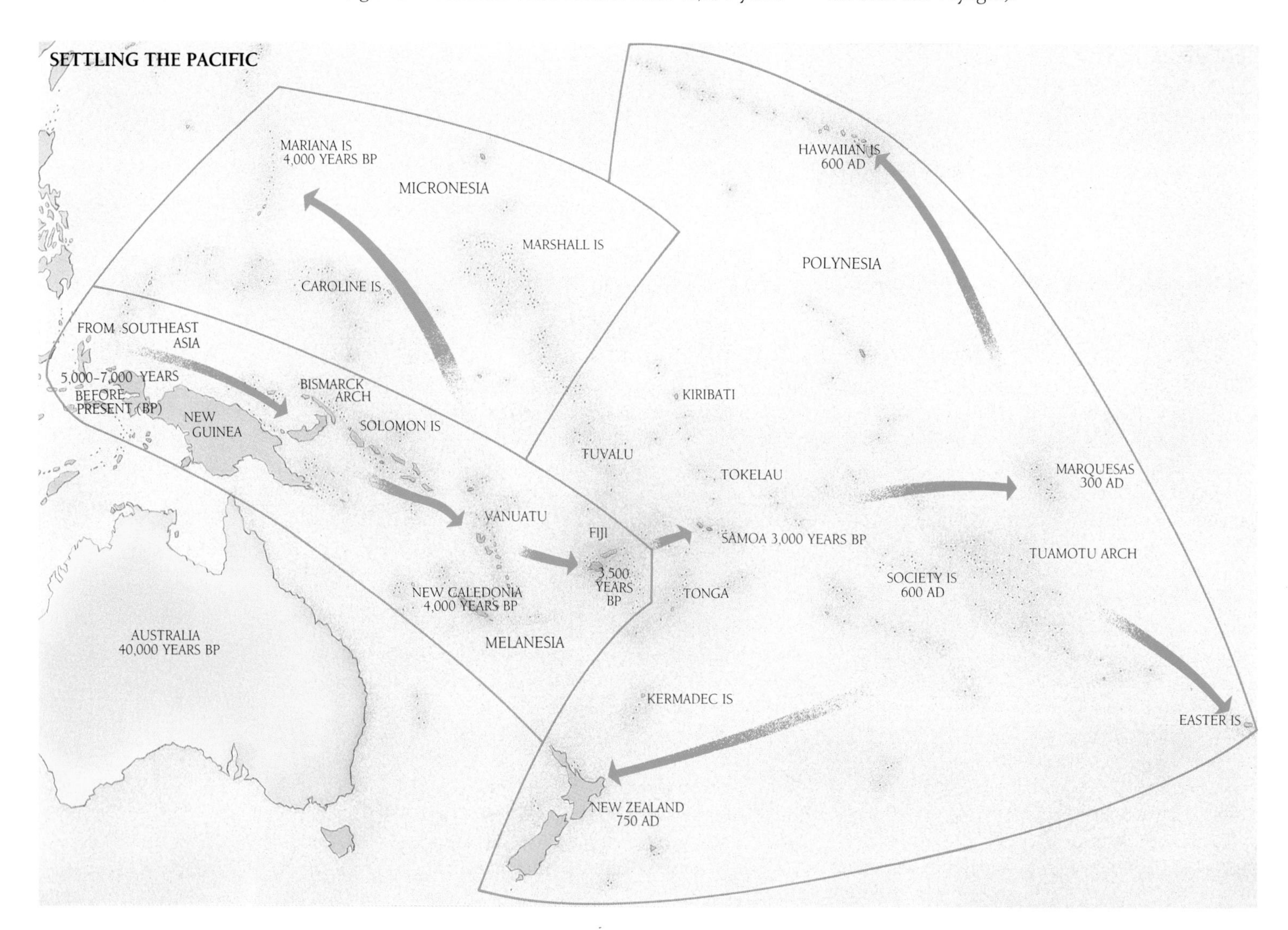

Mitchell Library, Sydney

▲ Captain James Cook's three voyages of exploration in the Pacific captured the imagination of the English public, and prints such as this one, which depicts the *Resolution* and *Adventure* in Papetoai Bay, Moorea, in 1777, were in demand, even though this artist, John Cleveley, was never in the Pacific. Cleveley's brother, James, was a carpenter on the *Resolution*.

## RACIAL GROUPS AND LANGUAGE

Melanesians, Micronesians, and Polynesians are still the main island racial groups, occupying three broad areas of the Pacific. Language, facial features, and skin pigment help to distinguish them. The pigmentation of the people of Melanesia (meaning " black islands") is darker than that of the Micronesians ("small islands") and Polynesians ("many islands"). Of the three, the Melanesians generally have the shortest stature and the darkest skin; the Micronesians are taller, their skin lighter; and the Polynesians are the tallest with the lightest skin.

However, for those not concerned with the specialized study of *Homo sapiens*, the division of Pacific island races into Melanesian, Micronesian, and Polynesian is probably no more than a scientific footnote of no great practical relevance. The geographical and racial divisions are blurred and there are many exceptions to the rule. Long-established racial pockets are found in islands thousands of kilometers from their "legitimate" groups. Islander characteristics are also being influenced by modern migrations from outside the Pacific, and, especially over the past century, by intermarriage.

Between them the islanders speak more than 1,200 distinct languages; more than 750 are spoken in Papua New Guinea alone. The Pacific's languages belong to either the Austronesian (Malaya-Polynesian) or the Papuan families, most being Austronesian. English is widely spoken in all island groups except the French possessions, where French is the official language. In Melanesia several versions of pidgin are also used, in some cases as the speaker's first language.

## EUROPEAN DISCOVERY AND COLONIZATION

The European discovery of the Pacific came only in comparatively recent times. Not until 1521 did Ferdinand Magellan make his first Pacific landfall, in eastern Polynesia's Tuamotu Archipelago. He sailed on westward without adding greatly to the world's knowledge of the Pacific islands or their people, for he saw few of either. Gradually, over the next three centuries, others gathered information about the Pacific: a notable body was recorded by Captain James Cook during his three voyages of exploration between 1768 and 1779.

Having charted the coasts and plotted the islands, from the beginning of the nineteenth century the Western powers began to extend their influence to the people themselves. Missionaries were in the vanguard of the invasion, introducing Christianity in place of the native religions that paid homage to many gods. The first Protestant missionaries, an inexperienced group of well-meaning people brought together in England by the newly formed London Missionary Society, established missions in Tahiti and Tonga in 1797.

Peter Hendrie

▲ With the arrival of the first missionaries as early as 1797, Christianity quickly spread through the Pacific islands, and religion remains a strong unifying influence. Fine churches (such as the Vaiusu Catholic church on Upolu, Western Samoa, shown here) are a feature of most island groups.

Christianity is today the major religion of the Pacific islanders.

Missionaries were followed by traders and by government. At the end of the nineteenth century, and the beginning of this one, the Western powers divided control of the island groups between themselves, often squabbling or horse-trading over particular islands, but rarely inviting the islanders to participate in the administration and economic exploitation of their lands. The European occupation of the islands was motivated by self-interest: either to control the islands' resources or to hold them as tactical bases against potential enemies and rivals.

France took possession of Tahiti and New Caledonia, and shared administration of the New Hebrides with Britain. Britain acquired Fiji and its islands, parts of New Guinea, the Solomon Islands, the Gilbert and Ellice Islands (including the phosphate-rich Ocean Island), Tonga and part control of the New Hebrides. Germany claimed parts of New Guinea, the major Samoan islands, Nauru and most of the Micronesian islands. The United States took Guam, the Hawaiian Islands, and those Samoan islands Germany had not occupied. New Zealand acquired the Cook Islands and Niue.

## PROBLEMS OF COLONIZATION

In the island carve-up, traditional tribal or even racial boundaries were often ignored for the sake of administrative or political convenience. The Gilbert and Ellice Islands, for example, were administered by Britain as a single colony although the Gilbert people were Micronesian and the Ellice people Polynesian. The great New Guinea mainland, the western half of which was already claimed by the Dutch, was further divided by drawing a line across the middle: everything south of the line went to Britain, and everything north to Germany. The Samoan islands were inhabited by one people but split between two powers. Britain and France formalized their joint claim on the New Hebrides with an odd treaty that gave them both control, yet enabled each to colonize the islands in its own way. The arrangement lasted for more than 70 years. Such arbitrary divisions were later to create many problems for the islanders.

More seeds of late-flowering dissension were sown when the powers introduced immigrant races: Europeans, Japanese, Chinese, and others from the Asian mainland and Indonesia. Many came as permanent settlers, others as imported laborers for agricultural and mining projects. For 40 years France used New Caledonia as a penal colony for convicts transported from France.

By the time Fiji became independent in 1970, more than half its permanent population was descended from the Indian workers who had been imported by Britain to work the canefields. Concern by the indigenous Fijians that they would lose control to the immigrant population resulted in a coup by the Fiji military in 1987, the overthrowing of the constitution, and restrictions on the political rights of Fiji-Indians.

## TOWARD INDEPENDENCE

The end of the First World War brought the first major change in the administration of the islands. Having lost the war, Germany was stripped of its Pacific island possessions, which were distributed among Australia, New Zealand, and Japan. More than half of New Guinea and its islands, most of the Micronesian islands, the phosphate island of Nauru, and the Western Samoa group became "mandated" territories, administered by one of the powers on a mandate from, and on behalf of, the League of Nations. In 1933 Japan withdrew from the League and thereafter treated its mandated islands as its possessions.

The end of the Second World War saw further significant changes in island rule, particularly the expulsion of Japanese settlers and administrators. The League of Nations was replaced by the United Nations Organization, and the mandates became "trusteeships" with the United States taking control of the Micronesian islands formerly administered by Japan. Australia made the separate entities of Papua and New Guinea into an administrative union and ruled them as the single territory of Papua New Guinea. France declared Tahiti and New Caledonia to be an integral part of the French republic, and populated New Caledonia with expatriates from Europe and people from other island groups, to the extent that the local Melanesians became a minority of the population.

The trusteeship system, which required the administering powers to make regular reports to the United Nations on conditions in the islands and to admit independent visiting missions on inspection tours, provided a climate in which the islanders could see that their future lay in taking control of their own affairs. This growing political awareness was assisted by a sense of fellowship shared by their membership of the South Pacific Commission, which the powers established in 1947 to enable them to discuss and perhaps resolve common economic, technical, and social concerns. Although the commission did not (and does not) involve itself in political affairs, selected islanders took part in its conferences, and many became leaders in their islands. From widely scattered groups of people isolated for thousands of years, speaking their own languages, the island populations had come together in the space of only a few years and saw that they had common interests and problems.

## INDEPENDENCE

The move to independence in the 1960s and 1970s was the most significant change in the Pacific islands since the nineteenth-century scramble for ownership. Decolonization occurred without the conflicts and bloodshed that typified Europe's rush for possession. This comparatively peaceful changeover was as much a tribute to the negotiating skills of the islanders as to the more enlightened policies of the colonial powers.

First to gain independence, in 1962, was Western Samoa, followed by Nauru in 1968, Fiji in 1970 (the year Britain also handed back full

▼ The sugar industry is a vital contributor to the economy of Fiji, and its success has largely depended on the work of farmers descended from indentured laborers brought from India by the British at the turn of the century.

Julia Brooke-White/Photo Researchers

control of its own affairs to Tonga), Papua New Guinea in 1975, the Solomon Islands and the Ellice Islands (renamed Tuvalu) in 1978, the Gilbert Islands (renamed Kiribati) in 1979, and the New Hebrides (renamed Vanuatu) in 1980. Other islands gained autonomy or internal self-government in varying degrees, and the political development of the Pacific continues today. All the independent states retained associations with their former political masters, and most continue to receive financial aid from them.

An important consequence of independence was that it enabled the new states to reclaim control of land that had been alienated by the colonial governments. On Vanuatu's independence day in 1980, for example, the constitution of the new republic automatically returned all land "to the indigenous custom owners and their descendants" and Vanuatu's land thereafter has been leased to users. Other newly independent states, such as Papua New Guinea and the Solomon Islands, set about strengthening their land legislation to protect it against alienation by non-islanders.

The independent and fully self-governing states, together with Australia and New Zealand, are members of the South Pacific Forum, which was created in 1971 by the island leaders to enable them to take concerted action on any political, economic, or social issue. The forum's agencies oversee for island countries offshore fishing rights, general trade, and a regional shipping service. Decisions are made by consensus, the chairmanship rotates among the leaders, and the organization has developed into an effective regional power bloc.

## POLITICAL ISSUES

Many island governments complain that they have inherited a swollen public service being paid comparatively high salaries for unproductive paperwork. The centralized bureaucratic system established by the colonial governments was designed as an instrument of administrative control rather than development. It is often too expensive to operate, too unwieldy to be effective. The islanders must find ways of dismantling the machinery with minimum disruption, and replacing it with decentralized decision making more in tune with the island style of consensus. Decentralization, however, has problems of its own: with the weak resource base typical of the island economies, capital expenditure and trained personnel may be spread so thinly as to be ineffective.

The political structures installed by the colonial powers have also brought problems for the newly independent states. The party system is perceived as a system of adversaries, unsuitable for small states where politicians come from close communities and heated political campaigns can cause deep personal divisions. Since independence in 1970 the Papua New Guinea parliament has had difficulty keeping any government in power for long, simply because its politicians do not have the same unyielding loyalty to parties that is common in many Western systems, and frequently change sides.

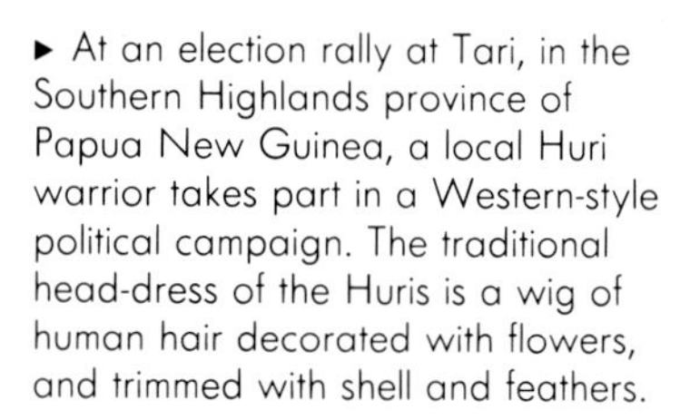

▸ At an election rally at Tari, in the Southern Highlands province of Papua New Guinea, a local Huri warrior takes part in a Western-style political campaign. The traditional head-dress of the Huris is a wig of human hair decorated with flowers, and trimmed with shell and feathers.

Jean-Paul Ferrero/Auscape

Nor do the voters offer them much reassurance, for at election time invariably about half of all sitting politicians lose their seats. The development of consensus style, coalition governments may be more suitable for the Pacific islands. The islands will, in time, find solutions in their own way, for many governments are only just emerging from their post-independence honeymoons and examining the alternatives.

ECONOMIC CONCERNS

The dilemma faced by the Pacific island states is that political independence does not necessarily equate with economic independence. What the forum leaders understand only too well is that while their islands have the same economic problems as larger, more developed countries, the challenges are greater because of their isolation, smaller populations, and smaller size. The islands are isolated from continents and remote from frequent shipping routes and world markets. They have fewer natural resources and are usually dependent on one mineral or agricultural product for their export income. They have to import many of their consumer goods, often much of their food, and all their fuel. Trained professional and technical people, and competent business and administrative staff, are in short supply.

Many Pacific islands are exposed to natural disasters such as typhoons, which not infrequently wipe out crops, destroy buildings, and cause major economic setbacks from which it is difficult to recover. Because the islands are small, their infrastructure and communications—shipping, air transport, and telecommunications—are expensive to establish and operate, and are often dominated by richer, developed countries with stronger resources and business networks and easier access to the global marketplace.

For the most part, Pacific island economies are agriculture-based, with minerals important only in New Caledonia and Papua New Guinea, and to a lesser extent in Fiji. Nauru's rich phosphate deposits, worked since the start of the century, are now nearing exhaustion; those on nearby Ocean Island, exploited by colonial powers for more than 70 years, were exhausted some years ago. Island agricultural exports include sugar, copra, timber, coffee, and cocoa—all of which are susceptible to storms and typhoons, disease, and the vagaries of the world market. Offshore fishing and tourism are of growing importance to many island economies.

Pacific populations are increasing faster than those of the developed countries. While it is impossible to generalize when discussing islands as disparate in size as Papua New Guinea (462,000 square kilometers/178,000 square miles) and Tokelau (12 square kilometers/4.6 square miles), the lack of work in the villages is creating pressures as people move to the towns. There they meet fierce competition for jobs and, without village support, often end up in crowded urban squatter settlements.

Claude Coirault

▲ Transportation on and between many Pacific islands can be primitive—partly because of the remoteness of the islands themselves, and partly because of economic hardship. At Taipivai on Nuku Hiva in the Marquesas an open boat transports copra (the main cash crop) down the river to the coast.

The pressures of island life, exacerbated by rising living costs and falling living standards, have led thousands of island citizens to migrate to those countries that will take them—notably New Zealand, Australia, the United States, and Canada. There are large Micronesian, Samoan, and Tongan populations in the United States; New Zealand is home to numbers of Samoans, Tongans, Cook Islanders, and Niueans. Inevitably, it is the islanders with skills who are accepted as migrants. The island economies, and the relatives left behind, benefit from the money sent back by the expatriates, but their skills and labor are lost. There are more American Samoans, Cook Islanders, and Niueans working abroad than there are left at home. Only the continued outflow keeps the populations in many islands constant.

The major struggle faced by island countries is to find ways of strengthening their economies. They are aware of the dilemma posed by the fact that, while politically independent, they require help to sustain their development. Large amounts of outside financial aid are accepted by island states every year, for without it their problems would be even greater. But they prefer trade to aid, and some are voluntarily reducing their aid packages in an effort to live within their means and be truly independent.

Although many have nothing to trade, not even destinations for the tourist dollar, all islanders cherish a spirit of self-reliance, and it is this that should carry them through the difficult times ahead. They know it is too late to go back to the village: that, like it or not, they have become part of the global economy.

STUART INDER

Nancy Durrell McKenna/The Hutchison Library

▲ Pulled from the water at low tide, these traditional open boats make a colorful addition to the shores of a Madeiran fishing village. Fishing plays an important role in the lives of those who inhabit coastal settlements such as this.

# THREE ATLANTIC ISLANDS

The Atlantic Ocean, in all its magnitude and majesty, did not become part of European consciousness until the Columbian voyages. Hence 1492 was a turning point in world and navigational history.

## THE EUROPEAN WORLD-VIEW

Europe had been accustomed to look southward and, even more, eastward; its marine center of gravity had been the Mediterranean Sea, not the Atlantic Ocean. Only after Columbus did Europe shift from what one writer has called a "thalassic" (sea-oriented) to an "oceanic" (ocean-oriented) view of the world. Only after 1492 can we speak of planetary empires, spanning whole oceans.

For Europe before Columbus, "the East" had always lain eastward, not westward; travel, whether overland by the Silk Route or by sea to the eastern Mediterranean, meant eastward travel. The Catalonians, the Amalfitans, the Genoese, the Venetians had all traded and sought to colonize in the eastern Mediterranean. Though many of Europe's scholar-sailors had conjured with the possibility of sailing westward to reach the East, before Columbus none had attempted it seriously. But massive alterations in the politics of empire to the east, cutting Europe off from prized substances, forcibly turned its attention afresh to the Atlantic.

The measure of the Atlantic was taken only in the course of several centuries. It turned out to be an ocean of unimagined vastness; not surprisingly, the islands within it were enormously varied. They are so many, and so scattered, that the histories and peopling of only a few can be described here; nor can these few be thought fully to represent all the others. Still, by talking of three island groups in particular, some sense of the grandness of the Atlantic itself may be conveyed.

## THREE ISLAND GROUPS

These three island groups are Madeira, Barbados, and St Pierre-Miquelon. All islands, they differ greatly otherwise. Madeira became an immensely important sugar producer for Europe, long before Columbus. Barbados, two centuries later, was Britain's first "sugar island" and a classic instance of British tropical colonialism. St Pierre and Miquelon, France's oldest and smallest colony (and its only "white colony"), still maintains ways of life of a more familiar north Atlantic sort.

The very location of these tiny islands, at enormous distances from each other, conveys some sense of the vast ocean in which they lie. The Madeira group, believed to be the same lands once identified by Pliny as the "Purple Islands", were uninhabited when discovered (or rediscovered; probably by the Portuguese mariners Vaz and Zarco in 1419). They were claimed by Prince Henry the Navigator, and their settlement was undertaken the following year. Lying about 645 kilometers (400 miles) northeast of Morocco and 965 kilometers (600 miles) southwest of Lisbon, at 32°40'N, these specks of land are at about the same latitude as the island of Bermuda, a British insular possession off the coast of the United States—and about 4,830 kilometers (3,000 miles) east of it. Other than several barren, uninhabited rocks, Madeira consists of two islands, Porto Santo (43 square kilometers/17 square miles) and Madeira proper (794 square kilometers/307 square miles). Today they constitute an autonomous region within the Portuguese state.

Barbados is among the most ancient of all overseas English colonies. It is a non-volcanic oceanic island, the easternmost of the Caribbean chain, somewhat isolated from the archipelago stretched north–south from eastern Puerto Rico to Trinidad and the South American mainland. Barbados' nearest neighbor, St Vincent, is 160 kilometers (100 miles) to the west. At approximately 13°N 59°W, Barbados is more than 4,845 kilometers (3,000 miles) west and south of Madeira, and nearly 3,508 kilometers (2,175 miles) south and east of St Pierre–Miquelon.

Unlike Madeira, uninhabited before Portugal's first settlement in 1420, Barbados had a lengthy pre-Columbian history. Though the Portuguese explorers who discovered it in 1536 found no one living there, archeology indicates that Barbados was populated at least as early as the start of the Christian era, by Amerindian settlers coming from the South American mainland. Around 1000 AD, Arawakan-speaking migrants, also from South

The Hutchinson Library

▲ Pushbikes and donkey-drawn carts fight for space among more modern vehicles in Bridgetown, the capital of Barbados. This is one of the world's most densely populated countries, and poverty and high unemployment make these cheap forms of transport a practical necessity.

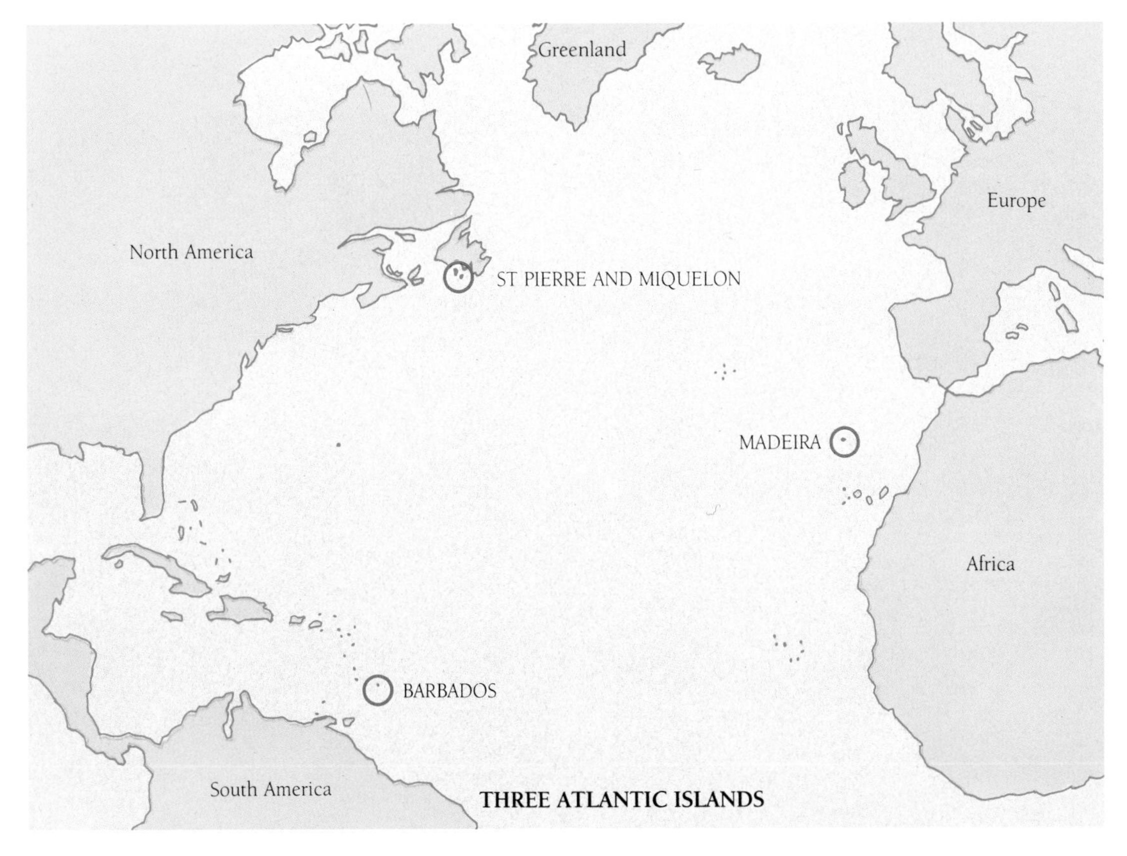


◀ The islands of the Atlantic are many and scattered. The three selected for discussion, geographically distant from each other and each with a quite distinct history, indicate something of the diversity of the Atlantic islands.

America, settled there. But in 1627, when claimed for the English crown, Barbados was uninhabited. A tiny island (430 square kilometers/166 square miles), by the time it achieved independence from Britain in 1966, Barbados had become one of the most densely populated nations in the entire world.

The colony of St Pierre-Miquelon consists of three islands: St Pierre, Grande Miquelon, and Langlade or Petite Miquelon, as well as some rocks and cays. Langlade and Petite Miquelon are joined by a sandbar. The total area of this ancient colony is about 240 square kilometers (93 square miles). It early won a sinister reputation, so dangerous was it to shipping: "the graveyard of the sea", many called it. The islands are separated from Newfoundland's south coast by a 24-kilometer (15-mile) channel near the Gulf of St Lawrence. As early as 1504, Breton and Norman fishermen seem to have frequented the islands. But they were not settled by the French until a century later. Located at 46°45'N 56°W, St Pierre-Miquelon forms the apex of the vast triangle we are drawing, and differs in almost every conceivable way from Madeira and Barbados.

One source of difference is physiographic and geological. Barbados, the vestige of a sunken continent, has a rolling surface, built up out of coral limestone, and heavily cultivated today. Mt Hillaby, the highest point in Barbados, is only 340 meters (1,115 feet) above sea level. Volcanic in origin, Madeira is strikingly different; there is hardly a level surface anywhere. The land is described as "crumpled" and its highest point (Pico Ruivo) towers to 1,830 meters (6,000 feet). Again different, St Pierre and Miquelon are the remnants of the Appalachian mountains, worn down by glacial movement. The coast is cliff-lined, the greenery is scant; there is fog nearly year-round.

## THE COLONIAL LEGACY

Not only do these islands differ in location, in terrain, in climate. They also represent different times and different imperial designs. Madeira's career as a sugar-producing Portuguese colony was brilliant but brief: initiated around 1450, it was diminishing, because of American (especially Brazilian) competition, a century later. Slave labor, which would figure so vitally in the new-world sugar industry, never acquired much importance in Madeira.

By the mid-sixteenth century, when Barbados had been launched upon a sugar career, the sugarcane in Madeira was being replaced by vines, and Madeira wine began to gain its

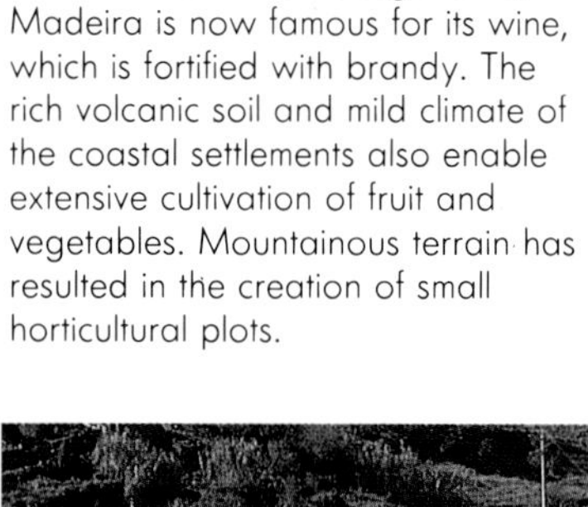

▼ Once known for its sugarcane, Madeira is now famous for its wine, which is fortified with brandy. The rich volcanic soil and mild climate of the coastal settlements also enable extensive cultivation of fruit and vegetables. Mountainous terrain has resulted in the creation of small horticultural plots.

A.G.E. Fotostock

E. de Malglaive/Explorer

▲ St Pierre is the administrative and commercial center of the stark, damp islands in the St Pierre and Miquelon group. Most of the forests were cleared long ago for fuel and much of the remaining landscape consists of peat bogs; consequently, fishing is the chief occupation of the people.

international fame. Never a vital link in Portuguese commerce or industry, in the twentieth century Madeira has become, most of all, a picturesque and balmy tourist haven.

Barbados, an important stepping-stone in Britain's American empire, was first settled as a small-farm colony. But in the subsequent two decades it changed—first gradually, then swiftly—into a premier plantation colony and sugar island. As more and more land was bought up to form large estates for producing sugar, the population changed from one of European colonists and smallholders to a few, mostly absentee, plantation owners and masses of enslaved Africans. As sugar mounted in importance in Europe and markets grew, so its production became commonplace in the Caribbean. Barbados soon was eclipsed by Jamaica and felt the competition of other islands. Since the end of the eighteenth century, all the Caribbean "sugar colonies" have lost ground, and Barbados may soon produce no sugar at all. Its economic alternatives, however, are meager: tourism will probably be most important in the island's future.

To these island societies, remnants of earlier imperial designs, St Pierre and Miquelon pose a striking contrast. Though some French fishermen had settled as early as 1670, it was the expulsion of the French from Acadia (Nova Scotia) that gave the colony its first stable settlement. Anglo-French rivalry repeatedly disturbed island life; taken and retaken in the ensuing half-century, the islands were finally left to the French in 1814, with the understanding that they would not be fortified as defence bases.

While whale fishing was important in the early years of the colony, cod fishing then emerged as the mainstay of the local economy. In the twentieth century, the fishing industry has had to move more and more in the direction of freezing the catch, preparing fishmeal from the leavings, and exporting both products. The larger part of the catch is landed by trawlers; an increasingly minor proportion is taken by small boats.

DISTINCTIVE IDENTITIES

From their differing locations and histories, these representative Atlantic islands are also quite distinctive unto themselves. Madeira may be thought of as an island of the first westward expansion: it became part of European awareness before the new world was discovered by Europeans. Barbados is an island of the second westward expansion: it marked the century when the north European states plunged into the world sugar trade. St Pierre and Miquelon were claimed by France only a few years before Barbados was claimed by Britain; yet their history has been strikingly different. So it is that each of these islands has its own story to tell.

SIDNEY W. MINTZ

Bibliotheque Nationale, Paris

▲ Before the maritime expansion that characterized the Renaissance, Europeans had little knowledge of societies beyond their immediate region, and imagination often filled the gaps. This medieval painting depicts the "dog-headed traders" of the Andaman Islands in the Indian Ocean bartering with each other.

# PEOPLE OF THE INDIAN OCEAN

The islands of the Indian Ocean lie on a great curve stretching from Africa to Southeast Asia around the vast emptiness of the ocean. Seaborne human settlement of the islands began at least 40,000 years ago and continued into the nineteenth century, when the last deserted islands were settled. Since earliest times, the history and lives of island people have been shaped by the unique monsoonal wind systems of the Indian Ocean, and by the rhythms and patterns of maritime trade.

## SETTLING THE ISLANDS

Between 40,000 and 50,000 years ago, in what may have been one of the first seaborne migrations, hunter-gatherers sailed across narrow seas in Southeast Asia to begin the settlement of New Guinea and Australia. Similar migrations occurred in the following millennia as seafarers unraveled the secrets of the monsoonal wind systems and were thus increasingly able to use the ocean for transport and trade.

At least 10,000 years ago skilled voyagers moved out of southern China into the Malay peninsula and the Indonesian archipelago. About 2,000 years ago some of their descendants sailed west to uninhabited Madagascar via Sri Lanka and the Maldives, both of which had been settled from India. By this time the largest Indian Ocean islands were inhabited. Some of the smaller islands had also been settled: the Andamans and Nicobars by hunter-gatherers who were to remain isolated until the nineteenth century; Suqutra from Arabia and Africa; and the coastal islands of eastern Africa by Bantu-speaking African peoples who also settled the west coast of Madagascar.

The importance of the ocean to these island settlers varied enormously. On large islands such as Madagascar, Sri Lanka, Sumatra, and Java, the lives of most people revolved around the rhythms of agriculture, and only small groups of fisherfolk and maritime merchants were affected by the ocean. For the inhabitants of many smaller islands, however, the ocean was an ever-present influence. Throughout much of the Indonesian archipelago, in the Maldives and Laccadives, and along the coast of Africa, island people were dependent upon the sea as a source of food and

▶ The saffron robes of these Sinhalese Buddhist monks in Sri Lanka are symbolic of the actions of the historic Buddha—a prince who gave away his clothes and possessions. To hide his nakedness he gathered up earth-colored rags from the rubbish; to the monks the saffron of their robes represents this earth color. In accordance with Buddhist teaching (monks are not allowed food or possessions for more than one day), these men must go out into the streets every morning, begging for alms.

Colour Australia

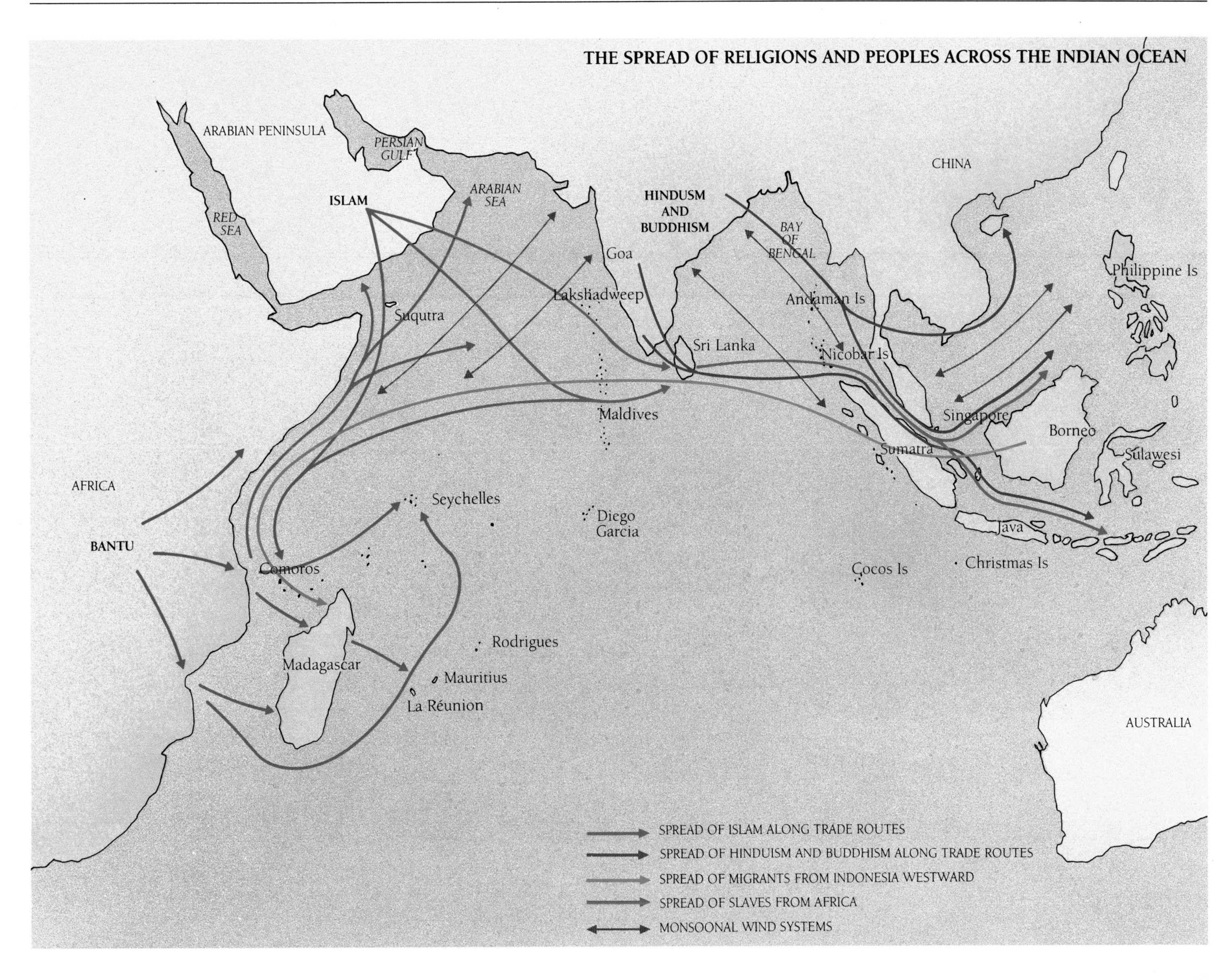


▲ The islands of the Indian Ocean have absorbed a diversity of religious and cultural influences as waves of settlement, conquest, and colonial exploitation merged with the patterns of maritime trade.

as the only highway linking scattered communities. The monsoon winds were vital to their existence. They enabled swift and regular passage by sea and underpinned a complex network of maritime trade which bound the islands into a larger world.

## MARITIME TRADE AND CULTURAL CROSSROADS

With the development of long-distance maritime trade, many of the islands of the Indian Ocean assumed a new importance as cosmopolitan cultural crossroads. During the first millennium AD Sri Lanka, for example, was a meeting ground for Buddhism and Hinduism from India, and Islam from the Middle East. Similarly the culture of the Indonesian islands was molded by the intermingling of indigenous animism with Hinduism, Buddhism, and finally Islam from south Asia. Such interaction was facilitated by seaborne trade.

Cultural blending also occurred in the Laccadive and Maldive islands, which were at the junction of maritime routes linking southern Asia, the Middle East, and eastern Africa. Some 2,000 years ago these islands were influenced by south Asian Buddhism and Hinduism. By the thirteenth century, however, their mercantile links with the Middle East had grown to such an extent that Islam displaced both of these earlier influences. During the same period many of the offshore African islands, from Suqutra to the Comoros, converted to Islam. Increasing trade contact with the Red Sea and Persian Gulf encouraged the spread of Islamic religious and cultural values to these islands and to trading settlements along the east African coast.

By the eleventh and twelfth centuries the peoples of coastal east Africa were evolving the unique Afro-Asian Swahili language and culture. The evolution of this culture was in part the result of settlement by Arabs and Persians, but more importantly it represented an African response to external stimulation. In eastern Africa, as in Indonesia, maritime trade opened up island and coastal communities to the outside world, brought prosperity, and changed traditional economic and cultural patterns.

Coo-ee Historical Picture Library

▲ Raffles Place, Singapore, in 1928—famous for its very British architecture and colonial atmosphere. The square was named after Sir Stamford Raffles, the British East Indian Administrator who in 1819 established the port settlement of Singapore. It rapidly became the center of British colonial activity in Southeast Asia.

THE EUROPEAN PRESENCE

By the sixteenth century Muslim seafarers dominated the Indian Ocean. From eastern Africa to the Indonesian archipelago, Islam had taken root among seagoing communities just as Arab and Persian words formed the *lingua franca* of commerce and navigation. From the sixteenth century, however, Europeans—Portuguese, Dutch, British, French, and Danes—began to trade in the Indian Ocean, and in the late seventeenth century the French began the colonization of the uninhabited islands of La Réunion, Mauritius, Rodrigues, the Seychelles, and Diego Garcia. These islands were settled both by European colonists and by their slaves from Mozambique, Madagascar, the Comoros, and India. The slaves, who formed the majority of the population by the mid-eighteenth century, adopted the religion of their masters and developed their own Afro-French language, Creole.

Elsewhere during these centuries, the Portuguese and Dutch left racial, linguistic, and cultural legacies in India, Sri Lanka, and throughout the Indonesian archipelago. However, the extent of their legacy should not be exaggerated. They were few in number; they did not displace local languages although they enriched them; and only in some islands and coastal areas did they add to the existing religious diversity. The greatest changes during this period were felt in Sri Lanka and places such as Goa and Kerala on the west coast of India where Christianity and Portuguese words, names, and music permeated coastal society.

During the late eighteenth and early nineteenth centuries the British became the dominant political and economic power in the Indian Ocean region. They conquered most of the French islands; Dutch Sri Lanka, Capetown, and Melaka; and much of the Indian subcontinent. But the full impact of this development was not felt until the Industrial Revolution of the nineteenth century when the lands of the Indian Ocean became suppliers of huge amounts of raw materials for the factories of Europe.

After slavery had been abolished in the British empire in 1833, plantations across the Indian Ocean were worked by cheap Chinese and Indian labor. Chinese and Indians worked on plantations in Malaya and on the island entrepot of Singapore, which the British founded in 1819 to dominate the maritime trade of Southeast Asia. Indian laborers migrated to the plantations and ricefields of the Caribbean, South America, Burma, Fiji, Kenya, Uganda, South Africa, Sri Lanka, Malaya, Singapore, Mauritius, and La Réunion.

They became the backbone of the tea industry in Sri Lanka and the sugar industry in Mauritius. In Singapore Indian immigrants ranked second only to the Chinese in commerce and industry. On deserted Christmas Island, Chinese and Malay laborers were imported to work the phosphate mine, while on the uninhabited Cocos Islands a semi-independent feudal plantation state of Malays was created by a Scots merchant family under the protection of the British.

## A MELTING POT OF CULTURES

On many Indian Ocean islands throughout the nineteenth century, these new streams of settlers profoundly altered cultural patterns. In the French-speaking Mascarenes, Hinduism and Islam became the religions of the majority. Singapore became a Chinese-dominated society, and in Sri Lanka the Tamil Hindu community grew rapidly in size and distribution. On the Cocos Islands a Malay Muslim community developed, entrapped in a feudal cocoon, while on Christmas Island Malay and Chinese communities coexisted under the rule of a private company.

Elsewhere in the Indian Ocean, island populations were remarkably unaffected by nineteenth-century developments. The northern coast of New Guinea became a protectorate of Germany; the Andamans and Nicobars were partially exploited by the British as penal settlements to serve India; the Laccadives, Maldives, and Suqutra slumbered on as remote Islamic corners of the British empire; under British rule the coastal islands of east Africa continued their age-old role as cultural and economic mediators between Africa and Asia.

In the twentieth century most of these islands have gained independence from colonial rule. Singapore, Sri Lanka, Mauritius, the Seychelles, the Maldives, Madagascar, Papua New Guinea, and most of the Comoros have achieved nationhood. The islands of coastal east Africa have been absorbed into larger mainland states. The Laccadive, Nicobar, and Andaman islands are part of India. The Indonesian archipelago has emerged as a wide-flung state comprising a range of cultures and languages. Christmas Island and the Cocos group are now part of Australia; Diego Garcia remains a British colony; and La Réunion and the Comorian island of Moroni have been absorbed politically into metropolitan France.

## COASTAL COMMUNITIES

Throughout the centuries the cultures of all the Indian Ocean islands have been profoundly affected by interactions stimulated by maritime trade and migration. The same generalization can be made of people living on the Indian Ocean coasts of Africa and Asia. The ocean had its maximum impact on these people where their contact with continental hinterlands was restricted. This was the case along much of the coast of western India and eastern Africa where mountains, climate, and inhospitable terrain impeded intensive movement between the coast and the interior. In these circumstances many coastal people were forced to look outward, and to earn their livelihood through maritime trade

KENNETH McPHERSON

Robbi Newman/The Image Bank

◂ Port Louis, the capital of Mauritius, cannot escape the twentieth-century advertising rivalries between international soft drink companies. The Mauritians however, are also historically recent imports—before the seventeenth century, there was no indigenous population. Now the island is one of the most densely populated countries in the world, inhabited by French, Creole, Indian, and Chinese people.

P. Plisson/Explorer

# PART THREE

Surf breaks on the coral fringing a thickly vegetated Pacific island.

# OF OCEANS AND ISLANDS

# 12 FOOD AND ENERGY FROM THE SEA

Emmanuel Valentin/Hoa-Qui

▲ Small-scale fishing for recreation or subsistence living is a common feature of tropical islands. This Maldivian fisherman is clearly delighted with the large trevally he has caught.

Although the seas and oceans cover about 71 percent of the surface of the earth, the marine contribution to human food supplies is much smaller than that from land. In tonnage, the world production of cereals is as much as 20 times the total of all fishery products; and the total of all agricultural products is probably 30 to 40 times as great. In effect, the production per unit area of the seas is much less than that of land. Nevertheless, the world's oceans are an important source of food—obtained either by harvesting from the wild or by farming. Moreover, the sea provides significant amounts of oil, minerals, and renewable energy, the extraction of which, at a time of increasing concern over depleted resources, is both a technological challenge and, perhaps, a pointer to the future of humanity.

## HARVESTING THE SEA

What is obtained from a resource, as distinct from what might be obtained, is measured in the first instance by the work done to effect the extraction. Therefore it is appropriate that this appraisal of the ocean harvest should begin with a survey of the methods of fishing.

### EQUIPMENT AND METHODS

Taking a panoramic view of the fishing equipment in use at present, and the way it is used, is a journey through time to observe the evolution of technology. Practically every kind of instrument ever devised for collecting, trapping, or catching fish can be found still in use, somewhere in the world: from primitive baskets for collecting shells, to spears, harpoons, and scoops, to dip nets and cast nets, to beach seines, Danish seines, and otter trawls, to steered midwater trawls. They are used by fishermen walking the shore, or standing on rocks, or diving to collect by hand or to drive fish into a net, or working from canoes, small boats, trawlers, and factory ships. In order to work, fishermen make use not only of a platform, namely a boat, but of equipment for navigation, for finding fish, for handling the catch and keeping it in good condition, and for communication.

The administration of these activities ranges from tribal rules through various levels of

◀ The ocean has always been an important and (if well managed) abundant source of food. After a successful day at sea, this commercial fishing boat returns to port in Maine.

Stephen Krasemann/NHPA

◀ Optimistic scavengers, these gulls follow a shrimp boat back to port in Georgia. Until some target fishing practices change, the birds will be well fed. The small meshed nets used to harvest shrimps catch many other fish species, which are often dead by the time they have been sorted and discarded overboard.

Doug Allen/Oxford Scientific Films

▲ Hauling in a cod trawl off Lofoten Island, Norway. There are several groups and hundreds of species in the cod family, all of which are demersal (bottom-dwelling) fish. *Gadus morhua* was the species most commonly fished in the north Atlantic until the fishery collapsed in 1989–90. This collapse was probably at least partly the result of the poor survival of young *Gadus* following the 1983 El Nino year.

governmental regulation to the international provisions of the United Nations Law of the Sea Treaty; these arrangements are based on knowledge ranging from local lore to the results of highly sophisticated scientific research. In addition to this bewildering array of equipment and methods, the fishing industry has reflected since the Second World War the pace of development observed in other fields, such as electronics. Thus, while retaining much of its traditional past it holds a place in the swift stream of modern technological evolution.

The equipment (or "gears") used for fishing are divided into three classes. Passive gears are those to which fish are brought by currents, or by their own movements, or by attraction (by light or bait). They include pots and traps, gill nets and drift nets. Active gears are those that move and seek out the fish: among them, seines, scoop nets and cast nets, trawls and dredges. The third class covers equipment directed against individual fish, including spears and harpoons, clubs and hooks.

## PRACTICAL ECOLOGY

To be a fisherman one must be a practical ecologist. One must know the distribution of each target species: where it will be at particular times (of day, lunar period, season, and year), and when it will be at particular places; one must know something of its life cycle; and one must know its behavior, in feeding, in moving about its ground and between grounds, in reproducing, and in schooling or remaining solitary. This knowledge relates to what we call "fishable stock", that is fish of a size and kind to be caught; but for some species it is necessary also to know something about the juvenile stock so as to avoid fishing on nursery grounds.

In primitive fisheries this knowledge was obtained by direct observation and applied by reading natural signs, such as weather, sea conditions, and bird behavior. In most modern fisheries this knowledge is made more particular and precise by science and technology, and its application is simplified by the use of instruments. But the strategy is unchanged: know the grounds; be on the grounds when the target species will be there at a fishable stage; locate the concentrations of fishable stock; operate the equipment that will take the fish.

Individual fisheries are named for the species

Bruce Coleman Limited

► A community effort—fishermen hauling in a beach seine in Kerala, India. To cast a beach seine, one end is held on the beach, while a boat carrying the net moves out beyond the waves, a short distance parallel to the shore, and then back to the beach, casting the net as it goes. The net is then brought to shore by pulling from both ends.

taken (such as herring or salmon); for the location of the grounds (such as Newfoundland Banks); for the kind of equipment used (such as drift net or longline); or for a combination of two or all three of these descriptors. For an appreciation of world fisheries, however, it is better to locate these individual names within a generalized classification based on dominant ecological characteristics of individual species which largely dictate the methods that can be used. Thus we distinguish surface (or pelagic), midwater, and bottom (or demersal) fisheries. Each of these classes is divided into those of the shoreline, those of the continental shelf, and those of the deep oceanic waters, except that there cannot be a shoreline midwater fishery.

That "all flesh is grass" is as true of the animals of the sea as it is of the animals of land. The "grass" of the seas are the microscopic plants known as phytoplankton. The seas have other vegetation, of course—seaweeds and seagrasses—but the greatest part of marine primary production is effected by phytoplankton. These minute plants are of little direct use to humans: they must be consumed, transformed, and passed on through various links in sometimes long food chains. Fed upon by herbivores, their material then passes to primary carnivores and on to secondary carnivores, from each of which humans take a harvest.

The species that constitute the living resources of the seas are numerous and varied. FAO's *Yearbook of Fisheries Statistics* has a list of more than a thousand species of plants and animals "taken for all purposes . . . except recreational" from "inland fresh and brackish water areas and [from] inshore, offshore and highsea fishing areas". The diversity of these organisms is indicated by the summary table in the Appendix of this publication.

## KINDS OF FISHERIES

There is one further complication in this picture of world fisheries, namely the distinctions between industrial, artisanal, subsistence, and recreational (or sport) fisheries. Once again we have the flavor of history: from the stone age to modern fisheries, from subsistence fishing to the great factory ships. Sport fishing is engaged in chiefly for recreation, even if its catch is taken home to eat, but this is not generally a sustained activity and the catch is not something on which any family depends. Subsistence fishing, in contrast, is a sustained activity carried out in order to obtain food for the fisherman and his family, and perhaps for other members of the community. All these people depend upon this activity, and any sale of a subsistence catch is fortuitous. Fisheries of this kind still operate in various parts of the world, especially from island communities.

J. L. Dugast/Hoa-Qui

▲ One of the more unusual methods of line fishing is employed by these Sri Lankan fishermen perched on stilts. Patience and a good sense of balance are required for success.

A fishing unit is a combination of boat, equipment, and manpower which can operate independently. Artisanal and industrial fisheries differ in the size and ownership of their fishing units. They both deliver their catch to markets, for the fishermen of these fisheries "don't fish for food, they fish for money"; but artisanal fishing is labor intensive, contrasting with the capital-intensive character of industrial fishing. The boat of an artisanal fishing unit is small, generally the equipment is simple and mostly it is hand operated, and the unit is owned by the person who works it; its operations are carried out pretty much according to the disposition of the owner-operator.

David Hiser/Photographers Aspen

◀ In polar regions such as Hudson Bay, where the ice shelf can extend a considerable distance across the sea, it is still possible to fish by dynamiting or drilling a hole through the ice. Polar fish have a sluggish metabolism to survive the extreme cold, so once they are hooked, they put up little fight and are easily reeled in.

Udo Hirsch/Bruce Coleman Limited

▲ In less well developed countries, fishing is labor intensive and can be a family business. Vessels are small and primitive, and the gear is usually simple—like the net used by these Turkish fishermen at Istanbul.

Although the catch is obtained with a view to cash sale, its disposal is arranged through loose arrangements with middlemen and local buyers. In the past most artisanal fishermen were in some sort of bondage to middlemen, who provided credit, sold to fishermen the nets and other materials they required, and fixed the prices at which they took the catch. This situation persists in some places, but with the spread of literacy, and market information available through radio and even television, the hold of middlemen over fishermen is steadily being eroded.

Industrial fishing is generally highly organized in terms of the fishing operations, disposal of the catch, and relations with processing plants and markets. The conduct of the fishing operations is dictated largely by the timetable of markets or the production schedules of processing establishments. The fishing units of industrial fisheries are highly modernized in motive power, navigational instruments, fish-finding equipment, winches, communications equipment, and the machinery needed to handle, store, and process the catch. The boats vary greatly in size, from 15-meter (50-feet) shrimp trawlers to great factory ships more than 100 meters (330 feet) long.

## THE WORLD CATCH

Almost 90 percent of the total reported catch is taken from marine areas. Of the 1987 reported catch of 80,501,200 tonnes, more than 61 percent came from the Pacific Ocean. While some part of the catch is accurately weighed and reported (this is true more especially of the industrial catch), the reports of artisanal and subsistence catch can only be estimates. It may be that exaggerations are cancelled out by omissions, leaving the total more or less correct, but all the figures should be treated as approximations.

The variation in the catch from area to area within each ocean is a result of important ecological phenomena. The size of the stocks of fish from which the catches are taken depends, to a large degree, on the food supply available to each stock, and at base that is the production of phytoplankton. The variation in the richness of phytoplankton depends upon the supply of nutrients to the places where the phytoplankton can live—notably the surface waters down to about 200 meters (650 feet) through which sunlight can penetrate. The reserves of nutrients

▶ Schools of fast pelagic fish such as tuna can be caught in purse seines. The net is released with a buoy or small boat on one end, and the other end is towed around the school to encircle the fish, as shown here. The bottom edge of the seine is pulled tight (pursed) to prevent fish from escaping by diving downward; then the net plus catch is winched on board the fishing vessel.

Planet Earth Pictures

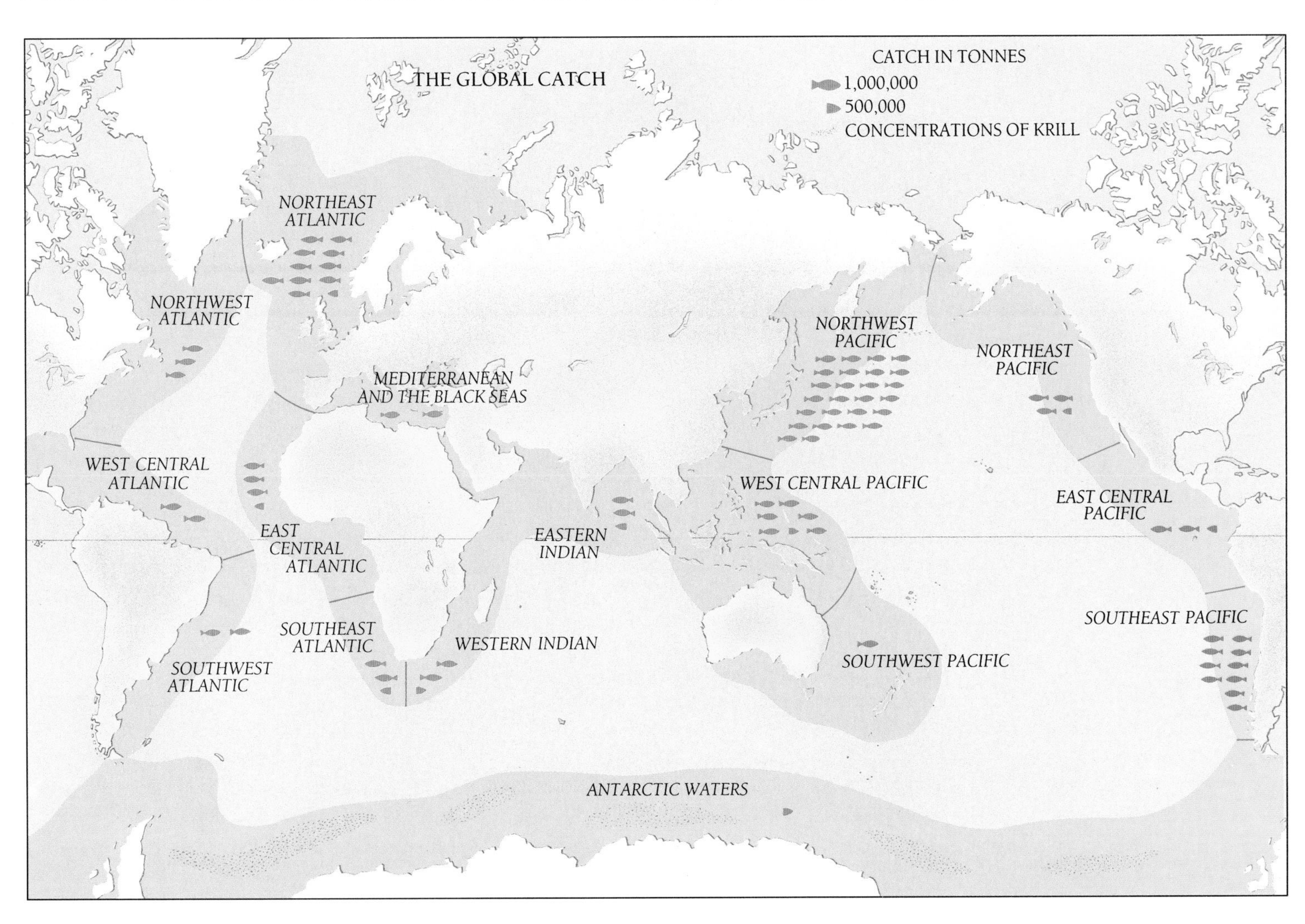


▲ The total global catch of fish, crustaceans, and molluscs (based on FAO data), here divided according to the established marine fishing areas, clearly indicates the importance of the Pacific Ocean in world fishing.

lie on the sea floor, an accumulation and breakdown of excrement and dead bodies cascaded down from overlying water. When material from these reserves is brought to the surface, as happens where deep-to-surface currents flow, there can be rich phytoplankton growth. It is this upwelling and the associated primary production that accounts for the very great stock of Peruvian anchovetta, from which as much as 10 million tonnes of catch has been taken in a single year.

Similar upwellings operate in other parts of the world, but phytoplankton growth is low in great areas of the ocean. In fact, most of the world's catch is taken from only the margins of the seas and continental shelves, many parts of which are narrow and almost desertic in character. The catch from the high seas, chiefly of the highly migratory species—tunas and billfishes—is only a very small proportion of the total. The catch of whales and other marine mammals is, of course, now very low.

It will now be clear that the comparison with agricultural production made at the beginning of this discussion is not as clear-cut as it seems. In the first place, comparison should be made between production from truly productive areas on both sides. Secondly, while some part of the total marine catch, probably less than 5 percent, is the product of aquaculture, the bulk is the result of harvesting. Thirdly, the efficiency of harvesting is still constrained by the inadequacy of the administrative machinery by which it is directed.

## THE PROSPECTS

Pessimistic forecasts of the future of the oceans are commonplace. One frequently hears warnings that the resources of the seas are being so heavily overfished that they are likely to be destroyed; that development of coastal areas, with destruction of mangrove stands, reclamation of shallow-water areas, and building of marinas and other structures, is destroying the habitat of

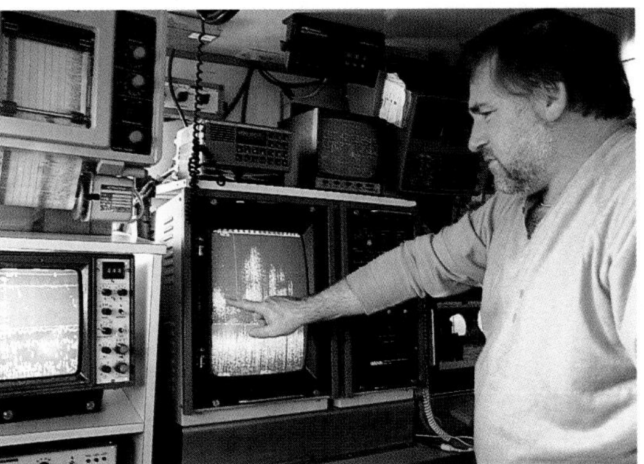

Kim Westerskov/Oxford Scientific Films

◄ On board a New Zealand deep-sea trawler, technology is the key to locating fishable resources. Here the skipper is indicating the large school of orange roughy *Hoplostethus atlanticus* he has picked up on his depth-sounder.

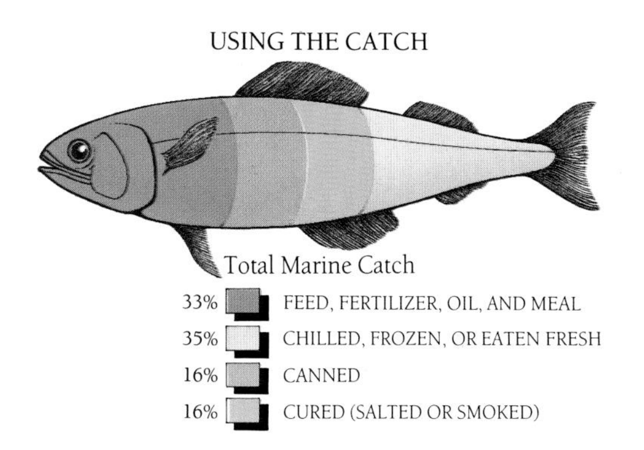


▸ The marine catch is used in a multiplicity of ways, depending on species, location, and local demand.

valuable species of fish, crustacea, and molluscs; that pollution is destroying habitats and organisms. And it is often declared that the result of all these activities has already reduced the yield from the seas.

However, that last claim is not supported by good evidence. Over the past 10 years the total reported catch has increased slightly, and the production achieved by the methods in use at present could be increased by rational management of fishing. Many stocks have been overfished and the catch from them is less than it might be. But if in each case procedures for monitoring the resource could be adopted, and if each year the catch could be limited appropriately to the stock-of-the-year, catch rates could be held at a level which, on average, would be higher than the levels of the past. Thus the total world catch could be increased just from the resources at present exploited.

Then there are the possibilities of extending the range of organisms, of nutritional or other value, to be caught. In this category could be included the use of the by-catch of species not sought in shrimp and other fisheries, which for the most part are thrown back dead into the sea. The effect that might be achieved by making use of the by-catch could be part of a set of interventionist techniques in the ecosystems from which catches are taken at present. These might include selective fishing of certain species in order to reduce predation on juvenile stages of sought-after species, and the plan referred to as "ocean ranching" in which broodstock will be held in breeding areas and the offspring fed and cared for until they are of a size to survive upon release into the open ocean where they will find naturally provided food. And, of course, there are the possibilities of aquaculture.

Just how far these possibilities can be realized will be dependent, to some degree, upon the scope and intensity of the effects of forces working against improved fish harvests. But it is

▾ Oysters have been farmed for centuries using a variety of methods. These oyster-farming boats in western Spain have their valuable crop growing on ropes suspended from the outriggers.

John & Gillian Lythgoe/Planet Earth Pictures

Mike Coltman/Planet Earth Pictures

Kathie Atkinson/Oxford Scientific Films

▲ On this clam farm in Baja California, Mexico (*left*), the clams are grown on open plastic mesh, which contains them and facilitates an easy harvest. One of the main areas for farming rock oysters in Australia is the Hawkesbury River near Sydney (*right*). On this farm, poles covered with young oysters are laid onto racks at low tide. In this traditional method, the oysters will remain on these maturing beds until they are a harvestable size.

important to take an informed view of those forces. For example, not all the materials deposited in the seas are noxious and damaging. A substantial proportion of the organic materials, particularly domestic sewage, can be rated as a return of nutrients to the sea. Similarly, not all fisheries are conducted in a highly competitive, prone-to-overfish style. Answers to problems in these two areas are not to be found in technology, nor even in natural sciences, but in social sciences—even ethics.

G. L. KESTEVEN

## FARMING THE SEA

The farming of fish and shellfish, popularly known as aquaculture, has been undertaken for centuries. As a means of animal production aquaculture has much in common with agriculture, given that the animals have aquatic rather than terrestrial life histories. In both cases the animals are cultured at much higher population densities than those that occur in natural habitats; they require optimal growing conditions including plentiful nutrition; and they are harvested from small enclosed areas or culture containers rather than from open waterways.

There are many advantages in farming fish and shellfish compared to catching the animals from the wild. It is possible to farm some species all year round and thus not be subject to seasonal variation. Individual animals of reasonably uniform size, weight, and color can be produced to suit the requirements of the retailer or consumer. Animals can also be supplied at regular times and in predictable quantities. All these considerations are of great commercial significance, and promote the increasing popularity of aquaculture products.

Species of fish and shellfish suitable for aquaculture occur in many varieties, sizes, and shapes. The most common group of farmed animals is the molluscan shellfish. These include the bivalve molluscs (which have twin shells) such as edible oysters, mussels, scallops, clams, and pearl oysters, and the monovalves (with single shells), principally the abalones. Prawns and shrimps, and to a lesser extent crabs and lobsters, are typical crustacean aquaculture species. Very many types of fish are grown, common species being salmon, trout, carp, tilapia, and milkfish.

Aquaculture species are successfully farmed in diverse tropical, subtropical, and temperate

Edward Parker/The Hutchison Library

▲ High-density fish farming requires considerable monitoring to ensure that water quality remains optimal, food supplies are adequate, and fish are healthy and free from parasites. On this farm in the Shetland Islands, fish are being returned to their tank after routine sampling.

regions around the world. Several countries have developed or sustained major aquaculture industries during the twentieth century, including Japan, Taiwan, Indonesia, Australia, France, Scotland, Norway, the United States of America, and Canada. Many other countries are now establishing significant aquaculture activities, among them the Asian countries of China and Malaysia, the South American countries of Ecuador and Chile, and various Pacific islands. As we approach the end of the twentieth century, aquaculture can be regarded as a boom industry.

## AQUACULTURE FARMS

Fish and shellfish are farmed in many ways, and four examples from different parts of the world are outlined here. A traditional example is the Asian village pond or dam stocked with tropical fish. These systems are community based, require water of only moderate quality, and often make efficient use of diverse nutritional sources including waste materials. The aquaculture species may be harvested at intervals between six months and two years, and will probably be consumed locally at low production cost.

A more technologically based example is the oyster farm in subtropical and temperate estuaries of Australia or New Zealand. These farms are deliberately located in areas of high water quality and high natural abundance of microscopic algae (or microalgae), the oysters' food source. Because the food costs nothing, the farms are often economically attractive. The animals are typically cultured in trays (or on sticks), graded at regular intervals to ensure vigorous growth rates, and harvested after 18 months to three years of farming. The product is usually sold for commercial gain.

Marine farms for fish, pioneered in Norway and Scotland, probably represent the most technologically advanced type of aquaculture grow-out activity. The fish are contained within pens or cages where their stocking density is carefully controlled. If the quality of the water or underlying sediment deteriorates, the pens are moved to a better area. The fish are given pelletized feeds with a defined content of protein, carbohydrate, lipid (including polyunsaturated fatty acids), vitamins, and minerals. The pellets may also have additives such as pigments to color the fish flesh or antibiotics to control possible

▶ In many underdeveloped countries, protein from the sea is an important part of the daily diet. Methods of preservation are often limited, so harvests of wild or cultivated prawns are sun-dried before being sold.

J. L. Mason/Ardea London

disease outbreaks. Automated devices can be used to deliver the feeds at specific times in specific quantities. The fish are typically harvested at 18 months to two years of age. In some instances they may be relatively costly for the consumer, reflecting the cost of the production technology.

Land-based ponds in Japan and Taiwan for prawns vary from simple low-cost operations to sophisticated systems. The quality of the bottom sediment is particularly important for sucessful prawn culture. In the intensive feeding style all the nutrients are provided, usually in the form of pellets. The diets are formulated according to the same principles as for fish. In the extensive feeding style a natural bloom of microalgae, zooplankton, and other small members of the food chain is allowed to develop. These small food items or their detritus form the food source for the prawns. In efficiently run tropical farms it is possible to harvest two crops of prawns per annum at moderate production cost.

## THE PRODUCTS OF AQUACULTURE

The most frequent use of aquaculture products is as food for human consumption. Examples range from tropical fish farmed by villagers for subsistence, to molluscs farmed in sheltered estuaries for sale in distant urban markets. The products can be delivered fresh or frozen, for sale or bartering. Alternatively they may be processed into a "value-added" state, such as smoked fillets or paté, before being sent to market.

Apart from their meat content, mass-cultured aquatic animals can fulfill other purposes. Pearl oysters are valued for their pearls and mother-of-pearl shell. Abalone shells are polished into a striking iridescence and used decoratively. Crocodile skin is esteemed by the clothing and accessories trades.

Harvested animals can also be used in the living state. In certain instances, millions of live animals have been released to replenish wild stocks in natural waterways which have been depleted. Such restocking can benefit either commercial or recreational fisheries. In this way regions in the northwest Pacific are being restocked with recreational species of salmon.

If we adopt a strict definition of aquaculture to include all mass-cultured fish and shellfish species, then we should also recognize the vast number of living animals (such as goldfish) which supply fish-hobbyists. Animals cultured as bait for fishermen are also aquaculture species.

Another aquaculture product to be noted involves the remnants of harvested animals. For some species it is economically feasible to incorporate the post-processing wastes into food such as pellets, which are then given to young growing animals back on the farm. Like all primary industries successful aquaculture thrives on efficiency!

Michael Freeman/Bruce Coleman Limited

▲ Crustaceans for sale at the Kowloon city market in Hong Kong. Accessible markets are an essential part of any fishing industry.

## FEATURES OF A SUCCESSFUL AQUACULTURE SPECIES

Several criteria must be satisfied for an aquaculture species to be farmed successfully in large numbers. Firstly, the developmental phases of the animal should be understood. Then the capacity to provide optimal growing conditions (such as oxygen level, temperature, salinity, stocking density, nutritional supply, and lack of predators) is important. Knowledge of techniques to maintain basic hygiene and control disease outbreaks is essential at all times. Opportunities for rapid spread of disease are an unavoidable consequence of the crowded growing conditions associated with aquaculture systems. This is similar to the spread of infectious disease among intensively reared pigs, sheep, and chickens.

From a commercial viewpoint the major

Gerald Cubitt/Bruce Coleman Limited

▲ De Kuilen trout hatchery near Lydenburg, South Africa, is a source of juvenile stock for local trout farms. Although the collection and cultivation of immature fish has been practiced since at least 500 BC, it was not until 1733 that trout became the first fish to be successfully raised from artificially collected and fertilized eggs.

indicator of a successful aquaculture species is strong and continual market demand. We should appreciate that different aquaculture products are produced to meet different needs in the community. These range from the subsistence or basic food item, through the readily affordable item on the supermarket shelf, to the gourmet or luxury product.

Over many years, and particularly in the twentieth century, hundreds of aquatic animals have been evaluated for their aquaculture potential. Few have been successful, because of the complex and changeable web of biological, industrial, and marketing factors that impinges on any aquaculture venture. The farming stage of aquaculture is the most time-consuming. It requires a substantial input of technical expertise and capital equipment. As mentioned earlier, a large supply of artificial food may also be needed for some species. But there is much more to a successful modern aquaculture venture than the farming stage. After harvest, the processing, post-harvest handling, and marketing stages must be undertaken. Beforehand, there are the hatchery and nursery phases which provide the essential juvenile stock.

## HATCHERIES AND NURSERIES

Traditionally fish and shellfish farmers obtained their juvenile stock from natural waterways. At the appropriate season bivalve spat, for example, were collected on sticks, or fish fingerlings were trapped in nets. However, these traditional methods have become increasingly unpredictable, often failing to yield the full quantity of young animals required by farmers.

Hatcheries, now the cornerstone of many communal and commercial aquaculture industries, were developed in the late nineteenth and twentieth centuries, often in response to the dwindling natural supplies of juveniles. Trout hatcheries were among the first. A modern aquaculture hatchery is a technologically advanced and compact operation. It is usually located on a few hectares of coastal land with access to high-quality sea water. Many thousand liters of water are likely to be needed every day.

Like terrestrial farming, the successful rearing of animals in hatcheries is no accident of fate. Many complex, skilled activities are undertaken there. Firstly, the breeding animals must be conditioned

▶ Modern aqauculture farms monitor and control their stock at all stages of production, from spawning to harvest. Good management of the hatchery and nursery stages is crucial to a successful industry.

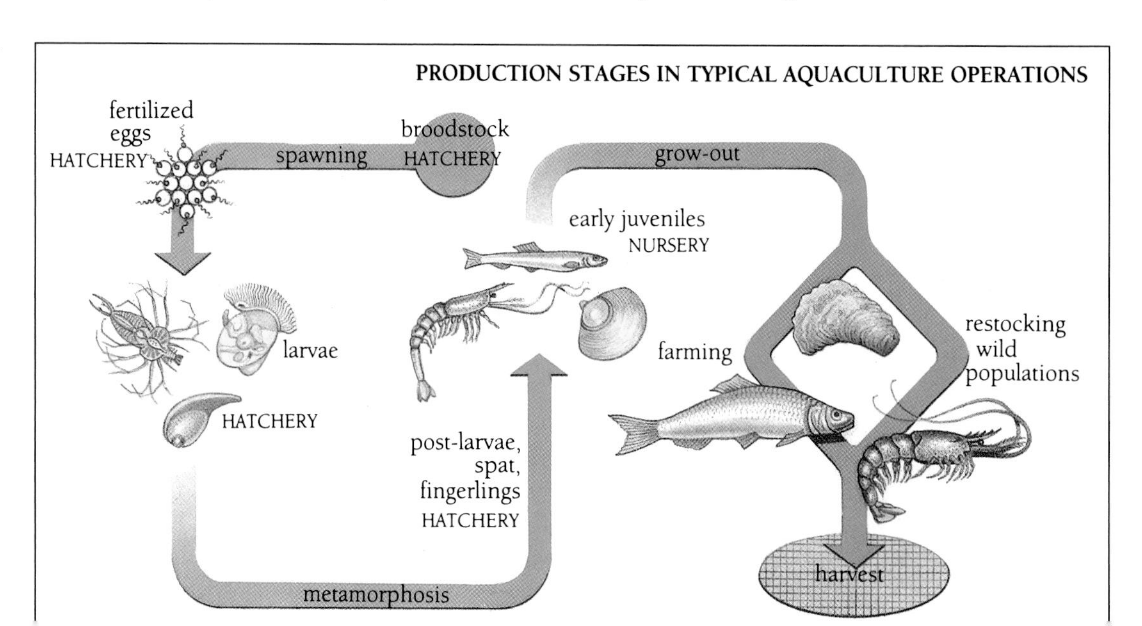

# PEARLS AND PEARLING

Coo-ee Historical Picture Library

*Pearling in Torres Straits,* an etching depicting the industry in 1875.

A natural pearl forms when a grain of sand or similar particle enters the shell of perhaps one mollusc (most commonly an oyster) in ten thousand. Over several years the mollusc covers the irritant with layer upon layer of iridescent nacre, or mother-of-pearl, to enclose the foreign body. Clams, conchs, mussels, and other molluscs can also produce pearls but, because they lack the iridescence of those of the pearl oyster, they are not as prized. The major natural pearl oyster beds are in the Persian Gulf, the Bay of Aden in the Red Sea, northwest Sri Lanka, off the northwest Australian coast, and around the south Pacific islands.

With cultured pearls, the procedure is not left to chance: an incision is made in the gonad of a pearl oyster, into which a tissue graft from the mantle of another oyster is implanted, together with a spherical mother-of-pearl nucleus, around which the pearl forms. The technique of modern pearl culture was developed at the end of the nineteenth century by Kokichi Mikimoto in Japan, which now produces up to 100 tonnes of pearls annually. The Japanese Akoya pearls—small pearls up to 10 millimeters (0.4 inch) in diameter—are the main pearls of international commerce.

China is probably the main producer by volume of pearls in the world. Chinese freshwater pearls are grown in mussel shells and are not as prestigious as marine pearls; they generally have a poor luster and are frequently artificially colored. Freshwater pearls are also obtained from the Mississippi River.

The most prestigious of all pearls are undoubtedly the white and black south sea pearls. White pearls are grown in the gold-lip oyster *Pinctada maxima* and so-called black pearls (in fact, they vary from dark gray to peacock green) in the black-lip oyster *P. margaritifera*. They can produce pearls up to 18 millimeters (0.7 inch) in diameter in half the time taken for the Akoya pearls to form. White south sea pearls are cultivated around the tropic shores of western and northern Australia and to a lesser extent in Indonesia, Burma, and the Philippines. The total annual production is around 1,000 kilograms (2,200 pounds). The production of black pearls began less than 20 years ago in Tahiti and has recently been introduced to the Cook Islands and elsewhere in the Pacific. Annual yield is about half that of the white pearls.

Commercial pearling began in northern Australia in the late nineteenth century, and workers from the Pacific islands and later Japan were recruited to supply labor for the industry. Today, divers working from luggers collect young pearl oysters from both offshore and inshore beds, and take them to floating "farms" where they are opened and the graft inserted. The oysters are kept in wire cages or netting baskets suspended from floating rafts or longlines for up to three years, when the pearls are extracted. It is sometimes possible to remove the pearl surgically without killing the oyster, and reoperate a second, or even a third, time. The pearlshell is a valuable byproduct of pearl farming, and is still in great demand for making buttons and jewelry.

Even with the most experienced seeding technicians and the best possible farm management, the chance of a flawless round pearl remains small. Round pearls are the most favored for necklaces, while irregularly shaped (or "baroque") pearls invite the creative designer to fashion works of jewelry art.

WILLIAM REED

Michael Freeman/Auscape

The pearl oyster *Pinctada maxima* is used in the Thai pearl industry.

Peter D. Capen/Planet Earth Pictures

▶ One way of managing fish intended for luxury markets is to farm them. Here harvested salmon are placed in tubs of ice at a farm in Washington, before being exported to epicurians around the world.

so that their gametes (eggs and sperm) are abundant and healthy, and bear the appropriate genetic characteristics. The animals are then induced to spawn their gametes, which are collected separately (when possible) by technical staff. The gametes are mixed to achieve fertilization. The young larvae hatch out, usually within one to two days.

Larvae are typically reared in tanks containing still or slowly replaced sea water. The culture conditions are carefully controlled, especially with respect to temperature, dissolved gases, pH levels, and stocking density. General hygiene is critical: infectious disease can spread rapidly and tens of millions of young animals may be put at risk.

Since natural sea water contains insufficient items for mass-cultured larvae, the food supply must also be carefully managed. Microalgae are the preferred food of bivalve larvae and spat, and of the early larvae of many fish and crustacean species. Consequently the mass culture of microalgae is often required. As the fish and crustacean larvae develop, they prefer larger live animal food items such as rotifers, copepods, and brine shrimps (which feed on microalgae) so these food sources must also be mass-cultured in the hatchery. Alternatively, the older larvae may take artificial feed such as pellets. To date these have been developed for relatively few species.

At regular intervals during the hatchery phase, technical staff catch larvae on sieves and examine them microscopically to check their size, shape, mobility, feeding habits, and organ development. Between one week and several months, and after one or more larval cycles, the animals are ready to metamorphose. Special habitat conditions are then provided so that, over several days, they can transform into juveniles.

The juvenile animals are usually produced in batches, and annual production by some hatcheries can amount to tens of millions of fish and crustaceans or thousands of millions of bivalve mollusc seed. These early juveniles may be despatched immediately to farmers or they may be acclimatized for several weeks in a nursery before being despatched. In either case, an efficiently operated modern hatchery or nursery can supply all the farmers located along several hundred kilometers of coastline.

▼ Salmon farming in the Outer Hebrides. Some farms collect their own eggs and raise the fish right through to maturity; others buy juveniles from a hatchery.

Liz & Tony Bomford/Ardea London

### THE FUTURE OF AQUACULTURE

A promising future awaits fish and shellfish farming. It is predicted that in some parts of the world aquaculture production will rival that of fisheries by the turn of the century. Certainly aquaculture is viewed by many fisheries scientists as an essential supplement to the fishing industry. To ensure that the prospects for aquaculture are maximized, substantial biotechnological research is being undertaken on the current range of species. Many new species are being evaluated for their potential. Marketing aspects are also receiving considerable attention from researchers, especially with regard to the nutritional value of products, together with safe and longer shelf-life.

The major concern for aquaculture is environmental pollution. Our world is increasingly at risk from the exotic materials accumulating in our waterways. Already there have been instances when pesticides and heavy metals (such as nickel, copper, and tributyl tin) have entered hatcheries via the intake sea water and profoundly affected larval survival. Growing animals have been weakened or disfigured, and occasionally killed, by toxic materials. Often their sources have been accidental spills or sediments which have become highly concentrated over the decades.

More sinister is the accumulation of pollutants by fish and shellfish which otherwise appear normal to the human eye and nose. The range of possible pollutants includes heavy metals, pesticides, organochlorines, and disease-causing bacteria and viruses from sewage. The consumer may well be the unwitting recipient of these compounds, and suffer acute or chronic illness.

Environmental pollution is of concern not just to aquaculture, but to the capture fisheries and to aquatic life in general. In the next decade society must work diligently to devise and implement strategies to control pollution. The methods will need to be simple and inexpensive so that they can be applied as readily in developing nations as in the technologically advanced countries.

CHRISTIAN D. GARLAND

## MINERALS AND ENERGY FROM THE SEA

For centuries, people have extracted useful minerals and energy from the sea. The most abundant of the ocean's mineral resources are its salts. On average, the salt content of the oceans is about 3.5 percent by weight, although it varies with the rates of evaporation and precipitation at

▼ The construction of this enormous self-contained oil-drilling rig is nearing completion, prior to being towed to a location in the North Sea. Lit up at night, it resembles a small city.

Wolfgang Volz/Bilderberg

Emmanuel Valentin/Hoa-Qui

▲ The human production of salt from the sea can be traced back to neolithic times. In Mauritius, traditional labor-intensive methods are still employed to harvest this commodity from shallow ponds.

the surface, with the freezing and thawing of sea ice at high latitudes, and nearshore with local runoff of fresh water from the land. Some 86 percent of the dissolved salt is sodium and chloride, but the salts of magnesium, calcium, and potassium, including sulfates, carbonates, and bromides, are economically and nutritionally important minor constituents of sea salt. Salt ponds, which sea water enters at high tide and which are then closed so that the water will evaporate and the salt precipitate, are common along coastal settlements, and international trade in salt has affected the course of history throughout the world.

The oceans and their beaches have long been exploited as a source of building materials. The rocks, sand, and gravel that have been winnowed, sorted, and concentrated by waves and currents are used worldwide for the construction of roads and buildings. Offshore from today's beaches are fossil beaches, the remnants of earlier geological times when the sea level was lower. These too preserve valuable resources, including submarine placer deposits of gold and other heavy metals, and along the Namibian coast of southwest Africa, diamonds, which are extracted by dredging at the mouth of the Orange River.

## PRECIOUS METALS

In recent times, attention has been paid to the minor dissolved constituents of sea water. Periodically, we read of the large amounts of precious metals—gold, silver, copper—that are dissolved in, say, a cubic kilometer of sea water. But a cubic kilometer is a great deal of water, and extracting trace metals that occur in nanograms per kilogram of water (a nanogram is one-billionth of a gram) is a costly process.

Instead of trying to extract dissolved metals directly from sea water, it is possible to recover them after time and the ocean have concentrated them on the sea floor. Perhaps the best known of these are the manganese nodules found on the ocean floor, particularly in the Pacific, south of Hawaii and northeast of Tahiti. These nodules contain significant amounts of manganese, iron, nickel, and copper, and minor amounts of cobalt, chromium, tin, and other metals. The world mining industry has expressed interest in recovering these nodules, but both recovery and refining raise environmental, economic, and

Anthony Bannister/NHPA

► Namibia provides about one-sixth of the world output of gem diamonds. Most of this supply is obtained by offshore dredging in the alluvial deposits between Oranjemund and Luderitz. Diamonds are removed from the sediment at this sorting plant at Oranjemund.

Robert Hessler/Planet Earth Pictures

technical questions that have, thus far, kept the nodules on the sea floor.

## MINERAL DEPOSITS

Oceanic mineral deposits result from the slow precipitation of dissolved metals from sea water or by the accumulation of tiny metallic particles carried in bottom currents. As a result, minerals form crusts that cover the bottom in many parts of the world ocean. Of special interest are the phosphorite deposits that precipitate in areas of high biological productivity, such as areas of coastal upwelling. These deposits represent an economically feasible source of phosphorus, whose main use is as a fertilizer. To date, however, they have not been exploited as there are many land deposits that are easily mined.

In the active centers of sea-floor spreading, ore deposits are precipitating at much faster rates. Where hot vents bring mineral-rich water from the interior of the Earth and eject it into the cold water of the ocean floor, the heated water cools rapidly and the minerals, including many metals and rare elements, are deposited as polymetallic sulfides near the vents. The mining industry is actively investigating the possibility of recovering these deposits. A major difficulty lies in developing techniques to extract the metals that are bound in very complex chemical compounds.

## OIL AND GAS

While the salt trade drove the politics of global wars and alliances in earlier times, the most influential economic resources of the ocean today are oil and natural gas. From Indonesia to the North Sea, from the Gulf of Mexico to the Sea of Okhotsk, and from southern California to the Persian Gulf, the recovery of oil and gas from offshore deposits is perhaps the single most important driving force in the global political arena. The value of offshore oil and gas deposits, worldwide, is still controversial, but new fields are being found, and as we learn more about the ocean floor, the discoveries are bound to continue. The production and transportation of offshore oil and gas are major features of the global economy.

Bob Thomason/Colorific!

Recovery of these resources is a subject of international debate, as offshore drilling accidents have damaged the ocean, at least temporarily. While one of the great strengths of the global ocean is its ability to absorb material, whether it comes from natural or human sources, the clearly better alternative is to extract resources without affecting either the ocean or its inhabitants. The technology used in the offshore oil and gas industry is among the world's most advanced.

## SOLAR ENERGY FROM THE SEA

Although the recovery of the mineral resources of the ocean is an ancient human endeavor, the recovery of useable energy from the ocean has a number of modern and futuristic dimensions. The ocean's energy comes from the sun, which moves the atmosphere and the great ocean currents, and

▲ Deep-sea manganese nodules (*left*) were discovered by the *Challenger* expedition in 1873–76. Their composition varies globally, but is predominantly manganese dioxide and iron oxide. These minerals appear to precipitate out of the surrounding water, gradually forming nodules over several million years. An offshore drilling rig, complete with helipad, in operation off Louisiana in the Gulf of Mexico (*above*). Over recent years, Louisiana has been one of the most important producers of oil and natural gas in America.

Larry J. Pierce/The Image Bank

▲ Waves such as this one at Oahu, Hawaii, could be harnessed to produce power, if only we had the technology. Such inexhaustible ocean energy sources have been largely ignored because of the relative cheapness of oil and coal.

from the gravitational energy of the tides. Recent work on the ocean floor reveals very large geothermal resources as well, but the recovery of geothermal energy is in its infancy even on land, and ocean floor geothermal power is years away.

The solar energy of the ocean is stored in the surface layer (the upper 200 meters/650 feet or so). The simplest form of energy to extract is that contained in the energy contrast between the surface and the deep layers of the ocean. Just as a refrigerator requires energy to maintain a temperature difference between its interior and exterior, energy is available wherever there is a temperature contrast in the ocean. The greatest contrast between the surface and deep layers occurs in the equatorial ocean, and between latitudes 20°N and 20°S the difference is large enough to recover energy through Ocean Thermal Energy Conversion (OTEC). The process has proven economically feasible at pilot plants such as the one at Keahole Point on the island of Hawaii. An improvement in the technology has also produced fresh water as a byproduct of the energy conversion process—a valuable resource for island nations in many parts of the world.

The second major source of solar energy in the ocean is from currents. These great rivers in the sea flow constantly along well-known routes. It may one day be possible to produce useful energy by mooring turbines in fast-moving currents such as the Kuroshio (literally "black current" but also known as the Japan Current), the Gulf Stream, and the East Australia Current.

Obviously, transmitting energy produced in mid-ocean, or for that matter on sparsely inhabited oceanic islands, through the OTEC process may introduce significant technological problems, but it is also possible to transmit energy in indirect ways. We now ship bauxite from the Caribbean to Scandinavia where cheap hydroelectric power makes the refining of aluminum economical. If the ocean energy is used near the production site to effect an energy-intensive process, such as bauxite refining, and the refined product is shipped, the energy is transmitted at a greatly reduced cost. Still, direct transmission is possible. Iceland is currently studying the possibility of transmitting inexpensive geothermal and hydroelectric power to Ireland and Britain by submarine cable, and similar studies have been done in other parts of the world ocean.

## WAVE AND TIDAL ENERGY

The atmosphere moves the ocean to create currents, but it also produces surface and internal waves in the water. Although these waves contain a great deal of energy, we have yet to develop a means of extracting it in an economical and environmentally safe manner. At present, wave energy remains a largely destructive rather than constructive resource.

The second major source of ocean energy is the gravitational energy embodied in the tides, and tidal energy has been used in many ways in human history. The simplest use is, perhaps, in salt ponds. The tides also lift ships into and out of drydocks around the world every day. Sea water flooding low-lying land areas in Maine and the Low Countries is drained back to the sea through turbines to produce hydroelectric power. Narrow straits around the world, such as the Bering Strait, the entrance to the Bay of Fundy, and the islands of the Sea of Cortez, may one day produce energy from tidal currents.

T. C. Middleton/Oxford Scientific Films

▲ Salt-producing industries such as this one at Lanzarote in the Canary Islands most commonly extract their product by solar evaporation of sea water or inland brines. Once the salt has crystallized in shallow pans, it is washed with a saturated brine solution, rewashed with fresh water, dried, and sold.

## OSMOTIC POTENTIAL

A potential source of ocean energy is available from the salt in the sea. As a desalination plant must use energy to remove salt from sea water, the contrast in salinity between the sea and freshwater river membranes would rise as the process of osmosis diluted the salt water. In a beaker, if an osmotic membrane separates fresh water from salt water, the fresh water moves through the membrane in an attempt to equalize the salinity on both sides of the membrane. The salt particles cannot move into the fresh water, so the osmotic flow is one-way. Over time, the surface of the side of the beaker containing salt water rises, and the surface of the side of the beaker containing fresh water drops. The difference between these two surfaces produces a useable hydrostatic pressure: a source of energy. At a site where a freshwater river enters an ocean, a pressure difference equivalent to a hydrostatic head of greater than 240 meters (790 feet) could be developed for typical ocean salinities. No practical system has yet been devised to utilize this energy source.

Many sources of ocean energy are available to us, but we have learned to use only a few. Unlike the mineral resources of the sea, its energy is renewed continuously by the sun and by the gravitational attraction of the moon, sun, and the other celestial bodies.

JAMES C. KELLEY

Ardea London

◄ Waves generated by freak weather conditions or geological events (such as submarine earthquakes) can have enormous destructive potential on coastlines and inappropriately designed buildings sited too close to the sea.

# 13 PLANNING FOR THE FUTURE

Jack Dermid/ Oxford Scientific Films

▲ The sea has always been a source of inspiration for humans, but we have not always found it easy to resolve the territorial disputes that transcend national boundaries and result from our claims on its mineral and biological wealth.

◄ Indigo blue of the ocean meets the intricate mosaic of coral reef, enclosing shallow turquoise waters on Australia's Great Barrier Reef. The framework of the reef is built by the limestone skeletons of living coral polyps and cemented together with coralline algae.

The oceans form a link between all who inhabit our planet, and their health is a matter of common concern. They challenge science and technology, national institutions and international resolve. Who owns the oceans and what rights do individual nations have to their resources? How serious is the threat of oceanic pollution, and what is being done about it? Has human intervention already destroyed the balance of nature? How are the environments of oceans and islands being protected? What is the likely future of Antarctica, a continent surrounded by ocean?

## THE LAW OF THE SEA

Before the Second World War most maritime claims were to territorial seas three nautical miles wide: a distance about the range of cannon fire in the eighteenth century. After the war, however, the extent of claims varied. Two United Nations conferences (in 1958 and 1960) failed to solve the problems and disputes that developed as a result of this ad hoc development.

In December 1982 119 delegations signed the Law of the Sea Convention which had been hammered out in the previous eight years at the Third United Nations Conference on the Law of the Sea. That Convention will come into force after 60 countries have ratified it and a year has elapsed after the sixtieth ratification. At the beginning of 1990 42 countries had ratified the Convention and it is expected to come into force in 1992 or 1993. Even though it is not yet in force, many countries appear to be applying the Convention's rules, and so it is appropriate to describe them in this analysis.

### THE FOUR ZONES

Under the 1982 Convention, coastal states are entitled to claim four main zones. Proceeding seaward from the low-water line, the first zone is called the territorial sea. With one exception the coastal state has complete authority over the territorial sea, and over the air-space above it and the seabed beneath it. The solitary exception is that foreign vessels are entitled to exercise the right of innocent passage through territorial seas. Passage is deemed to be innocent if it is not prejudicial to the peace, good order, and security of the coastal state. The maximum permitted width of the territorial sea is 12 nautical miles from the baseline used by the coastal state. Normally that baseline is a low-water line. One nautical mile is equal to one second of latitude, which is 1,852 meters (6,075 feet).

The contiguous zone lies beyond the territoral sea, and in it the coastal state may exercise controls necessary to prevent infringements of its customs, fiscal, immigration, and sanitary laws. The outer limit of the continguous zone may not be more than 24 nautical miles from the baseline.

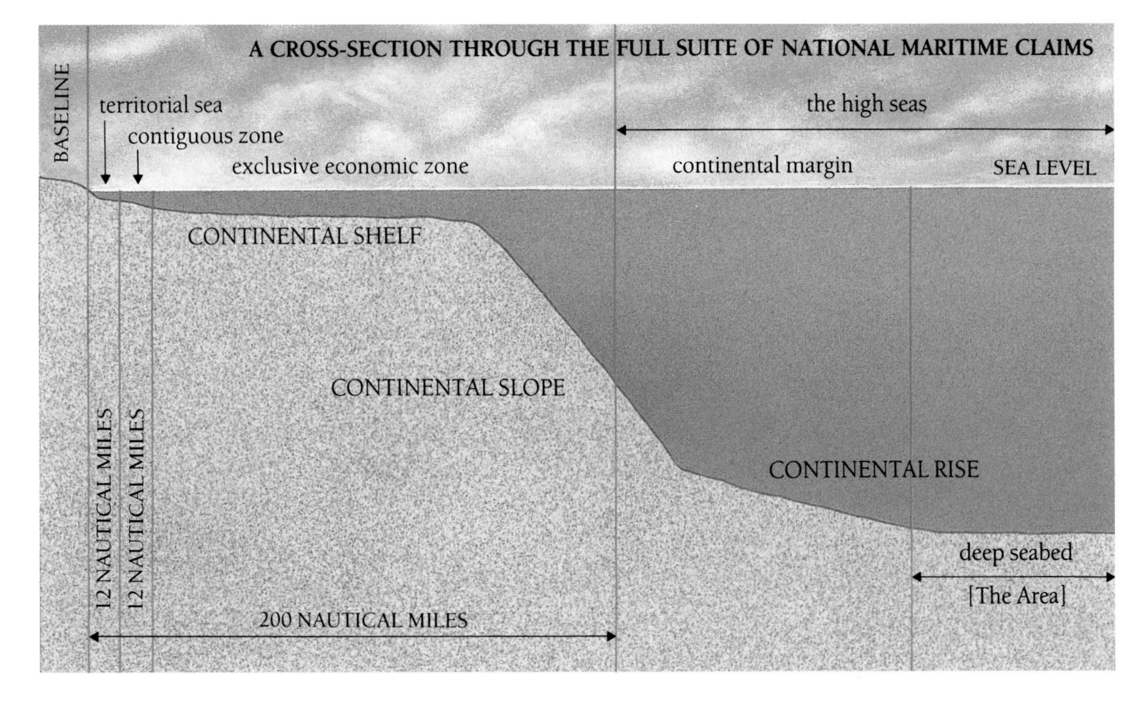


◄ This schematic representation indicates the full range of national maritime claims as agreed at the 1982 Law of the Sea Convention.

Colin Monteath/Hedgehog House New Zealand

▲ Tabular icebergs in Hope Bay, Antarctica. Ice flows from the high central dome of the south pole, outwards in all directions, eventually falling from oceanfront glaciers to form these huge icebergs. Antarctica is the only continent around which maritime claims have not been made.

▼ For centuries the sea has been the sump of the world. Waste drains into the sea via river systems and terrestrial runoff. This is especially true of enclosed waterways such as the Bosphorus, the link between the Black Sea and the Aegean.

The next zone is the exclusive economic zone. This zone may have a maximum width of 200 nautical miles measured from the baseline. In it the coastal state has the sovereign right to manage and exploit the living and non-living resources of the waters and the seabed. This right includes the generation of power from waves or tides. The coastal state also has the sole right to establish artificial islands in the zone, and jurisdiction over research and protection of the marine environment. If artificial islands or other installations are established within the zone, the coastal state may create a safety zone of 500 meters (1,640 feet) around them, within which it can make rules for navigation.

If the continental margin, which consists usually of a continental shelf, continental slope, and continental rise, extends more than 200 nautical miles from the baseline, the coastal state may claim it. The claim is only to the seabed and is not to the waters above that seabed. On the continental shelf claim, as in the seabed of the exclusive economic zone, the coastal state has the sole right to manage the resources of the seabed whether they be lobsters or oilfields. Countries that possess shelves wider than 200 nautical miles include Australia, Brazil, Canada, the Soviet Union, and the United States.

R.F. Porter/Ardea London

BOUNDARY AGREEMENTS

There is no country in the world so distant from its neighbors that it can make the full suite of maritime claims without overlapping with similar claims from a neighbor's coast. In such cases the countries involved must agree on how to divide the waters and seabed which they could both claim. At the beginning of 1990 about 120 marine boundary agreements had been concluded. Most of them were in enclosed or semi-enclosed seas such as the Caribbean, the North Sea, the Baltic Sea, the Mediterranean Sea, the Persian Gulf, and the Andaman Sea. In some cases countries have asked the International Court of Justice or an arbitral tribunal to select the appropriate boundary or the means of defining any boundary. Significant cases have included those between France and the United Kingdom over the boundaries in the Channel; Canada and the United States in the Gulf of Maine; and Libya and Malta in the Mediterranean.

INTERNATIONAL ZONES

There are two international zones in the oceans. First, the high seas lie beyond national claims to territorial seas and the exclusive economic zone. All states, whether coastal or landlocked, have equal rights to make peaceful use of the high seas. Second, the deep seabed, called The Area in the Convention, is the seabed beyond national claims to the seabed of the exclusive economic zone or the continental margin where it is wider than 200 nautical miles. Where states claim margins beyond 200 nautical miles the waters above that distant margin will be part of the high seas. The Convention lays down very detailed rules for mining The Area; an international body provided for in the Convention will oversee any such mining. These rules were developed to control the mining of manganese nodules. To date, no such commercial mining has occurred. Today, polymetallic deposits within the exclusive economic zones of some states offer more attractive mining ventures outside the authority of the international body.

ISLANDS AND ARCHIPELAGOS

Countries can claim maritime zones from islands as well as from the coasts of mainlands. So the United States can claim an exclusive economic zone around the Hawaiian Islands, as India can around the Nicobars and Britain around the Falkland Islands. There is a rule by which claims to territorial seas are allowed only from rocks. It is imprecise because it provides no way of distinguishing a small island from a large rock.

Archipelagic states enjoy special provisions under the Convention. If a country is constituted wholly by one or more archipelagos, it can draw straight lines connecting the outermost islands and measure its maritime claims from those straight lines. There are rules which prevent some widely scattered archipelagic states such as Kiribati or some very compact ones such as New Zealand from drawing such lines, but archipelagic baselines have been drawn by countries such as Indonesia, Sao Tome and Principe, and Fiji.

The only continent around which maritime claims have not been made is Antarctica. The members of the Antarctic Treaty have established conventions to protect marine animals and fish, and are trying to do the same for non-living resources, such as minerals and fuels on Antarctica's continental margin.

VICTOR PRESCOTT

# OCEANIC POLLUTION

For most of human history, the sea has been regarded as a vast sink into which anything could be dumped with impunity. Those who thought about it at all, assumed that dissolved substances rapidly became diluted in the sea until their presence could not be detected, while just about everyone believed that sea water killed all "germs". The single event that served to change this attitude and to stimulate marine pollution research was the outbreak of Minamata disease on Japan's most southerly island in the 1950s. After many false trails had been followed, the origin of the disease was traced to the release of mercury into Minamata Bay by a local factory—but by then scores of people had died from eating mercury-polluted fish and a still greater number, mostly children, were permanently paralyzed.

## THE EFFECTS OF POLLUTION

Research undertaken since then has shown that virtually all previously held concepts were erroneous. Many pollutants do not become progressively diluted but are trapped and even concentrated in marine sediments, from where they may enter the food chain. The pollutant is further concentrated as animal eats animal, the concentration in the final predator—usually a fish, bird, or mammal—being up to a million times what it was in the sea water. Humans, at the top of the food chain, therefore run the greatest risk of all. It was also discovered that, once in the sea, pollutants may be converted to even more toxic forms; an example was the conversion of inorganic to organic mercury in Minamata Bay.

By 1970, when at last the United Nations began to take an interest in marine pollution, the situation had become very serious indeed. The Mediterranean was dying; the Bosphorus resembled a giant sewer; many European estuaries were completely fouled; cities were having problems disposing of their sewage; and pesticides such as DDT could be found in the tissues of most marine animals, even in the remote Antarctic. It was also discovered that, far from the sea killing all pathogenic organisms in sewage, the causative agents of such diseases as cholera, typhoid, and hepatitis could survive for long periods in sea water and were accumulated in shellfish such as mussels and oysters. In 1973 the United States Environmental Protection Agency proposed that even the slightest contamination of coastal waters with sewage was unacceptable, although the majority of countries still do not take this advice seriously and there are still areas where sewage contamination makes swimming and shellfish consumption hazardous.

## POLLUTION CONTROL

There is no question that we will always have to endure some marine pollution; what we *can* hope

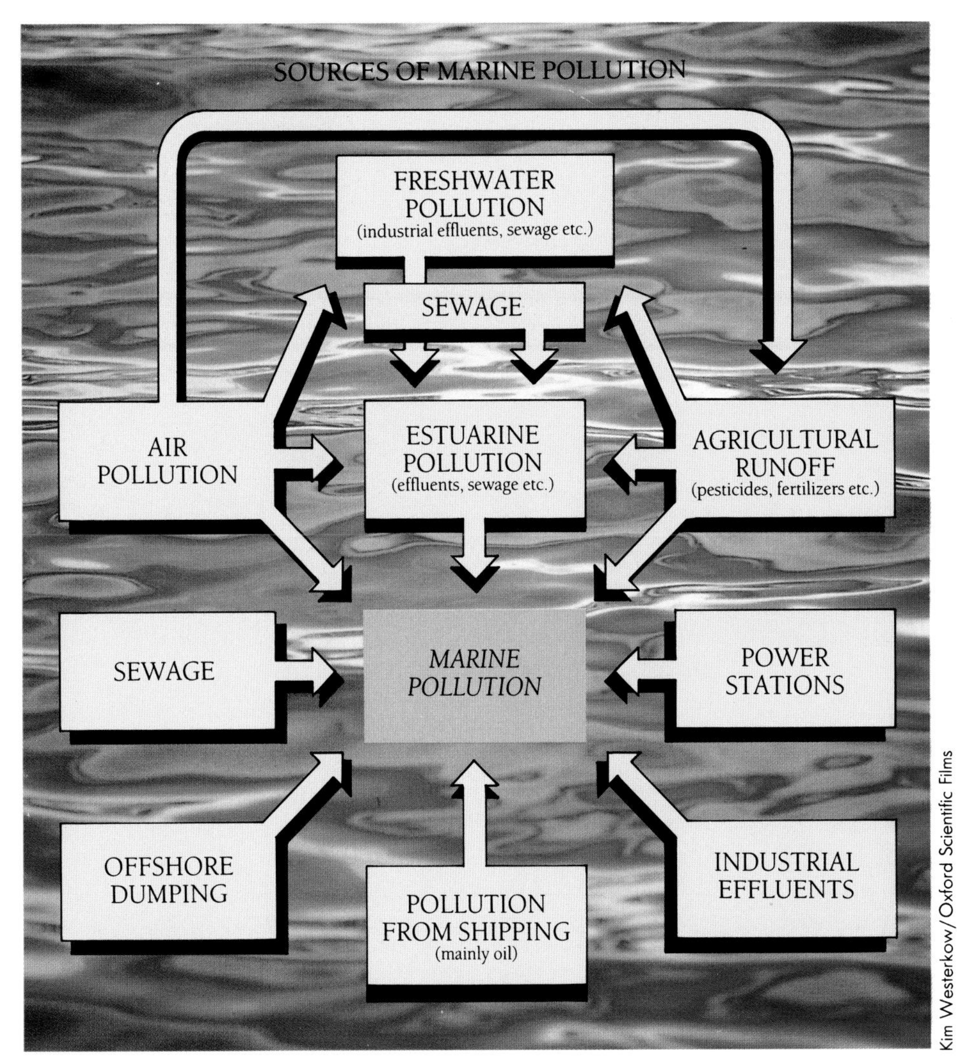


Kim Westerkow/Oxford Scientific Films

▲ Pollution of the air, land, fresh water, and estuaries eventually reaches the sea to contribute to marine pollution.

Martin Coleman/Planet Earth Pictures

▸ Beach pollution, Sydney, Australia. Pollution of the coastline in heavily populated areas is a combination of debris washed in by the tide and rubbish left on the beach.

▾ In 1973 the International Convention for Prevention of Pollution from Ships (MARPOL) was established. It sets minimum distances from the coastline for the discharge and dumping of treated and untreated garbage and sewage, noxious chemicals, and oil. Its provisions state that oil cannot be discharged from tankers in heavily polluted and vulnerable areas. To date the provisions of MARPOL have not been fully adopted by the international community.

for is to reduce it to an acceptable level, a level which the sea can assimilate without grave ecological effects and without health hazards to humans. Unnecessary pollution cannot be tolerated—and it must be stated that a good deal of marine pollution is quite unnecessary and easily prevented. For example, the mercury spilled by the factory in Minamata Bay was not a waste product but a valuable catalyst: it escaped because of a faulty recycling plant. When the factory was forced to rectify the fault, not only did the pollution cease but profits increased as the factory's demand for mercury decreased.

Pollution by plastic materials, a comparatively recent but now very widespread and severe form of pollution, is largely preventable by public education coupled with appropriate legislation. Most pollution by crude oil is also preventable and in fact has decreased quite dramatically since courts began to impose heavy fines for avoidable spills. Sewage disposal continues to present a problem, although many countries have resolved that no new permits will be granted for the discharge of raw or partially treated sewage to sea. However, a great many pipelines discharging raw sewage remain in operation and while the

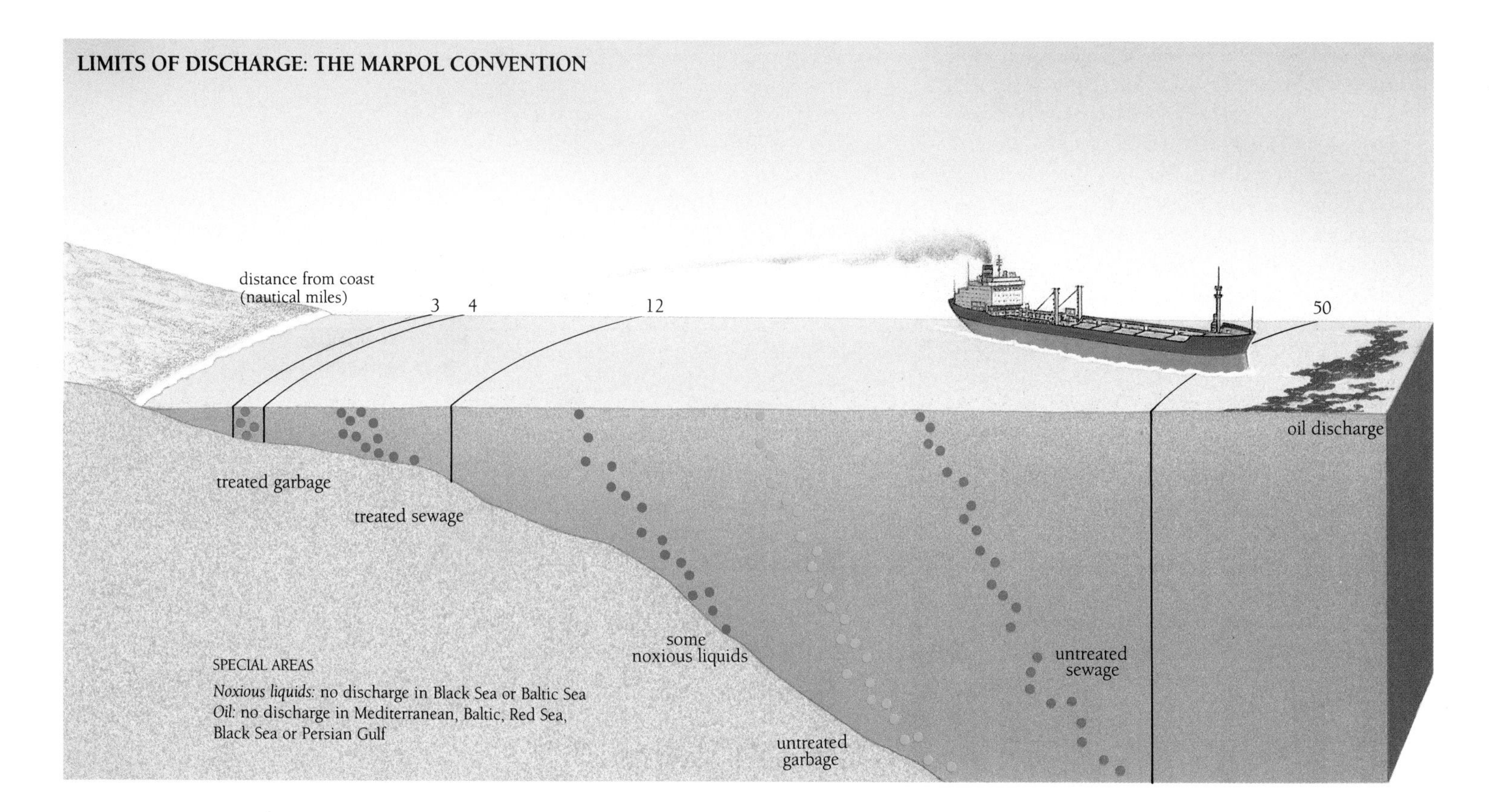

Hans-Jurgen Burkard/Bilderberg

tendency in developed countries has been to increase the length of these pipes, this can never be regarded as the final answer; neither can it be considered an appropriate solution to transport sewage sludge offshore and dump it, as is done in New York.

In most countries, a permit is now required for the discharge of an industrial effluent to sea, although in many cases the legislation governing the granting of permits is unnecessarily complex or open to abuse. For many years it was considered that once "safe" discharge levels had been established for one factory, these could be applied to all other factories with similar effluents. This was poor thinking, as clearly a factory discharging into strong currents in the open sea can afford to have higher levels of toxicants than one discharging into an estuary or a sheltered bay, where mixing and turnover of the water body are slow. Also, other discharges into the same area must be taken into account, so that acceptable seawater quality is maintained. The question is, what is acceptable?

## MEASURING POLLUTION LEVELS

For a long time the toxicity of an effluent was tested by discovering the concentration that would kill 50 percent of a given species (usually a fish or a crustacean) in a given period (usually 48 or 96 hours). A "safe" discharge level was then considered to be a tenth, a fiftieth, or a hundredth of that value, depending on the country concerned. Such simplistic tests have now largely been abandoned in favor of much more sensitive sublethal tests, in the knowledge that almost any change in the physiology or behavior of a population will reduce its reproductive potential and is thus likely to eliminate it in the long term. Sublethal testing has reached a high level of sophistication and is constantly being refined. What one would really like to know, of course, is the effect of a pollutant on the whole ecosystem, in order to evaluate the assimilative capacity of the system, but this is difficult and remains something for the future.

The concepts of assimilative capacity and seawater quality (rather than effluent quality) have been combined with the notion of "beneficial seawater uses", the idea being that sea water used to cool a power station need not be of very high quality, while at the other end of the scale areas used for aquaculture or recreation must have

▲ Marine organisms are particularly vulnerable to oil spills. Inundation, or the dissolution of the organism's natural protective fatty barrier by the oil, usually results in a high death toll. A further "cost" of oil spills is the expensive, labor-intensive effort required to clean up the mess. Here a crew works to clean up a despoiled beach in Prince William Sound, Alaska.

Frans Lanting/Minden Pictures

▲ Early visitors to the Galapagos Islands introduced horses as a means of transport, but they had little regard for the effect of these animals on the unique flora and fauna of the islands.

water of the highest purity. The concept is a useful one, although it has obvious dangers for the marine environment.

An enormous amount of research has been undertaken on marine pollution, and it continues with ever-increasing momentum. It has already borne fruit by way of application and this, combined with improved river-water quality and reduced pollution from runoff from the land, might be expected to alleviate the situation. So it would were it not for the continuing increase in the world's population. More effort is therefore needed, coupled with greater understanding of marine ecosystems, if the oceans are to survive.

ALEC C. BROWN

# INTRODUCED SPECIES

Alien species have provided problems in virtually every type of ecosystem on earth—and the oceans and their islands are no exceptions. Not all alien species have been introduced by humans, of course, and many, both marine and terrestrial, have been spread by ocean currents, on the feet or in the alimentary canals of birds, or attached to fish. However, we are concerned here with species introduced by humans, either inadvertently or deliberately, and which have established themselves at the expense of the local flora or fauna.

## CHANGING THE ISLAND BALANCE

Islands commonly have high proportions of endemic species or subspecies, differing slightly from those of even nearby islands. Because of this, and their relative isolation, islands are natural "evolution laboratories" of great importance to our understanding of biological processes. They are also highly susceptible to alien invasion, and unique species may rapidly become extinct if care is not taken to preserve the ecosystem. An extreme example is provided by the intensive cultivation of sugar cane on some Caribbean islands, where this plant has largely replaced the natural flora.

The islands of the Southern Ocean—and particularly those that are or have been inhabited—present a sorry tale of the destruction of vegetation and wildlife. These islands cannot provide large populations with enough to eat, so that over the years visitors and settlers have taken

▶ Goats were commonly introduced to islands such as the Galapagos as a source of food for visiting or shipwrecked sailors and early settlers. As herds grew, their trampling and grazing decimated many local habitats.

Udo Hirsch/Bruce Coleman Limited

Frans Lanting/Minden Pictures

▲ Reindeer were introduced to the subantarctic island of South Georgia in the early 1900s. These hard-hoofed grazers now number in their thousands and have caused widespread destruction of the island habitats.

much of their food with them, in the form of cattle, sheep, goats, pigs, and rabbits. The effect of these domestic animals on the indigenous biota has frequently been devastating. As long ago as 1874 sealers acquainted with the Southern Ocean gave it as their opinion that "it would not be well to introduce pigs to the southern islands as they would destroy the birds, the main support of chance castaway mariners". Such foresight has been extremely rare, however; mammals have been introduced deliberately to 15 islands or island groups in the Antarctic alone and are at present established on 13 of these.

In the early years of the present century, Norwegian whalers introduced 11 reindeer onto the island of South Georgia; since then they have multiplied into a herd of several thousand head, to the great destruction of the island's natural vegetation. More common has been the accidental introduction of mice and rats, sometimes followed by the deliberate introduction of cats in an effort to control the rodents. Classic examples are provided by Marion Island and Macquarie Island, both breeding places for subantarctic birds.

INTRODUCED MARINE SPECIES

The introduction of non-indigenous marine species along the world's coastlines is much less well documented than the introduction of terrestrial species onto islands. The earliest introductions were undoubtedly accidental and arose from invertebrates being carried on the wooden hulls of sailing vessels. Thus the present distribution of barnacles and some molluscs may be partly due to early introductions by humans. More recent modes of transport, such as the water used as ballast by oil tankers, may also be responsible for introducing alien species.

A number of recent introductions have been documented. For example, *Carcinus*, a crab common in Europe but not in the Southern Hemisphere, was discovered in Table Bay harbor, at the southern tip of Africa, in 1972, having possibly been introduced from oil rigs. At first the crab was quite rare but it has since colonized the harbor in large numbers, replacing much of the natural fauna, and it has also been found at Bloubergstrand, 15 kilometers (9 miles) north of the harbor. The situation is being monitored with some apprehension because, as *Carcinus* is a very robust and highly aggressive predator, the spread of the crab along the coast would probably have a very considerable effect on the indigenous biota. It has certainly become a pest in some other countries to which it has been introduced and in the United States was regarded as a serious threat

Kathie Atkinson

▲ The whelk *Morula marginalba* has been accidentally introduced to many parts of the world. Commonly known as the oyster drill because of its habit of boring holes through the shells of oysters and other bivalves, it is now a pest in some aquaculture operations.

▼ The worldwide distribution of the rocky shore mussel *Mytilus* results partly from deliberate introduction for aquaculture; partly from its planktonic larvae which enable it to spread away from the point of introduction; and partly from its ability to out-compete the indigenous marine fauna.

to the soft clam industry. Regrettably, tackling the problem of an introduced marine species is even more difficult than for a terrestrial one. In the case of *Carcinus* there is simply no way that the entire population can be eradicated once it is established, without adversely affecting the naturally occurring biota.

In other cases the situation is somewhat more hopeful. The whelks known as oyster drills because they feed by boring holes through the shells of oysters and other bivalve molluscs, travel well on ships' hulls and have been accidentally introduced to the shores of many parts of the world. In some areas they have become a serious pest, and a threat to commercial oyster and mussel farms. However, unlike most molluscs and unlike the crab *Carcinus*, oyster drills do not have free-swimming larvae. They can thus spread only slowly along the shore. Moreover, populations can be eradicated by collecting their egg capsules and allowing them to dry out in the sun, a method which has been used successfully in Britain.

Introduced species of the winkle *Littorina* are easier to deal with, for not only do they have no free larval stages, but they occur near the top of the shore, from which they can be removed by hand. Molluscicide poisons have sometimes been used in an attempt to eradicate discrete populations of introduced molluscs. However, they are seldom completely effective and their benefits must be weighed against the damage they are almost certain to do to the indigenous molluscs in the area.

Jeff Foott/Survival Anglia

## DELIBERATE INTRODUCTIONS

For about four centuries alien species of molluscs have been deliberately introduced for cultivation as food. For instance, many countries cultivate the Japanese oyster in preference to local species, because it grows faster and is thus marketable sooner. As a result of such introductions, the Pacific coast of the United States, for example, now supports some 40 non-native molluscan species. As most molluscs have planktonic larvae, there is no way of preventing their spread to adjacent areas along the coast. Although in many cases these aliens cannot compete with the indigenous fauna, there are exceptions. The rocky shore mussel *Mytilus*, cultivated in South Africa, has established itself along the coast, and is proving a problem in many areas. It was probably introduced long before it was cultivated. No sound method of eradicating this introduced mussel has been proposed.

A problem seldom accorded the attention it deserves is that of parasites and pathogens introduced with the animals to be cultured. Such parasites may spread to indigenous species which are likely to be ill adapted to deal with them. The effects are potentially devastating. The growth of molluscan aquaculture makes it essential that reliable methods for eliminating parasites from transported stock be explored. Furthermore, the biology of disease-causing organisms and their interactions with host organisms need urgent investigation. Too little attention has also been given to introduced algae.

The problem of introduced species and their parasites in the marine environment is not being faced squarely, and it is not possible to report any great success in the control of such species. For there to be any hope of eradication, the life history and behavior of the alien must be fully understood, as well as its reproductive patterns, growth, and rate of dispersal in its new habitat. Even then there is at present little chance of controlling it without damaging the ecosystem. Prevention is therefore not just better than cure—it is essential in the absence of any cure.

ALEC C. BROWN

# PROTECTING THE MARINE ENVIRONMENT

More than 300 marine and estuarine sites around the world have been reserved as protected areas for their outstanding features. They are known variously as marine parks, sanctuaries, or reserves; aquatic, fisheries, habitat, wetland, or nature reserves; historic or shipwreck sites. Marine protected areas have been reserved to protect endangered species, marine and estuarine environments, aesthetic and cultural values; to manage and improve fisheries; and to provide opportunities for conservation, education, and scientific or historical research. They vary in size from less than 3 hectares (7.5 acres), as for example at the Shiprock Aquatic Reserve in Port Hacking near Sydney, Australia, to the approximately 35 million hectares (87 million acres) of the Great Barrier Reef Marine Park.

## PROTECTING MARINE SPECIES AND HABITATS

Initially, areas were set aside to protect the beauty of their marine features for tourists. The first marine protected areas, established in the 1930s, were located at the Dry Tortugas off southern Florida in the United States, and Green Island in the Great Barrier Reef off Cairns in Queensland, Australia. Marine areas have since been reserved in many countries specifically to protect endangered marine species and their habitats. For example, over 19,800 square kilometers (7,640 square miles) in the southwest Pacific Ocean (Lihou Reef and the Coringa-Herald national nature reserves in the western Coral Sea) and the northwest shelf of Australia (Ashmore Reef National Nature Reserve) have been set aside by the Australian government primarily to protect the feeding and breeding habitats of migratory seabirds and turtles.

By the mid-1980s coastal populations of the black cod *Epinephelus daemelli*, a large, sedentary reef fish once common on the southeast Australian coast, had declined to the extent that it was declared a protected species in Australian waters. Its protection has been helped by the establishment in 1987 of the Elizabeth and Middleton Reefs Marine National Nature Reserve (east of the New South Wales coast). Each reef supports a large population of the black cod, and commercial and recreational fishing are restricted within the reserve.

Whales are a significant resource in many protected areas. The St Lawrence River in Canada has been home to the beluga whale *Delphinapterus leucas* since the last ice age. Today only several hundred belugas remain of the 5,000 strong population reported at the beginning of the century. Despite total bans on hunting in recent years, a continuing decline appears to be related to high levels of toxic industrial compounds in the environment. Concern in the community has led to a multimillion dollar Canadian government project to establish a marine park to protect the whale's prime habitat in a 96-kilometer (60-mile) section of the St Lawrence River. Simultaneously, the government will be acting to reduce levels of toxic materials entering the habitat from land-based industries.

The Channel Islands National Marine Sanctuary off southern California, established in 1980, is the largest marine protected area in the United States. Many of the world's whale species, such as California gray, humpback, blue, fin, and sei whales, regularly migrate through it. Southern humpback whales swim along the east coast of Australia from Antarctica to breeding grounds in the Coral Sea, and many stop over in Hervey Bay

▼ The map shows part of the Great Barrier Reef and two protected areas in the Coral Sea. Between them they cover nearly 360,000 square kilometers (139,000 square miles) of continental, offshore, and oceanic coral reef habitats.

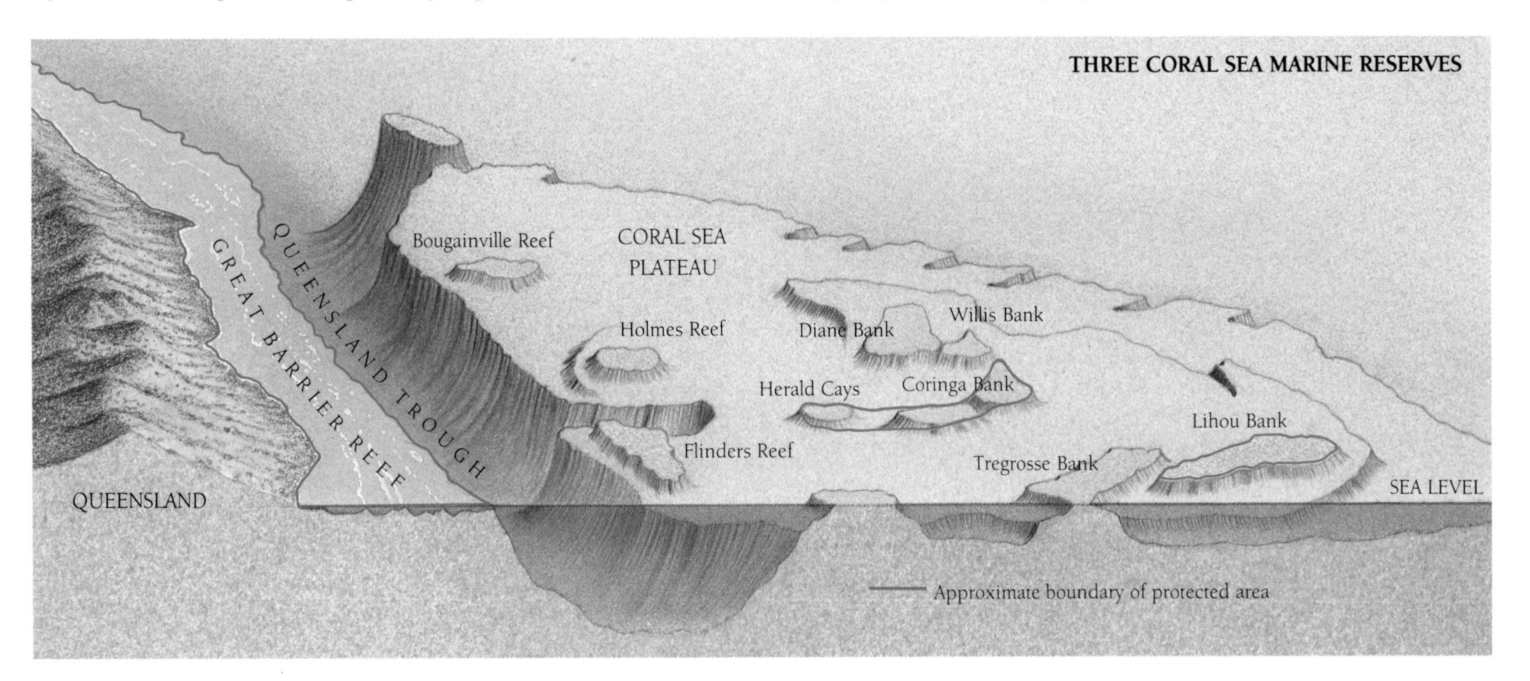

Al Grotell

▲ Although the John Pennekamp Coral Reef State Park was established as a marine protected area off Florida, these snappers and grunts within its boundaries are not protected from recreational or commercial fishing, and their numbers are continuing to decline.

(south of the Great Barrier Reef Marine Park). The Hervey Bay Marine Park was established in 1989, primarily to control a rapidly growing whale-watching industry in the bay and to minimize possible adverse effects on the whales.

HOW MUCH PROTECTION?

Because many communities have feared the possible or perceived effects of restrictions, most marine protected areas have had to allow continued access to the natural marine resources for recreational and commercial purposes. However, many features for which the areas were protected are beginning to degrade in quality. Lush coral growths in parts of the Florida Reef Tract off Key Largo which were protected in 1960 (John Pennekamp Coral Reef State Park) and 1975 (Key Largo National Marine Sanctuary) have been damaged or have died. Fish and lobsters are becoming harder to catch as numbers decrease. These declines are attributed to too many visitors and too much pollution resulting from increases in nearby coastal residential developments lacking appropriate controls. Kelp forests and populations of the black abalone in the Channel Islands National Marine Sanctuary have declined sharply in the past five years. Pressures imposed by continuing human activities are intensifying the effects of natural environmental stresses on these species.

Some marine protected areas do exclude exploitative activities. Studies in the Philippines and New Zealand are finding important differences between populations of commercial and recreational species inside and outside totally protected areas, which raises the possibility of replenishing fish stocks in exploited waters from those in the protected areas.

The Cape Rodney-to-Okakari Point Marine Reserve (north of Auckland, New Zealand), established in 1975, excludes all forms of fishing and collecting, and access to visitors is limited. Research conducted there by the University of Auckland Marine Biological Station indicates that local snapper and lobster are larger and more numerous inside the reserve than in exploited areas beside it. The reserve was originally considered a threat by the local commercial fishing community, but its beneficial effects are now recognized and the industry contains some of the most ardent defenders against illegal fishing activity.

## THE GREAT BARRIER REEF MARINE PARK

The Great Barrier Reef Marine Park, which extends for some 2,000 kilometers (1,240 miles) along the northeast coast of Australia, is the world's largest and best-known marine protected area. Inscribed on the World Heritage List in 1981, it includes a vast array of tropical habitats including fringing coastal reefs, coral cays, barrier reefs, and deeper areas between the reefs and offshore. Its four regional management areas are zoned through extensive community consultation and regular review to accommodate commercial and recreational activities.

General use zones (75–85 percent of the total area) provide for most activities; more protective zones (14–24 percent of the total area) allow only non-exploitative activities such as underwater photography and observation; and reference and research zones (1 percent or less of the total area) totally exclude the public. Significant bird and turtle breeding areas are closed seasonally, as are "reef appreciation areas" and "replenishment areas". Oil drilling and mining, spearfishing with scuba-diving equipment, capture of the potato cod, giant groper, and other nominated species of fish, and littering are totally prohibited throughout the marine park.

## THE FUTURE

Major challenges to the management of this marine park, and many other marine protected areas, include an exploding tourism industry, damage to reefs by crown-of-thorns starfish and other species, and excess nitrogen, phosphorus, and sediments from land-based activities, and the decline of commercial and recreational fisheries.

The survival of healthy oceans to provide a wealth of natural resources now and in the future will depend on extensive protection and integrated management of the marine environment through the reservation of significant or representative coastal and offshore areas. All lovers and users of the sea have a responsibility to promote the protection of our rich marine heritage by encouraging cooperation at all levels of government and within our own communities.

ANGELA MARIA IVANOVICI

Al Grotell

▲ This nassau grouper *Epinephelus striatus* waits patiently at a "cleaning station" while cleaner gobies (a species of small fish) remove parasites from its body.

▼ Although marine sponges are among the world's most primitive multicellular organisms, they exhibit an infinite variety of colors and growth forms, attaching themselves to most hard substrates. One of the more spectacular species is the "elephants ear" sponge, which can be seen in the Great Barrier Reef Marine Park.

Kev Deacon/Dive 2000

# ANTARCTICA: A WORLD APART

Antarctica is a special place. It is the coldest continent, the driest, the windiest, and the most isolated. It is the only continent without an indigenous population. Because of its isolation, and its own oceanic surface circulation system, it has developed a unique biota, dominated by krill, whales, seals, and penguins.

In recent years, Antarctica has become a focus of heightened international interest for three major reasons: the recognition there of a major stratospheric ozone depletion in the austral spring; the value of the ice core record as a source of baseline information on the natural variation of the earth's atmosphere and oceans over hundreds of thousands of years; and the debate over protection of the Antarctic environment, occasioned particularly by proposals to regulate mineral exploitation, but also in relation to other resource developments such as fishery, tourism, and iceberg utilization.

Until the signing of the Antarctic Treaty in 1961, management of the Antarctic was a matter of concern principally to the seven nations—Argentina, Australia, Chile, France, Great Britain, New Zealand, and Norway—that claim Antarctic territory. This was not a simple matter, as the claims of three nations overlap and not all the claimant nations recognize each other's claims. The success of the International Geophysical Year of 1957-58, when 12 nations—the claimants plus Belgium, Japan, South Africa, the United States, and the Soviet Union—agreed to cooperate in a major scientific program, led to the international negotiations that established the Antarctic Treaty to cover the surface south of 60°S. Careful formulation of Article 4 of the treaty enabled territorial disputes to be put to one side so that Antarctica could be internationalized as a continent for science.

The other important element of international Antarctic affairs is the Scientific Committee on Antarctic Research (SCAR), established in 1958 to guide the planning for the International Geophysical Year, and thus predating the Antarctic Treaty. Membership of this organization is through national academies of science. Both SCAR and the Antarctic Treaty consultative meeting are held every two years, but in alternating years.

▼ Seven countries have laid claim to parts of the continent of Antarctica, and others have established sovereignty over the subantarctic islands. The future of Antarctica depends upon international cooperation and the successful working of the Antarctic Treaty.

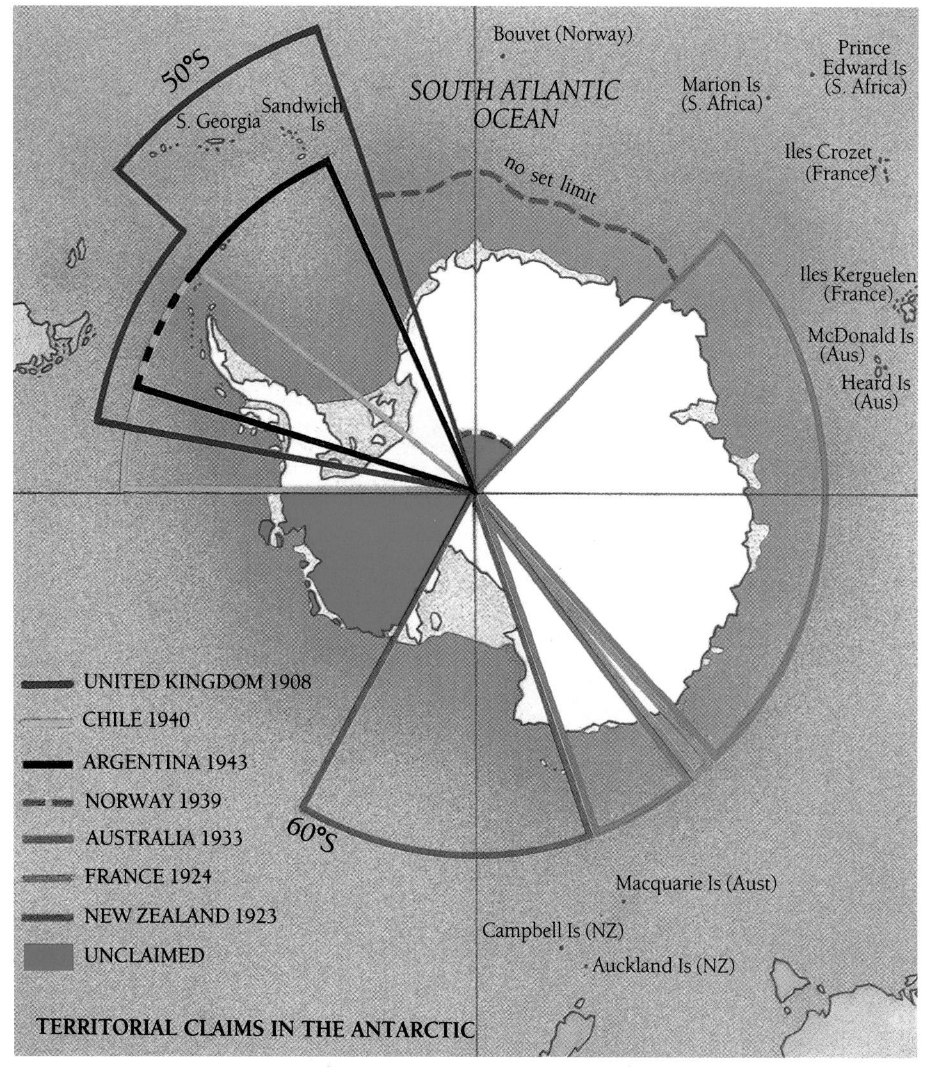


## PROTECTED AREAS IN THE ANTARCTIC

The Antarctic Treaty system has recognized the need for protected areas in the Antarctic for a variety of historical, conservation, environmental, and scientific reasons. There are four principal categories. Specially Protected Areas (SPA) are designated for the protection of fauna and flora. These are largely based on islands, some including a small marine environment, to protect bird or seal breeding sites. Sites of Special Scientific Interest (SSSI) are identified for the duration of specific experiments. Among them for marine research are Chile (or Discovery) Bay on Greenwich Island in the South Shetland Islands (for studies of its benthos), and two small sites in the crater of Deception Island (for colonization studies). A new category is the Specially Reserved Area (SRA) that can be designated for a variety of reasons, including protection of non-biological features such as geologically significant sites. Another new category is the Multiple Use Area. Although none has yet been designated, the concept is a broad one and includes the possibility of identifying areas of particular significance for tourism.

There is also provision for an ecosystem monitoring program to operate from land-based sites, which will record changes in the marine ecosystem. At present, only one site has been nominated, at Magnetic Island near Australia's Davis Station. Other sites are being examined in East Antarctica and on the Antarctic Peninsula.

Colin Monteath/Hedgehog House New Zealand

▲ Antarctica is a wild and beautiful continent. In recent years it has become an increasingly popular destination for tourists aboard cruise-liners such as this one, seen here off Cape Hallet, North Victoria Land.

## THE FUTURE

The Antarctic Treaty came into force for 30 years in the first instance, allowing the possibility of renegotiation in 1991, or in later years, if called for. At present there is no evidence of such a call because the treaty has proved to be a very flexible document and hundreds of recommendations from treaty consultative meetings have been accepted by member governments. In addition, many other instruments have been negotiated, including agreements for the conservation of fauna and flora.

Future developments include a consideration of the impact of tourism on Antarctica. At present, some 6,000 tourists per year visit the continent, mainly on well-organized ship-based tours. There are also a few aircraft-based tourism ventures, especially from South America. New ships are being built for Antarctic tour operators, and pressure will undoubtedly grow for increasing numbers of tourists and associated infrastructure.

Antarctic waters are already the target of a major fishery concentrating on krill and finfish such as *Notothenia rossii* and *Champsocephalus gunnari*. The level of finfishery is a cause for concern, and some species in particular localities have been fished nearly to extinction. It is possible that the new environmental measures may address this concern. Although the krill catch is now relatively small compared with its potential, it is already the largest single-species crustacean fishery on the globe. Because of its high protein content and potential as a source of chitin (an organic material in the skeletons of crustaceans of value in the medical industry and as a purifier), krill is seen in some quarters as the growing fishery product of the 1990s. This is made more likely by the collapse of several traditional Northern Hemisphere fisheries, leaving a large trawler fleet available for use in the Southern Ocean.

There has been considerable discussion in the past of the potential use of Antarctic ice as a water source and as refrigerated storage for excess production from elsewhere on Earth. To date no firm proposals for development have come forward but pressure can be expected to build as world population increases. Scientists have predicted that some traditional grain-producing areas, such as southwestern Australia, may dry out due to the impact of human-induced global climate change, and this may provoke a call for new sources of water. Already possibilities for transporting ice to cope with this eventuality are being discussed. At any time, there are some 200,000 icebergs floating around Antarctica and the total volume of ice shed annually as ice exceeds the total world water usage.

Antarctica is still remote, but decreasingly so. As communications improve the continent will become less isolated. The Antarctic Treaty system will evolve and change to meet new challenges, but Antarctica is unlikely to become heavily populated or subject to major local sources of pollution. It will probably remain a continent for science into the foreseeable future.

PATRICK G. QUILTY

# THE IMPLICATIONS OF GLOBAL CLIMATE CHANGE

Surrounding the Earth is the atmosphere—a sphere of gases 500 kilometers (300 miles) thick. Although the lower atmosphere—the troposphere—contains most of the gases, the next layer—the stratosphere—contains the ozone that provides protection against ultraviolet radiation. Thus the Earth and all that exists upon it are protected from the direct path of the sun's rays by this mixture of atmospheric gases, of which the most important are nitrogen (about 78 percent), oxygen (about 20 percent), water vapor, and carbon dioxide.

## THE GREENHOUSE GASES

Some of the energy of the sun that falls on the Earth is reflected back into space; the rest is absorbed, warming the Earth's surface, which itself emits heat upward. The absorption of some of this heat by trace gases in the atmosphere—the so-called greenhouse gases—makes the Earth an inhabitable planet.

▼ Tropical areas like Hawaii are already known for their seasonal convectional storms. With global warming, this phenomenon is likely to increase. A greater level of irradiation could result in increased evaporation and therefore an increase in cloud cover, eventually changing local weather patterns.

Michael S. Yamashita/Colorific!

The most important of the greenhouse gases is carbon dioxide. It is absorbed by the oceans, taken up by plants in the presence of sunlight, and given off to the atmosphere when combustion takes place. Thus active volcanoes and the burning of vegetation by wildfire, together with the respiration of plants and animals, put carbon dioxide into the atmosphere. Methane, sometimes called "marsh gas" or "rotten egg gas", is another significant greenhouse gas. It is produced naturally from marshes and boglands, and from the digestive processes of animals. Water vapor, nitrous oxide, chlorofluorocarbons (CFCs), and ozone are other greenhouse gases.

As well as occurring naturally, carbon dioxide, methane, and nitrous oxide are also produced by human activities such as running cars, trucks, buses, and trains, and generating electricity from coal or petroleum. Methane is generated in rice paddyfields and its production has also grown as a result of a worldwide increase in cattle herds. Chlorofluorocarbons are produced only by human activities—they were invented in the 1930s and are used to replace ammonia in refrigerators. They have two separate effects on climate change: they contribute both to greenhouse gases and to the depletion of stratospheric ozone.

## THE EFFECTS OF GLOBAL WARMING

Over the last 30 years, increasing carbon dioxide concentrations have been measured in the atmosphere. This, together with data from analyses of air trapped in polar ice, indicates an increase in concentration of about 5 percent per decade, particularly since the Industrial Revolution. Based on similar data, methane concentrations indicate a 110 percent increase since pre-industrial times. The concentrations of nitrous oxide have climbed by about 8 percent in the same period. CFCs contribute more to the problems of climate change: they are long lasting; they play a major part in greenhouse gas build-up; and they are the primary contributor to atmospheric ozone destruction.

The accumulations of greenhouse gases will give rise to an increasing average temperature level in the atmosphere, the extent of which will depend upon a number of factors that are not yet well understood. In particular, although carbon dioxide is absorbed by the oceans, little is known about the processes involved. The role of phytoplankton—the microscopic green plants of the ocean—is not yet understood well enough to allow accurate predictions. Another unknown is the response of global cloud systems to the new conditions. Cloudiness may increase in some regions, but the effect on energy transfers within the atmosphere is still not resolved. Observations of temperatures around the world indicate a real but irregular increase in global surface temperature since the late nineteenth century.

Depending upon the extent of warming, two major effects will be observable. The initial heating will lead to an expansion of the oceans, which must lead to a rise in sea levels around the world. Only after some centuries of increased temperatures might the glacier ice begin to melt and contribute to further sea-level rise. The second major effect will be an intensification of the global hydrological cycle—more evaporation, more cloudiness, and more precipitation. At a regional level, there will be changes in the distribution of rainfall and storms. Tropical climates will expand toward the poles, bringing cyclones or hurricanes further south and north in the respective hemispheres.

Global sea level rose between 1 and 2 centimeters (0.4–0.8 inch) over the last century. There is no firm evidence of an acceleration this century, although some measurements suggest that the sea level has risen faster this century than in the previous two centuries. The most recent predictions indicate a rise in sea level of between 16 and 51 centimeters (6–20 inches) by the year 2050 if emissions of greenhouse gases continue at a high rate. Even if very strict controls on emissions are introduced, the prediction is of a rise of about 10 centimeters (4 inches) by the year 2050 because of the gases already in the atmosphere.

## THE IMPACT ON ISLANDS

A rising sea level could have quite catastrophic impacts in many countries even if the most conservative predictions are used. Low-lying countries such as Bangladesh are likely to be the first affected. Between 18 and 34 percent of the habitable lands of that country could be lost and a more extensive area affected by saltwater intrusion into the groundwater.

Islands fringed by coral reefs may survive better than those without reefs if coral growth and deposition can keep up with the increased rate of sea-level rise. A change in the number and intensity of storms may complicate this situation by causing more erosion of the cays and more damage to the protecting coral reefs. Islands such as Kiribati or the Maldives may be eroded by higher seas and their small freshwater lenses destroyed by saltwater intrusion. Should they be in a region that experiences an increase in rainfall, the groundwater lens may be protected to some degree.

High islands in regions with good rainfall, such as Hawaii and Guadalcanal, may experience an increased growth rate in vegetation because of the fertilizing effect of increased carbon dioxide. This assumes that soil nutrients and moisture are sufficient to support enhanced growth. Increased storminess and a widening of the extremes in climate may, however, overshadow such potentially beneficial effects.

Both high and low islands will lose low-lying land to the sea. In high islands this may be the most arable land; its loss would thus have severe impacts on food production. These low-lying lands may also support nursery areas or prime habitat for fish, shellfish, and crustaceans. In the initial phases their erosion may limit the capacity of nearby reefs to grow and respond to increasing sea levels.

In non-tropical areas, an increase in rainfall and humidity may lead to increased incidence of pathogens. Bacterial and fungal pathogens of humans, other animals, and plants may limit the ability of a community to adjust to new circumstances. Other social impacts may include a greater propensity for wildfire and a loss of tourist beaches.

The potential impacts of global climate change induced by an enhanced greenhouse effect will disrupt island communities more seriously than large continental nations. The smaller landmass and proportionately longer coastline indicates that a large component of their most productive land could be adversely affected. Moreover, the environmental management strategies of small island communities, including resource and social planning capacities, are less well developed than those of larger nations.

ALISTAIR J. GILMOUR

Paul Steel/Stock Photos Pty Ltd

▲ A tropical palm tree beach in the Maldives. Such exotic scenes from present-day low islands may be something of the past if global warming produces the radical effects some scientists are predicting. A rise in sea level could inundate low-lying areas, and changing weather patterns could mean that the "tropics" are expanded to include areas that are currently temperate.

# FACTS ABOUT OCEANS AND ISLANDS

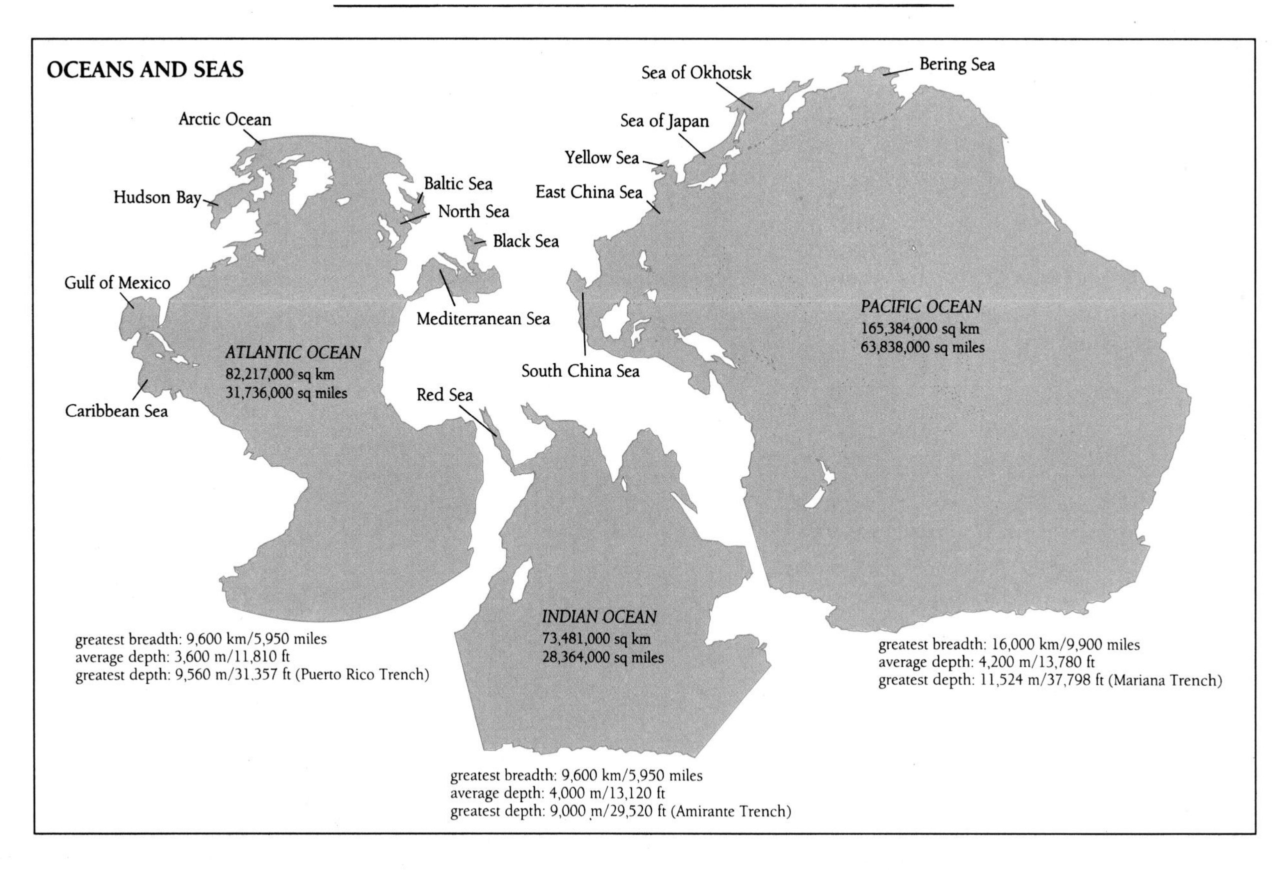


## LARGEST ISLANDS

Australia is regarded as a continental landmass rather than an island. The total area of Australia (including Tasmania) is 7,682,300 sq km/ 2,965,370 sq miles.

Greenland, Denmark
2,175,610 SQ KM/ 839,790 SQ MILES

New Guinea
820,660 SQ KM/316,770 SQ MILES

Borneo
746,550 SQ KM/ 288,170 SQ MILES

Madagascar
587,040 SQ KM/226,560 SQ MILES

Baffin Island, Canada
476,070 SQ KM/183,760 SQ MILES

Sumatra, Indonesia
473,600 SQ KM/182,810 SQ MILES

Honshu, Japan
230,330 SQ KM/88,910 SQ MILES

Great Britian
229,960 SQ KM/88,765 SQ MILES

Ellesmere Island, Canada
212,690 SQ KM/82,100 SQ MILES

Victoria Island, Canada
212,200 SQ KM/81,910 SQ MILES

Sulawesi, Indonesia
189,220 SQ KM/73,040 SQ MILES

South Island, New Zealand
150,460 SQ KM/58,080 SQ MILES

Java, Indonesia
131,430 SQ KM/50,730 SQ MILES

North Island, New Zealand
114,690 SQ KM/44,270 SQ MILES

Cuba
110,920 SQ KM/42,810 SQ MILES

Newfoundland, Canada
112,300 SQ KM/43,350 SQ MILES

Luzon, Philippines
105,710 SQ KM/40,800 SQ MILES

Iceland
102,850 SQ KM/39,700 SQ MILES

Mindanao, Philippines
95,590 SQ KM/36,900 SQ MILES

Ireland
84,400 SQ KM/32,580 SQ MILES

## ENDANGERED MARINE SPECIES

The following marine species have been listed as endangered by the International Union for Conservation of Nature (IUCN). Their habitats are shown on the accompanying map.

In addition to the species listed here, a number of seabird, fish, and invertebrate species are also endangered.

### MAMMALS

| | |
|---|---|
| fin whale | *Balaenoptera musculus* |
| humpback whale | *Megaptera novaeangliae* |
| bowhead whale | *Balaena mysticetus* |
| northern right whale | *Eubalaena glacialis* |
| Japanese sea lion | *Zalophus californianus japonicus* |
| Mediterranean monk seal | *Monachus monachus* |
| Hawaiian monk seal | *Monachus schauinslandi* |
| Caribbean monk seal | *Monachus tropicalis* (possibly extinct) |
| saimaa seal | *Phoca hispida saimensis* |
| Steller's sea cow | *Hydrodamalis gigas* (extinct) |

### REPTILES

| | |
|---|---|
| green turtle | *Chelonia mydas* |
| hawksbill turtle | *Eretmochelys imbricata* |
| Kemp's ridley | *Lepidochelys kempii* |
| olive ridley | *Lepidochelys olivacea* |
| leatherback turtle | *Dermochelys coriacea* |

Source: *1988 IUCN Red List of Theatened Animals* (IUCN Conservation Monitoring Center, Cambridge, 1989)

**LIFE IN THE OCEAN** The International Standard Statistical Classification of Aquatic Animals and Plants (ISSCAAP)

**FRESHWATER FISHES**
carps, barbels, and other cyprinids
tilapias and other cichlids
miscellaneous freshwater fishes

**DIADROMOUS FISHES**
sturgeons, paddlefishes etc.
river eels
salmons, trouts, smelt etc.
shads, milkfishes etc.
miscellaneous diadromous fishes

**MARINE FISHES**
flounders, halibuts, soles etc.
cods, hakes, haddocks etc.
redfishes, basses, congers etc.
jacks, mullets, sauries etc.
herrings, sardines, anchovies etc.
tunas, bonitos, billfishes etc.
mackerel, snocks, cutlassfishes etc.
sharks, rays, chimaeras etc.
miscellaneous marine fishes

**CRUSTACEANS**
freshwater crustaceans
sea-spiders, crabs etc.
lobsters, spiny-rock lobsters etc.
squat-lobsters, nephrops etc.
shrimps, prawns etc.
krill, planktonic crustaceans etc
miscellaneous marine crustaceans

**MOLLUSCS**
freshwater molluscs
abalones, winkles, conchs etc.
oysters
mussels
scallops, pectens etc.
clams, cockles, arkshells etc.
squids, cuttlefishes, octopuses etc.
miscellaneous marine molluscs

**WHALES, SEALS AND OTHER AQUATIC MAMMALS**
blue whales, fin whales etc.
minke whales, pilot whales etc.
porpoises, dolphins etc.
eared seals, hair seals, walruses etc.
miscellaneous aquatic mammals

**MISCELLANEOUS AQUATIC ANIMALS**
frogs and other amphibians
turtles and other reptiles
sea-squirts and other tunicates
horseshoe crabs and other arachnoids
sea-urchins and other echinoderms
miscellaneous aquatic invertebrates

**MISCELLANEOUS AQUATIC ANIMAL PRODUCTS**
pearls, mother-of-pearl, shells etc.
corals
sponges
aquatic bird guano, eggs etc.

**AQUATIC PLANTS**
brown seaweed
red seaweed
green seaweed and other algae
miscellaneous aquatic plants

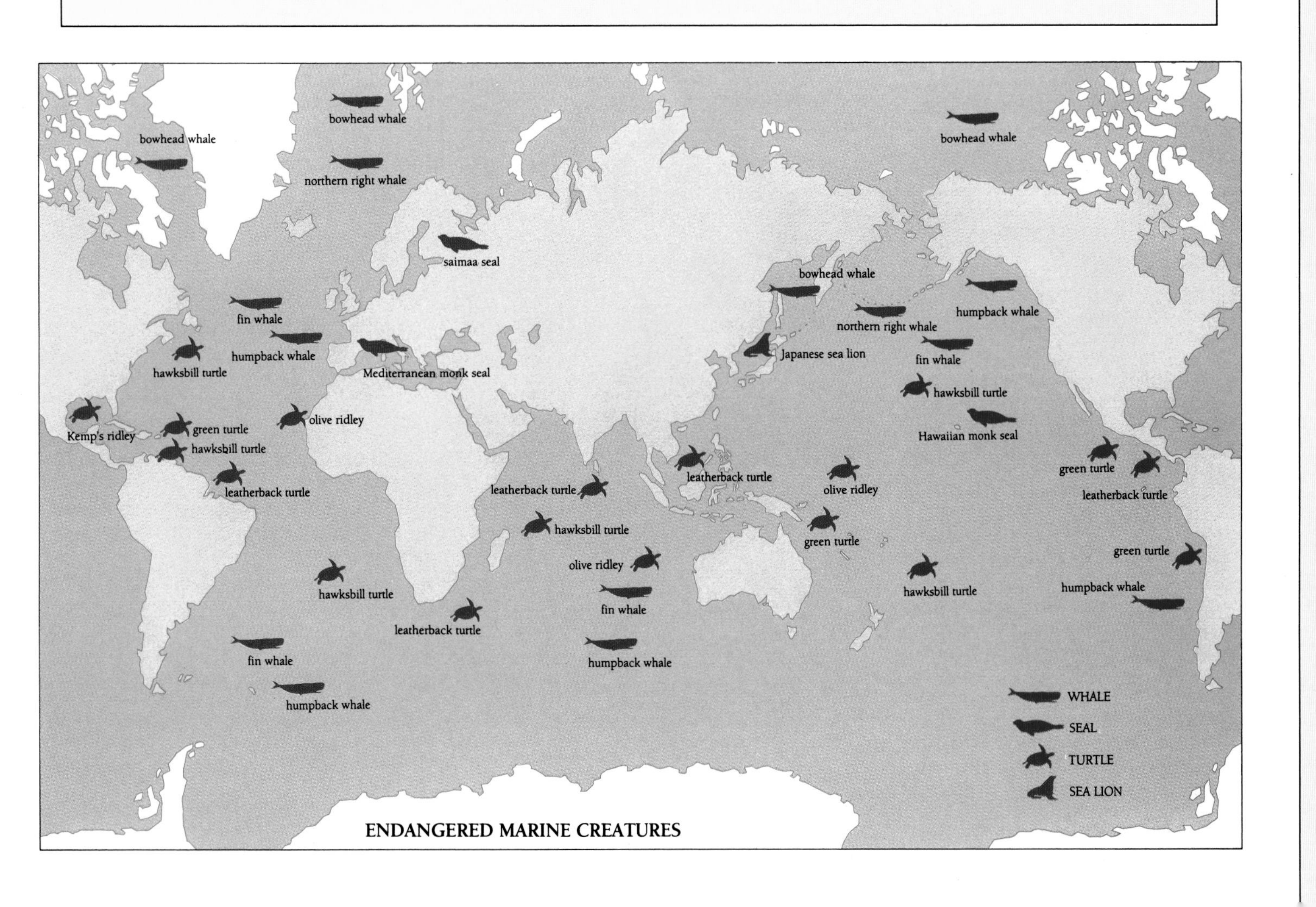


ENDANGERED MARINE CREATURES

# GLOSSARY

**A**

ALGAE Aquatic plants; unlike terrestrial plants, algae are a nonvascular (unable to transmit or circulate fluid) division of the plant kingdom. They range in size from unicellular phytoplankton to giant kelp.

AMPHIPOD A marine or freshwater crustacean which usually has a laterally compressed body e.g. sandhopper.

ANADROMOUS Describes fish which migrate from salt water to fresh water to spawn e.g. salmon.

ANNELID A ringed or segmented worm, including bristle worms, earthworms, and leeches. The annelid body is more complex than that of round worms or flatworms.

ANTARCTIC CONVERGENCE Important biogeographical boundary that marks the meeting between warmer subantarctic waters and colder Antarctic waters; coincides roughly with the northern limit of Antarctic icebergs.

AQUACULTURE The farming of aquatic or marine plants and animals, as opposed to their exploitation.

ASTHENOSPHERE A shell of uncertain thickness, tens of kilometers beneath the surface of the Earth. In this shell, the Earth's crust is very plastic, enabling a more even distribution of the Earth's mass.

ASTROLABE An instrument used by astronomers to measure the altitude of planets and stars.

ATOLL A ring-like island or series of islands or reefs, which encircle a lagoon.

**B**

BALEEN The horny plates attached to the upper jaws of true whales, and used to sieve zooplankton from the ocean.

BARRIER REEF A coral reef separated from the coast by a stretch of water that is too deep for coral growth.

BATHYMETRY Measurement of depths, especially the ocean.

BATHYPELAGOS Deep sea.

BENTHIC Describes organisms living on the sea floor; also applied to all the bottom terrain from the shoreline to the greatest depths.

BERM A terrace-like, nearly flat point of a beach formed by sediment deposited through wave action. Some beaches have no berms; others have one or more.

BIOLUMINESCENCE The production of light by any living organism, either aquatic or terrestrial.

BRITTLESTAR Common name for the Ophiuroidea—a class of echinoderms (related to starfish and sea urchins) that have narrow arms radiating from a central disk, to form a star-shaped body.

BRYOZOAN A member of the phylum Bryozoa, these are small colonial marine animals called sea mosses because they slightly resemble some seaweeds.

**C**

CALDERA A large basin-shaped crater usually formed by the subsidence of the top of a volcanic mountain. The crater is sometimes occupied by a lake, and may have one or more active cones if the volcano is not extinct.

CARRION Dead and decaying flesh, which is eaten by scavengers.

CATADROMOUS Describes fish which migrate from fresh water to salt water to spawn.

CAY A low flat mound of sediment (usually coral fragments and sand) which builds up on a reef flat, slightly above the high tide level. Eventually, these mounds may develop into low vegetated islands.

CEPHALOPOD A mollusc such as octopus and squid which has a large well-developed head surrounded by tentacles, and a large mantle cavity. Water entering this cavity via a siphon is used to create jet propulsion—the means by which this animal moves.

CHEMOSYNTHESIS The process of building up organic compounds in some organisms (e.g. bacteria) via the oxidation of inorganic compounds (e.g. hydrogen sulfide). This process is similar to plant photosynthesis but does not require light.

CONTINENTAL ISLAND An island that is close to and geologically related to a continent. Usually such islands are detached fragments of nearby continents, from which they are separated by shallow water.

CONTINENTAL MARGIN Consists of the continental shelf, continental slope, and continental rise; together these make up a zone which separates the deep-sea floor from continental landmasses.

CONTINENTAL RISE Underwater surface beyond the continental slope. Generally occurs at depths between 1,370 meters (4,500 feet) and 5,200 meters (17,000 feet), and slopes gently to the abyssal plains.

CONTINENTAL SHELF Shallow underwater extension from the edge of a continent (shoreline), sloping gently down to an abrupt increase in bottom gradient (continental slope). The shelf around some countries is very narrow, but can extend more than 320 kilometers (200 miles) around others.

CONTINENTAL SLOPE Begins at the edge of the continental shelf and slopes rapidly to about 1,370 meters–3,000 meters (4,500–10,000 feet) to the beginning of the continental rise.

COPEPOD A group of free living or parasitic crustaceans, generally microscopic, that are found in marine or freshwater environments.

COPRA The dried flesh of the coconut, from which coconut oil is pressed. Copra is the main cash crop of many tropical nations.

CORIOLIS EFFECT The directional force resulting from Earth's rotation, causing a moving body (e.g. water) on the Earth's surface to be deflected to the left in the Southern Hemisphere and to the right in the north.

CRUSTACEAN A subphylum of mostly aquatic arthropods that have hard shell or crust covering their bodies. They include lobsters, prawns, crabs, and barnacles.

CTENOPHORE A small marine invertebrate (phylum Ctenophora), known as a comb jelly because of its resemblance to jellyfish and the way it swims using comb-like bands of hairs (cilia).

CURRAGH A type of small rounded boat that has a wickerwork frame covered with skin or canvas. Used in Ireland, Wales and some parts of northern England.

CUSP Low mound of sediment formed by wave action along the foreshore of a beach. Cusps are usually separated from each other by crescent-shaped hollows.

**D**

DECOMPRESSION The gradual return of people who have been diving or working in deep water to conditions of normal atmospheric pressure.

DEMERSAL Describes organisms that live on or near the bottom of lakes or seas.

DEPOSIT FEEDER A bottom-living (demersal) animal which feeds on the organic material that falls from the water column and collects on the bottom.

DIADROMOUS Describes animals that spend part of their lives in fresh water and part in salt water.

**E**

ECHINODERM A phylum of marine animals that are organised into a five segment pattern of body symmetry as adults, with a calcareous internal skeleton and a water vascular system. The latter is used like a hydraulic pump to enable the organism to move its body parts. Includes sea urchins, starfish, and sea cucumbers.

ECOSYSTEM The complex community of organisms within any defined environment; their interactions among themselves and with their physical environment.

EDDY A usually circular current of air or water running contrary to the main current, as in a whirlpool.

ELASMOBRANCH Subset of the cartilaginous fishes (Chondrichthyes) which have platelike scales, a spiracle (air hole) and multiple gill slits e.g. sharks and rays.

ENDEMISM The confinement of some species to a particular area or region.

**F**

FILTER FEEDER An animal that filters planktonic food from the water column.

FORAMINIFERA Protozoans which have a hard shell and are members of the order Sarcodina.

FRINGING REEF A coral reef partially or completely surrounding a landmass or island; unlike a barrier reef, a fringing reef is attached directly to the shore.

**G**

GESTATION The period of time during which an embryo develops inside its mother's uterus.

GRABEN A geological term to describe a long piece of rock that has moved downwards along fault lines relative to the rocks on either side of it.

GUANO Seabird excrement found in large deposits on offshore islands. Rich in nitrogen and phosphorus, it is a valuable fertilizer.

GUYOT A generally flat-topped (usually circular) seamount.

GYRE A form of oceanographic current which has a circular motion.

**H**

HERMATYPIC Describes corals that have zooxanthellae (symbiotic algae) in their subsurface (endothermal) tissues.

HIGH SEAS Usually refers to the open waters of any sea or ocean, beyond the territorial waters of any country.

HOLOTHURIAN A class of echinoderms. Because of their characteristic sausage shape they are commonly called sea cucumbers.

HOT SPOT A region of existing volcanic activity, often on a plate boundary.

**I**

INTERSTITIAL Refers to the flora and fauna living between sand particles.

ITEROPARITY Describes animals which reproduce more than once.

ISOPOD An order of crustaceans that inhabit freshwater, marine, or terrestrial environments. They have a dorsoventrally (top to bottom) flattened body and seven pairs of legs e.g. sea lice.

**K**

KRILL Large planktonic marine crustaceans; they provide food for baleen whales, fish, and seabirds.

**L**

LARVAE Immature form of insects, aquatic invertebrates, and some fish. Usually describes embryonic stages that are independent but do not have the characteristic features of their parents. During the larval development of aquatic organisms, nutrition is obtained from a yolk sac and/or feeding on plankton.

LITHOSPHERE The Earth's crust; the outer shell of our planet as distinguished from the layers above and beneath it.

LIVERWORT Common name for plants of the class Hepaticae; mosslike plants that grow mostly on damp ground, on water, or on tree trunks.

**M**

MAGMA The molten material within or beneath the Earth's crust. The extrusion and subsequent solidification of magma is the source of volcanic (igneous) rocks.

MEDUSAE Commonly called jellyfish; free-swimming sexual stages of Hydrozoan and Scyphozoan coelenterates. Some of these coelenterates spend their entire lives as jellyfish, but most undergo a bottom-living, polyp-like asexual stage.

METHANE A colorless, odorless, flammable gas with a simple chemical structure composed of carbon and hydrogen. Methane is commercially obtained from natural gas, but is also found in marsh gas and associated with coal seams.

MICROALGAE Microscopic algae; includes phytoplankton and some juvenile stages of macroalgae.

MID-ATLANTIC RIDGE A swelling of the sea floor several hundred kilometers wide extending the length of the Atlantic Basin and lying parallel to the edges of the continents. An area of active sea-floor spreading and volcanism that often forms islands e.g. the Azores.

MOLLUSC A member of the phylum of invertebrates that includes chitons, snails, bivalves, squid, and octopus. The body usually consists of a head, a foot, and digestive and reproductive organs all covered by a soft skin (mantle). The body is then partially or completely enclosed by a calcareous shell.

MORPHOLOGY The form and structure of a whole organism; describes the scientific study of form and structure in plants and animals.

**N**

NACRE Mother of pearl; an iridescent layer on the inside of mollusc shells, composed mainly of calcium carbonate.

NEAP TIDE Tides that fall midway between spring tides and reach the least height, with the narrowest range, in a tidal cycle.

NEMATODE Commonly called roundworms, these worms are unsegmented and have a thick skin. Includes free-living species, soil-dwelling worms, and parasitic worms.

NUDIBRANCH Sluglike marine mollusc which has a naked gill on its upper (dorsal) surface.

**O**

OCEANIC ISLAND These islands often occur in close groups, rising from deep water a long way from continents e.g. the Azores, the Hawaiian Islands.

ORDER A taxonomic group of related organisms, ranking between family and class on a hierarchical scale of description.

**P**

PANTEMPERATE Temperate waters of the world.

PANTROPICAL Tropical waters of the world.

PATCH REEF A small isolated patch or "bommie" of one or more species of coral, usually within a lagoon but separated from other coral growth by sand.

PELAGIC Found in the water column of the sea or ocean; organisms which are not bottom dwelling.

PHYLUM A primary division of the animal or vegetable kingdom.

PHYTOPLANKTON Plant plankton, most of which are microscopic.

PINNIPED A subset of carnivorous mammals which are marine and have a long gestation period e.g. seals, sea lions, and walruses.

PLATE TECTONICS A theory proposed in the 1960s which has led to the conclusion that the Earth's lithosphere consists of a number of rigid plates that appear to be in continual motion, sliding slowly on the asthenosphere below. Movement at the boundaries of plates results in: crumpling and formation of mountains and/or subduction, as one plate slides beneath the other; spreading, as two plates separate—this can result in the formation or extension of ocean basins; earthquakes, when two plates slide past each other.

PNEUMATOPHORE Usually refers to the aerial roots of estuarine plants such as mangroves. Can also describe the air bladder of some marsh or shore plants; the float or air sac of siphonophores.

POLYP Generally refers to an individual or zooid of a colonial animal. More specifically in coelenterates, an individual that has a tubular body and a ring of tentacles around its mouth.

PRODUCTIVITY Generally, the amount of organic (carbon-based) material incorporated into an ecosystem per unit time; frequently used to describe the seasonal or latitudinal fluctuations in phytoplankton growth.

**R**

RIDGE A long, steep-sided elevation of the ocean floor, with a frequently rough topography.

RIFT VALLEY Valley produced when a strip of land, bounded by two parallel rifts, subsides. Rifts are special kinds of fault lines that run parallel to the dominant geological structure in a region e.g. African Rift Valley, San Andreas Rift in California.

ROTIFER Tiny, planktonic single-celled aquatic animals. Known as wheel animals because of an anterior circle of hairs (cilia).

**S**

SALP Common name for the urochordate class Thaliacea (urochordates are in the same phylum as humans but have chordate features only during their larval phase). Salps are free swimming and have a transparent gelatinous body, and two distinct reproductive phases—sexual and asexual.

SCUBA Self-contained underwater breathing apparatus. Consists of a cylinder or cylinders containing compressed air and attached to a breathing apparatus. This enables a person to dive freely without being attached to the surface via an umbilical air hose.

SEA-FLOOR SPREADING Magma from the Earth's mantle wells up continuously along the crests of the mid-ocean ridges and as it cools, spreads out laterally away from them. This process creates a successively older ocean floor away from the ridge and also increases the width of the ocean basin.

SEAGRASS Flowering marine plant closely related to terrestrial flowering plants but also related to seaweed. Seagrasses grow in sand or muddy sediment in shallow estuarine or sheltered waters. Because of their complex root system they play an important role in stabilizing sediment; in addition, their grass-like blades provide shelter for many animals.

SEAMOUNT A submarine mountain (usually a volcanic cone) rising more than 900 meters (3,000 feet) above the ocean floor.

SEMELPARITY Describes animals which reproduce only once.

SIPHONOPHORE A planktonic jellyfish-like creature made up of groups of cells which are specialized for feeding, reproducing, or defense.

SPREADING CENTER A mid-ocean ridge, the site at which sea-floor spreading is generated.

SPRING BLOOM The massive increase in growth and reproduction of plant and animal plankton typically observed in spring and early summer. This cycle is closely linked to increased water temperature and nutrient availability and is most obvious at high altitudes.

SPRING TIDE A tide that occurs at or close to the time of new and full moon. In contrast to a neap tide, it rises highest and falls lowest from the mean (average) tide level.

SUBMARINE CANYON Generally a steep, valley-like depression that crosses the continental shelf and slope—sometimes continuing out over the continental rise.

SWELL The undulating (rolling) movement of the surface of the open sea.

SYMBIOSIS Individuals of two different species living together; this association does not imply that there is a mutual benefit for both species, although it is often (incorrectly) used as a synonym for mutualism.

SYNERGY The working together of two or more substances e.g. hormones (or objects) to produce a result or effect greater than the sum of the two individual effects.

**T**

TALUS Usually coarse or angular rock fragments (of any size or shape) derived from and lying at the base of a cliff or steep rocky slope.

TARDIGRADE A small, usually microscopic animal commonly found in mosses. The small phylum Tardigrada has morphological similarities with animals like worms and insects.

THERMOCLINE The narrow transition zone between warm (upper) and cold (lower) water masses; especially common in summer.

TRANSFORM FAULT A fault that characterizes mid-ocean ridges.

TRENCH Long, narrow depression of the deep-sea floor; usually has steep sides and lies parallel to the direction of the nearest continent, between the continental margin and the abyssal hills.

TSUNAMI A gravitational marine wave produced by any large-scale, short-duration disturbance of the ocean floor.

**U-Z**

UPWELLING Nutrient-rich bottom waters rising to the surface as a result of convection, thereby creating localized and often seasonal regions of high productivity.

WHELK Synonym for a predatory marine gastropod mollusc that has a strong snail-like shell.

ZOOAXANTHELLAE Symbiotic unicellular algae that live in the tissues of various animals (usually marine) including corals and jellyfish.

ZOOPLANKTON Animal plankton, most of which are microscopic.

# THE CONTRIBUTORS

MARGARET ATKINSON
Margaret Atkinson is a research assistant at the University of Sydney, and is currently coordinating a biological survey of the epifauna and algae of rocky reefs in Jervis Bay, New South Wales. She completed a BSc and MSc (Hons) at the University of Auckland before moving to Sydney, Australia, in 1987. Her fascination with the sea and most things marine has resulted in work on various projects at the University of Auckland's marine laboratory at Leigh, the Australian Museum, and the University of Sydney. Much of this work has been field based and has consequently taken her on many memorable trips to some of the most interesting coastal regions of New Zealand and the east coast of Australia. In addition to her work she enjoys sailing, bushwalking, painting, and people.

MICHEL A. BOUDRIAS
Michel A. Boudrias is a PhD candidate at Scripps Institution of Oceanography, University of California. He became interested in marine biology in high school and pursued his interests by working on barnacle functional morphology as an undergraduate at McGill University in Montreal, Canada. His marine biology degree led him to Oregon State University where he studied the feeding and life history of Arctic amphipods. At both Oregon State and Scripps, he became involved in deep-sea photographic analysis and participated in the generation of a biological scale map of the Rose Garden hydrothermal vent. At Scripps, he has devoted his time to the study of crustacean locomotion, particularly the fluid dynamics and functional morphology of swimming deep-sea amphipods.

ALEC C. BROWN
Alec C. Brown is Professor of Marine Biology and member of the Marine Biology Research Institute at the University of Cape Town, South Africa. Born in Cape Town, he attended Rhodes University and then worked on water treatment for the CSIR before joining the staff of the University of Cape Town. He was appointed Professor and Head of the Department of Zoology in 1975. He has published five books, several chapters in books, and about 150 research articles, has worked in Chile and Antarctica, has been visiting Professor to the University of Manchester, and has worked at the University of Cambridge and University College, London, as well as at several marine institutes in the United States. He is a Life Fellow of the University of Cape Town, past president of the Royal Zoological Society of South Africa, and holder of the gold medal of the Zoological Society of Southern Africa.

MICHAEL BRYDEN
Michael Bryden has spent the past 26 years dedicated to the study of marine mammals. In 1963-64 he worked as a veterinarian in Tasmania, then spent three years as a biologist with the Antarctic Division of the Australian government, which included 16 months at Macquarie Island as a member of the Australian National Antarctic Research Expedition, studying growth and development of the southern elephant seal. Professor Bryden has held academic posts at the universities of Queensland and Sydney, Australia, and Cornell University in the United States. He has studied the growth and adaptation of seals on four summer research trips to Antarctica and one to the Norwegian Sea, and carried out research on the reproductive biology of cetaceans at the University of Cambridge in 1978 and 1981. He took up the Chair of Veterinary Anatomy in the University of Sydney in 1988. Since the mid-1970s he has supervised surveys of bottlenose dolphins and humpback whales, and in 1988 began a project to study humpback whales in Hervey Bay, Queensland. He is joint editor of three books, and author or joint author of more than 80 research papers and eight government reports.

M. G. CHAPMAN
M. G. Chapman has a BSc (Hons) degree from the University of Natal and an MSc from the University of Sydney. She is a professional officer at the Institute of Marine Ecology at the University of Sydney, where her work involves experimental ecology and environmental assessment. She is also a writer and scientific advisor for publishing companies.

SYLVIA A.EARLE
Sylvia A. Earle is President and Chief Executive Officer of Deep Ocean Engineering Inc. A marine scientist with a PhD from Duke University, she has held research positions in several universities. In 1982 she co-founded Deep Ocean Engineering Inc. to to design, develop, manufacture, and operate equipment in the ocean. She has extensive field experience as a deep-diving aquanaut, has led more than 50 expeditions using a variety of submersibles and equipment, and holds the depth record for solo deep diving. She has written and lectured widely on marine science and technology, serves on many boards and committees related to oceanographic research and conservation, and is the recipient of numerous prestigious awards to honor her pioneering work.

RICHARD S. FISKE
Richard S. Fiske is a research geologist at the National Museum of Natural History, Smithsonian Institution. A volcanologist, he received his undergraduate training at Princeton University and a PhD in geology from the Johns Hopkins University in 1960. He was a postdoctoral fellow at the University of Tokyo before joining the US Geological Survey, where he carried out research on the active volcanoes of Hawaii and ancient volcanic rocks of the Sierra Nevada, California. He joined the Smithsonian Institution as a geologist in 1976, served as Director of the Smithsonian's National Museum of Natural History from 1980 to 1985, and has now returned to the position of geologist. He maintains an active research program that includes the study of submarine explosive volcanoes in Japan, the potentially dangerous volcanoes of the eastern Caribbean, and the mobile south flank of Kilauea volcano, Hawaii.

SCOTT C. FRANCE
Scott C. France is a PhD candidate at Scripps Institution of Oceanography, University of California, and National Science Foundation Research Assistant, San Diego Museum of Natural History. Before coming to Scripps in 1986, he studied the ecology of freshwater copepod populations at Concordia University in Montreal, Canada. There he became fascinated with the problem of how vent populations propagate from one location to another. To prepare for work in deep-sea biology, he spent a summer in the laboratory of Dr J. Frederick Grassle at the Woods Hole Oceanographic Institution, where he analyzed thousands of photographs of hard-bottom communities. His research at Scripps attempts to address dispersal of deep-sea invertebrates through studies of the genetics and morphometrics of amphipod populations.

CHRISTIAN D. GARLAND
Christian D. Garland is a senior lecturer in the Department of Geography and Environmental Studies/Department of Agricultural Science, University of Tasmania, Australia. He received his doctorate from the University of New South Wales (Sydney) and has worked as a marine scientist on the island state of Tasmania for the past 10 years. His initial contact with aquaculture was as a microbiologist investigating disease outbreaks in a commercial oyster hatchery. His interests then broadened to the general importance of bacteria and microalgae to mass-cultured marine animals, fish, and shellfish. Dr Garland has visited numerous aquaculture ventures in Australia and the Indo-Pacific area, and has witnessed the industry increase dramatically in scope and financial value since the mid-1980s. He leads an innovative research program on bivalve aquaculture, and is also involved in assessment of environmental pollution in aquatic habitats, especially urbanized estuaries. His recent studies have concerned the impact of organic enrichment on the biota of the highly degraded Derwent estuary.

STEPHEN GARNETT
Stephen Garnett has always been drawn to seabirds. As a child he worked with the legendary Dom Serventy as he unraveled the story of the short-tailed shearwater in Tasmania, and he has since led or participated in expeditions to study albatrosses, penguins, frigatebirds, and boobies. His research on birds, crocodiles, turtles, and goats has taken him to islands all around the Pacific. More recently he has been involved in the production of a comprehensive handbook that includes all the seabirds of Australia

and Antarctica. He is the editor of *Search* magazine, and a consultant to the Food and Agriculture Organization, and the Queensland and Northern Territory governments.

ALISTAIR J. GILMOUR
Alistair J. Gilmour is Professor of Environmental Studies and Director of the Graduate School of the Environment at Macquarie University in Sydney. His research interests are in coastal management, environmental policy, and the environmental issues of the south Pacific. He was elected Fellow of the Australian Academy of Technological Science and Engineering in 1977 and to the Institute of Biology in 1986. Before his appointment to Macquarie University in 1985, he was Executive Officer of the Great Barrier Reef Marine Park Authority, and before that, Director of Marine Studies, Ministry for Conservation, Victoria.

RICHARD W. GRIGG
Richard W. Grigg is a Professor of Oceanography at the University of Hawaii. He received his PhD in oceanography from the Scripps Institution of Oceanography in 1970, and is now recognized as a world authority on the ecology of both deep-water precious corals and shallow-water reef corals. His pioneering work on precious corals has led to the development and management of a major industry in Hawaii which utilizes precious coral for jewelry. Dr Grigg's research on shallow-water coral reefs includes extensive work on colonization and successional processes as well as the evolutionary history of fossil reefs in the Hawaiian Archipelago. His recent article on the paleoceanography of coral reefs in Hawaii, published in *Science* magazine, is considered a major milestone in the field. Dr Grigg is presently the chairman of the Pacific Science Association Committee on Coral Reefs and serves on numerous other scientific committees. He has published over 75 scientific papers and is the author or coauthor of three books.

RICHARD HARBISON
Born in Miami, Florida, Richard Harbison received his AB from Columbia University in 1966 and his PhD from Florida State University in 1971. In 1972 he joined the scientific staff of the Woods Hole Oceanographic Institution, where he is presently a senior scientist. Between 1980 and 1982, he was a principal research scientist at the Australian Institute of Marine Science, and also worked as Director of the Division of Marine Sciences at the Harbor Branch Oceanographic Institution from 1987 to 1989. His primary research involves the study of large planktonic animals, such as medusae, siphonophores, ctenophores, salps, and pteropods. He is interested in their systematics, physiology, behavior, and distribution, with the goal of understanding their roles in the oceanic ecosystem. Although he is mainly interested in animals that inhabit the open sea, he has also studied the gelatinous macroplankton of the Arctic and Antarctic.

HAROLD HEATWOLE
Harold Heatwole is Associate Professor in the Department of Zoology, University of New England, Australia. Born in the United States, he has taught in Puerto Rico as well as Australia (from 1966). He has been president of the Australian Society of Herpetologists, the Great Barrier Reef Committee, and the Australian Coral Reef Society. He is a Fellow of the Explorers Club, the Institute of Biology, and the Australian Institute of Biology. He is the author of more than 200 scientific reports and books on coral islands, reptiles, and eucalyptus dieback, and serves on the editorial board of *Fauna of Australia*.

ROBERT R. HESSLER
Robert R. Hessler is Professor of Biological Oceanography, Scripps Institution of Oceanography, University of California. His career in science stemmed from a childhood fascination with arthropods and fossils. This led him to study trilobites at the University of Chicago, leading to a PhD in paleobiology. He went to Woods Hole Oceanographic Institution in 1960, to collaborate with Howard Sanders in describing the anatomy of the Cephalocarida, the most primitive living crustacean, which he had discovered in 1955. Sanders, a benthic ecologist, began work on the ecology of deep-sea bottoms in 1960, and invited Hessler to join him. Since then he has pursued interests in arthropods and the deep sea concurrently, often intertwining them. In 1969, he was invited to join the faculty at Scripps Institution. When hydrothermal vents were discovered, he was part of the first biological expedition in 1979 and has been studying them ever since.

STUART INDER
Stuart Inder, MBE, is a specialist writer on Pacific affairs. Born in 1926 in Sydney, where he lives with his wife, and a professional journalist since 1944, he is a regular contributor to journals, books and encyclopedias, and a radio commentator on Pacific islands matters. He is a former director of the daily *Fiji Times*, Suva, and for nearly 25 years was editor and later publisher of the *Pacific Islands Monthly*, the *Pacific Islands Yearbook* and many Pacific handbooks, histories, biographies, and guides. He was Pacific affairs writer for the Australian weekly news-magazine, the *Bulletin*, from 1982 until 1987. He has been a frequent visitor to all the Pacific island groups and has lived in Japan, Papua New Guinea, Fiji, and Hawaii. Since 1987 he has been a writer with *Australian Geographic*, journal of the Australian Geographic Society.

ANGELA MARIA IVANOVICI
Angela Maria Ivanovici is Senior Project Officer with the Australian National Parks and Wildlife Service, with responsibilities for the selection and proclamation of marine and estuarine protected areas. Born in Italy and educated in Australia, she pursued postdoctoral studies in the United States (on a Harkness Fellowship) and the United Kingdom. She is vice-president of the Coast and Wetland Society and secretary to the Australian committee of the International Union for the Conservation of Nature's subcommittee on marine reserves. She has been involved in scuba diving since 1968, winning several national titles, and is a keen flyer.

DAVID JOHNSON
David Johnson studied for his BSc degree at the University of Sydney, and after industry experience returned to complete his PhD at the University of Western Australia in 1974. Following a period overseas he joined James Cook University in 1978, where he is now Associate Professor in Sedimentology. His major research interest has been the nature of, and processes operating on, the seabeds of the Great Barrier Reef, Australia, and off Vanuatu in the southwest Pacific. He is a member of several geological societies and currently serves as chairman of the Consortium of Ocean Geoscientists of Australian Universities.

GRAHAM JOYNER
Graham Joyner is a lecturer in history and curator of the ancient history teaching collection at Macquarie University in Sydney, Australia. He studied in Sydney and Athens, and has taken part in excavations in Cyprus, Crete, and Andros. He was a member of the Australian team excavating at Torone in Greece from 1975 to 1985. His interests are divided between Greek language and archeology, and he teaches in both areas, as well as courses in Greek pottery and sculpture, and Greek and Roman numismatics. He has published works on Mycenean Greece and classical Greek, and has written and presented a television program on writing in antiquity.

E. ALISON KAY
E. Alison Kay is Professor of Zoology at the University of Hawaii and editor in chief of *Pacific Science*. Born on Kauai in the Hawaiian Islands, she was educated in Hawaii and California before attending Cambridge University on a Fullbright scholarship. She is the author of *A Natural History of the Hawaiian Islands: Selected Readings* (1972) and *Hawaiian Marine Shells* (1979). She has served in several administrative positions at the University of Hawaii, including that of Acting Vice Chancellor. She is an honorary associate in malacology of the B. P. Bishop Museum and is currently chair of the Mollusc Specialist Group, Species Survival Commission of the International Union for the Conservation of Nature.

JAMES C. KELLEY
James C. Kelley was educated in geology and joined the faculty in oceanography at the University of Washington in Seattle in 1966. His research interests have been in biomathematics and applied statistics in the analysis of marine productivity data measured in coastal upwelling systems. He has led and participated in many research cruises in the world's oceans. In 1975 he became Dean of Science and Engineering at San Francisco State University. Since 1986 he has been president of the California Academy of Sciences.

KNOWLES KERRY
Knowles Kerry is a senior research scientist with the Australian Antarctic Division. Since joining the Australian National Antarctic Research Expeditions in 1966 as a biologist, he has wintered on Macquarie Island and spent many summers there and in Antarctica conducting biological research, and has led numerous expeditions to Antarctica, including his work as chief scientist on marine science cruises. As a permanent member of the scientific staff of the Australian Antarctic Division he established in 1972 the Australian biological research program in Antarctica and in 1979 the marine biology program. He served on the Australian delegation which negotiated the Convention for the Conservation of Antarctic Marine Living Resources and represented Australia at meetings of the scientific committee of the commission. He has recently edited *Antarctic Ecosystems: Ecological Change and Conservation*.

G. L. KESTEVEN
G. L. Kesteven was educated at the University of Sydney, and has worked as a fisheries investigation officer with the New South Wales Department of Fisheries; a fisheries biologist with the CSIR; Deputy Controller of Fisheries in the Department of War Organization of Industry; and an advisor on fisheries to the South Pacific office of UNRRA. In 1947 he joined the Food and Agriculture Organization and was stationed in Singapore, Bangkok, and Rome unil 1960 when he returned to Australia to become assistant chief of the CSIRO Division of Fisheries and Oceanography. In 1967 he rejoined FAO, undertaking fieldwork in Mexico, Peru, and other Latin American countries. He has since retired, but continues to act as a consultant to fishery companies.

M. J. KINGSFORD
M. J. Kingsford was born in Hastings, New Zealand. After a first degree at Canterbury University, he undertook research for MSc and PhD projects from the University of Auckland's Leigh Marine Laboratory. He then carried out contract work for New Zealand fisheries and took up a postdoctoral fellowship at the University of Sydney, where in 1988 he obtained a lecturing position in the School of Biological Sciences. Other teaching activities include presenting adult education lectures on fish and marine ecology. His research has focused on reef-associated juvenile and adult fish as well as the early life history stages of fish. He has a major research interest in the importance of oceanographic features for influencing the distribution, movements, and survival of small fish and plankton. He has published 15 research articles on fish and plankton, and has contributed to educational books.

JOHN E. McCOSKER
John E. McCosker has been Director of the Steinhart Aquarium, a division of the California Academy of Sciences, since 1973. Although trained as an evolutionary biologist, his research activities have subsequently broadened to include such diverse topics as the symbiotic behavior of bioluminescent fishes, the behavior of venomous sea snakes, the predatory behavior of the great white shark, the biology of penguins, the biology of the coelacanth, and dispersed and renewable energy sources as alternatives to national vulnerability and war. His shark research has been summarized in BBC, NOVA, and *National Geographic* television specials.

KENNETH McPHERSON
Kenneth McPherson is Executive Director of the Center for Indian Ocean Peace Studies at Curtin University and the University of Western Australia. He established the *Indian Ocean Review* in 1980, the Center for Indian Ocean Regional Studies in 1987, and the Peace Studies Center in 1990. He is the author of numerous articles on Indian Ocean history and is currently working on a major history of the region.

ALEXANDER MALAHOFF
Alexander Malahoff is Professor of Geological Oceanography, Department of Oceanography, University of Hawaii; and Director of NOAA's National Undersea Research Center at the University of Hawaii (Hawaii Undersea Research Laboratory). His research over the past 25 years has been focused on the geology and geophysics of the ocean floor and of volcanoes and volcanic islands, using ships, airplanes, and submarines in order to conduct these studies. He is the author of more than 70 scientific papers and has undertaken over 150 dives in a number of submersibles including those that are nuclear powered. Much of his recent work has centered around studies on hydrothermal vent processes and cobalt-rich ferromanganese crusts around the Hawaiian Islands and other areas.

COLIN MARTIN
Colin Martin is Reader in the Department of Scottish History at the University of St Andrews, Scotland. He worked as a flying instructor and army officer before becoming a freelance journalist specializing in archeological subjects. Between 1968 and 1983 he was Archeological Director on the excavation of three Spanish Armada wrecks in Irish and Scottish waters. He established the Scottish Institute of Maritime Studies at St Andrews in 1973. He is a frequent broadcaster on radio and television, president of the Nautical Archeology Society, and author of two books on the Spanish Armada and numerous papers and chapters on maritime topics.

SIDNEY W. MINTZ
Sidney W. Mintz is William L. Strauss Jr Professor of Anthropology at the Johns Hopkins University, Baltimore. He is a regional specialist in the anthropology of the Caribbean region and topical specialist in economic anthropology, the history of peasantries and agrarian proletariats, and the anthropology of food. His publications include *The People of Puerto Rico* (coauthor), *Worker in the Cane, Caribbean Transformations, Sweetness and Power*, six other edited and coauthored volumes, and about 200 articles and reviews.

STORRS L. OLSON
Storrs L. Olson is Curator of Birds, Department of Vertebrate Zoology, National Museum of Natural History, Smithsonian Institution, with research interests in systematics and the evolution of recent and fossil birds of the world, particularly human-caused extinctions of birds on oceanic islands, especially the Hawaiian Islands, West Indies, and Atlantic. Dr Olson has been associated with the Smithsonian Institution since 1968 and has maintained an office in the National Museum of Natural History since 1971. An authority on fossil birds in general, he has for many years specialized in extinctions of birds on islands, beginning with his first expedition to Ascension Island in 1970. He has collected fossils on four other islands in the Atlantic, and on several islands in the West Indies. With his wife, Helen James, he has also discovered and described dozens of new species of fossil birds in the Hawaiian Islands.

JOHN R. PAXTON
John R. Paxton is a senior research scientist at the Australian Museum, Sydney. He was born in Hollywood and received his degrees from the University of Southern California, working on deep-sea fishes at the Allan Hancock Foundation. He joined the Australian Museum in 1968 as curator of fishes. He has conducted fieldwork in many south Pacific countries and has taken part in numerous deep-sea expeditions. He and coauthors have recently published the first fish volume of the *Zoological Catalogue of Australia*. His current work involves attempts to unravel the mysteries of the biology and evolutionary relationships of the deep-sea whalefishes. He has served as foundation secretary and president of the Australian Society for Fish Biology, and has written more than 60 scientific papers and popular articles.

VICTOR PRESCOTT
Victor Prescott is a political geographer whose first appointment was in Nigeria, at University College Ibadan. While lecturing there for five years he completed his doctoral thesis on Nigeria's international and regional boundaries. Since 1961 he has been on the staff of the Department of Geography at the University of Melbourne. He was appointed to a personal chair in that department in 1986. He is author of *Maritime Political Boundaries of the World* (1986) and *Political Frontiers and Boundaries* (1987).

PATRICK G. QUILTY
Patrick G. Quilty is Assistant Director, Science, with the Australian Antarctic Division. Born in Western Australia, he held positions in Australian universities before establishing his career with the Antarctic Division in 1980. Since 1965 he has made 10 visits to Antarctica, and has the distinction of having landmarks there named in his honor. He has been awarded the United States Antarctic Services medal, has written over 100 scientific papers about the region, and comments regularly on Antarctic affairs. A geologist by training, his special interests lie in microfossils.

WILLIAM REED
William Reed is part-owner and director of several companies pearl farming in Western Australia and the Cook Islands. Born in Queensland, he worked for eight years with the Food and Agriculture Organization developing methods of pearl oyster farming in the Red Sea. From 1968 to 1975 he was consultant to the first black pearl farm in Tahiti. He returned to Australia in 1975 to establish pearl farms in the Broome area of Western Australia. He is a consultant to a number of pearl production, jewelry design, manufacturing, and retail companies. He is a Member of the Institute of Biology, London.

PAUL SCULLY-POWER
Paul Scully-Power is a research associate at Scripps Institution of Oceanography, California. He holds the Distinguished Chair of Environmental Acoustics at the Naval Underwater Systems Center, New London, Connecticut, where he is additionally Chief Scientist of the Warfare Systems Architecture and Engineering Office in the Surface ASW Directorate. He is also a flight crew instructor in the Astronaut Office at the Johnson Space Center, Houston, Texas. He has an international reputation in physical oceanography, underwater acoustics, naval systems, and space operations. He was the first Navy civilian and the first oceanographer in space as a crew member of the *Challenger* mission in 1984. He has received the Navy's highest civilian honor, the Distinguished Civilian Service Medal, for his achievements in naval research, and holds over 40 Navy special achievement awards.

J. R. SIMONS
J. R. Simons is a former Associate Professor of Biology and Dean of the Faculty of Science at the University of Sydney. Educated in Sydney and London, where he received his PhD, he began his academic career as a lecturer in zoology at the University of Sydney. Following his retirement in 1983, he has pursued his interest in the history of Australian biology.

ROBERT E. STEVENSON
Robert E. Stevenson is Secretary-General of the International Association for the Physical Sciences of the Ocean and a consultant to NASA for space oceanography. His distinguished academic career has included teaching and research positions in several universities and consultancies to institutions such as the National Academy of Sciences and NATO. From 1985 to 1988 he was scientific liaison officer and Deputy Director, Space Oceanography, in the Office of Naval Research, Scripps Institution of Oceanography. He is recognized as a leading world authority on space oceanography and continues to work with NASA to develop space oceanography programs.

FRANK H. TALBOT
Frank H. Talbot is a marine biologist who has specialized in coral reef fish ecology. He has studied reef fish communities off the east coast of Africa, and over the past 25 years has researched mainly off Australia's Great Barrier Reef, where he was involved in setting up two island research stations. He lived and studied on a Caribbean coral reef in the Tektite II underwater habitat program. He has sailed a small yacht to islands in the Pacific, and with his wife and youngest child has voyaged from Australia to Africa across the Indian Ocean, returning via the Southern Ocean and calling at a subantarctic island group along the way. He has been Professor of Biology and Environmental Studies at Macquarie University, Sydney; Director of the Australian Museum; Executive Director of the California Academy of Sciences; and is presently Director of the National Museum of Natural History/National Museum of Man, Smithsonian Institution, Washington.

A.J.UNDERWOOD
A. J. Underwood is Director of the Institute of Marine Ecology and Reader in Experimental Ecology at the University of Sydney. He is experienced in marine ecological investigations in several fields of study. His main interests are the experimental analysis of interactions in complex assemblages of species. He is an authority on the design of sampling and experimental programs and has published numerous scientific papers.

DIANA WALKER
Diana Walker is a lecturer in marine botany at the University of Western Australia with particular interests in seagrasses and macroalgae. She was born in the United Kingdom and studied marine biology at the University of Liverpool's Marine Biological Station on the Isle of Man. She carried out her PhD research on coral reef algae in the Red Sea (Jordan) while based at the University of York in England. From there she was appointed to a postdoctoral fellowship and then a lectureship at the University of Western Australia, where she has carried out research on seagrasses and macroalgae from the northwest of Western Australia to the south coast, but concentrating on Shark Bay. Her main interests are on the factors influencing the distribution of marine macrophytes covering ecophysiology to biogeography.

G. M. WELLINGTON
G. M. Wellington is Associate Professor of Biology, University of Houston, Texas. He spent two years (1973–75) living and working in the Galapagos Islands for the Charles Darwin Foundation and the Galapagos National Park Service (Ecuador) on a survey of marine environments, and the development of a comprehensive plan for their protection. Following graduate studies at the University of California (1976–81), he took a position at the University of Houston where he has pursued studies on the ecology and evolution of coral reef organisms. He is currently working on a project to evaluate the relationship between planktonic larval duration and gene flow in marine shorefishes of the tropical eastern Pacific. The goal of this study is to understand the evolution of endemic species found on offshore islands such as the Galapagos.

## FURTHER READING


Bannister, K. and Campbell, A. *eds.* 1985, *The Encyclopedia of Aquatic Life*, Facts on File, New York.

Brown, A. C. and McLachlan, A. 1990, *Ecology of Sandy Shores*, Elsevier, Amsterdam.

Clark, R. B. 1986, *Marine Pollution*, Clarendon Press, Oxford.

Couper, A. *ed.* 1983, *The Times Atlas of the Oceans*, Van Nostrand Reinhold, New York.

Decker, R. W. *et al eds.* 1987, *Volcanism in Hawaii*, US Government Printing Office, Washington.

George, J. D. and George, J. J. 1979, *Marine Life: An Illustrated Encyclopedia of Invertebrates in the Sea*, Harrap, London.

*Island Life*, 1978 (Wild, Wild World of Animals), Time-Life, New York.

*Life in the Coral Reef*, 1977 (Wild, Wild World of Animals), Time-Life, New York.

Kennett, J. P. 1983, *Marine Geology*, Prentice-Hall, Englewood Cliffs, NJ.

Macdonald, G. A. *et al.* 1970, *Volcanoes in the Sea*, University of Hawaii Press, Honolulu.

*The Mitchell Beazley Atlas of the Oceans*, 1977, Mitchell Beazley, London.

Myers, N. *ed.* 1985, *The Gaia Atlas of Planet Management*, Pan Books, London.

*Oceanography: Readings from Scientific American*, 1971, W. H. Freeman, San Francisco.

*Ocean Science: Readings from Scientific American*, 1977, W. H. Freeman, San Francisco.

Ranwell, D. S. 1972, *Ecology of Salt Marshes and Sand Dunes*, Chapman & Hall, London.

Teal, J. and Teal, M. 1975, *The Sargasso Sea*, Little Brown, New York.

Thurman, H. V. 1981, *Introductory Oceanography*, 3rd edn, Charles E. Merrill, Toronto.

Veron, J. E. N. 1986, *Corals of Australia and the Indo-Pacific*, Angus & Robertson, Sydney.

*Volcano*, 1982, Time-Life, Amsterdam.

# INDEX

# ACKNOWLEDGEMENTS

The publishers would like to thank the following for their assistance in the production of this book: Dr K. Radway Allen, Christine Deacon, Sharon Freed, Macquarie Library Pty Ltd, Tristan Phillips, and Annette Riddell.

Many of the illustrations prepared for this publication were based on original references provided by the contributors. Other sources of illustrations are listed below.

**Page 16** : *Stages in continental drift* and **Page 19**: *Major tectonic plates* are from *The Macquarie Illustrated World Atlas* 1984, Macquarie Library, Sydney, pp. 46—7. They are reproduced by permission of the publishers. **Pages 21-2**: The maps of the ocean basins are reproduced by permission of Verlag Das Beste GmbH, Stuttgart. **Page 27**: *The ocean currents* is adapted from *The Macquarie Illustrated World Atlas*, p. 74. **Page 34**: *The annual cycle of plankton production* is adapted from Couper, A. *ed.* 1983, *The Times Atlas of the Oceans*, Van Nostrand Reinhold, New York, p. 71. **Page 38**: *The skipjack tuna: a Pacific Ocean wanderer* is adapted from *The Times Atlas of the Oceans*, p. 101. **Page 47**: *The Arctic tern: a long distance wanderer* is adapted from *The Atlas of the Living World* 1989, Weidenfeld & Nicolson, London, p115. **Pages 48-9**:*How seabirds catch their prey* is adapted from Ashmole, N.P. and Ashmole, M.J. 1967, *Comparative Feeding Ecology of Seabirds on a Tropical Oceanic Island*, Yale Peabody Museum Natural History Bulletin 24. **Page 55**: *Faunal regions of the Atlantic Ocean* is adapted from Backus, R.H. *et al* 1977, "Atlantic mesopelagic zoogeography", *Fishes of the Western North Atlantic. Order Iniomi (Myctophiformes)*, Yale University, NY, pp.266-87. **Page 62**: *Mangrove distribution in the Indian Ocean* is adapted from Lear, R. and Turner, T, 1977, *Mangroves of Australia*, University of Queensland Press, Brisbane. **Page 68**: *Divisions of the marine environment* is adapted from Thurman, H.V. 1981, *Introductory Oceanography*, 3rd edn, Charles E. Merrill, Toronto, p. 307. **Page 82**: *The Bering Canyon* is adapted from a block diagram by T. R. Alpha, US Geological Survey. **Page 85**: *Tidal influences on time out of water* is adapted from Underwood, A.J. and Hutchings, P. A. 1987, *Australia's Seashores*, Collins, Sydney. **Page 87**: *A typical sandy coastline*, **Page 89**: *How sandy beaches act as filters*, and **Page 90**: *Some characteristic interstitial animals found on sandy beaches* are adapted from Brown, A. C. and McLachlan, A. 1990, *Ecology of Sandy Shores*. Elsevier, Amsterdam. **Page 92**: *An estuarine food web* is adapted from Underwood, A. J. and Hutchings, P. A. *Australia's Seashores*. **Page 127**: *The making of oceanic volcanoes* is adapted from *Volcano* 1982, Time-Life Books, Amsterdam, pp. 62-3. **Page 132**: *The Hawaiian seamount chain* is adapted from *Airone* 108, April 1990. **Page 142**: *The formation of a coral atoll* is adapted from Davis, W. M. 1928, *The Coral Reef Problem*, American Geographical Society. **Page 145**: *Changes in sea level over the last 140,000 years* is adapted from Chappell, J. and Polach, H. A. 1976, *Geological Society of America Bulletin* 87, 235. **Page 157**: *Diversity and adaptation: the finches of Galapagos* is adapted from Grant, P. R. 1986, *Ecology and Evolution of Darwin's Finches*, Princeton University Press. **Page 158**: *The converging currents* is adapted from Glynn, P. W. and Wellington, G. M. 1983, *Corals and Coral Reefs of the Galapagos Islands*, University of California Press. **Page 160**: *The limits of distribution of freshwater fishes, amphibians, and fruit bats* is adapted from Kay, E. A. 1980, *Little Worlds of the Pacific*, Lyon Arboretium Lecture 9, University of Hawaii. **Page 176**: *Minoan trade routes* is adapted from Judge, J. 1978, "Greece's brilliant Bronze Age", *National Geographic* 153, 2, February, p. 154. **Page 178**: *Pacific island settlement* is adapted from Howe. K. R. 1984, *Where the Waves Fall*, Allen & Unwin, Sydney, p. 14. **Page 199**: *The global catch* is based on statistics from FAO *Yearbook* 1987. **Page 200**: *How the catch is used* is adapted from Myers, N. *ed.* 1985, *The Gaia Atlas of Planet Management*, Pan Books, London, p. 82. **Page 204**: *Production stages in typical aquaculture operations* is adapted from Garland, C. D. 1988, *Australian Mariculture: the Role of Hatcheries in Animal Production*, University of Tasmania, Hobart. **Page 216**: *Limits of discharge: the MARPOL convention* is adapted from *The Gaia Atlas of Planet Management*, p. 92. **Page 221**: *Three Coral Sea marine reserves* is adapted from Australian National Parks & Wildlife Authority 1990, *Coral Sea National Nature Reserves*, information brochure. **Page 228**: *Largest islands* is from *The Macquarie Illustrated World Atlas*, p. 391. It is reproduced by permission of the publishers. **Page 229**: *Endangered marine creatures* is based on information in *1988 IUCN Red List of Threatened Animals* 1988, IUCN Conservation Monitoring Centre, Cambridge.

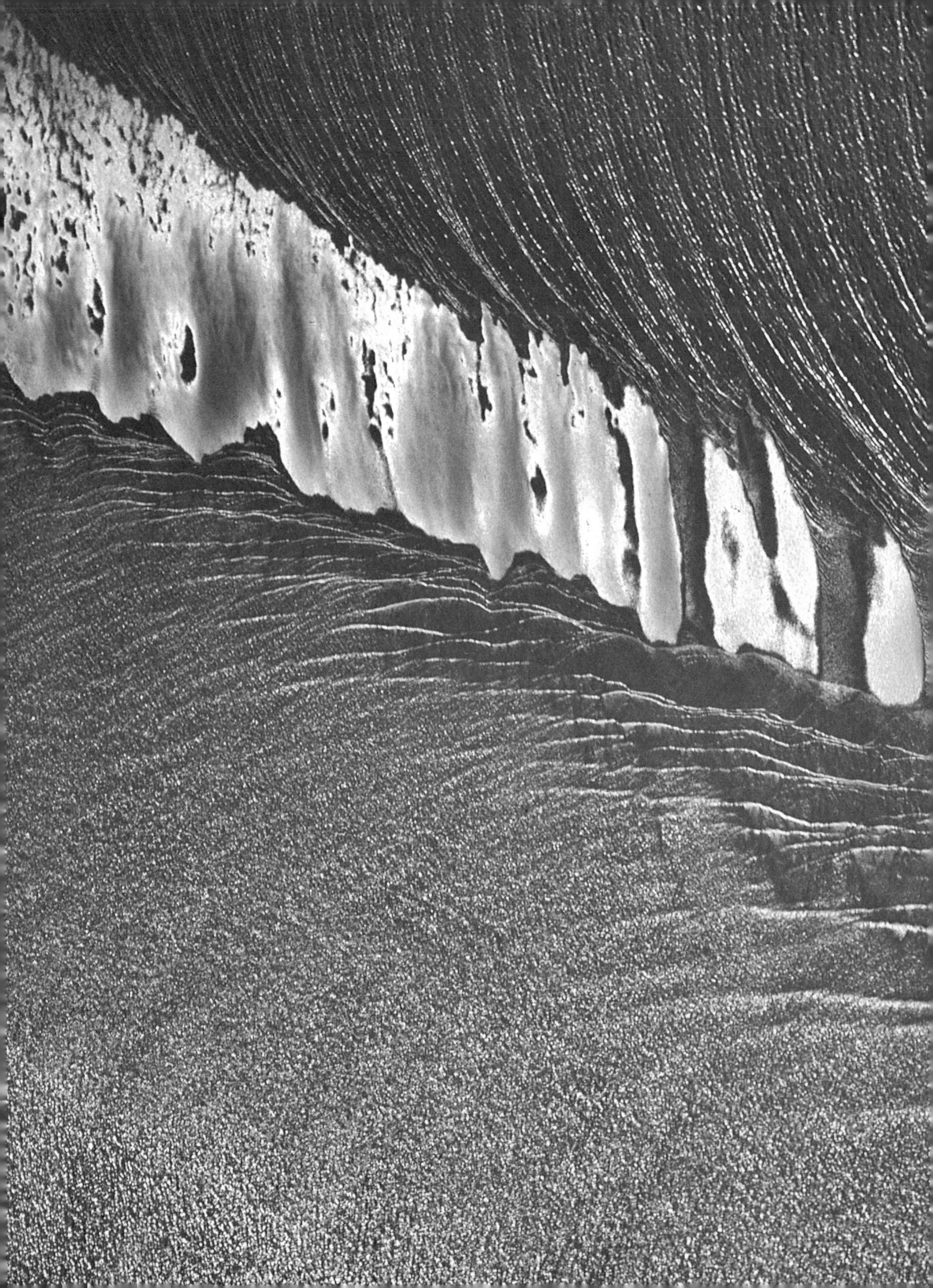